The Stablecoin Book

By

Christopher W. Smithmyer

Elite exclusivity

Dysart Detroit Dallas

First Edition 2023

Send All Inquiries to: Info@EliteExclusivity.com

ISBN Paperback: 978-1-952651-28-1

Printed in the United States of America

"If I had a world of my own, everything would be nonsense. Nothing would be what it is because everything would be what it isn't. And contrary wise, what is, it wouldn't be. And what it wouldn't be, it would. You see?"

The Mad Hatter

Lewis Carroll, Through the Looking Glass.

This was meant to be a fictitious quote, not the basis for a system of government.

To My Niece and Nephews,

If we cannot build a better world for future generations,

De we even deserve the pitiful world that past generations left us?

Table of Contents

Forward

This book was written in two days in response to the unprecedented attacks by the SEC against the cryptocurrency community. Not only are these attacks unconstitutional, but also seditionary against the economy and sovereignty of the United States. Cryptocurrency is a viable system for the replacement of the fiat currency for international exchange. It is not a direct replacement for the dollar, or any fiat currency in the world. Because the cryptoverse (and the world in general) is so large, there will always be room for both currencies to exist side by side. Whether the action is through greed or through utter incompetence we, the citizens of the United States who use and work in the cryptocurrency field these complaints at the feet of our elected representatives (who are to represent us) and the world community in general. People around the world have a right to their wealth, and if no other right in this world is self-evident, then that should be the starting point to all property law. Therefore, we submit the complaints:

1.) The Securities and Exchange Commission, and to a lesser extent the Commodities and Futures Trading Commission, are acting illegally and beyond the scope of their organizational charter in attempting, not to regulate, but to destroy the cryptocurrency industry in the United States.[1] This attack is coming in the form of unconstitutional ex post facto

---

[1] Levy, Ari (2023). "Crypto tokens plunge this week after Gensler stepped up SEC crackdown." CNBC at https://www.cnbc.com/2023/06/09/crypto-tokens-plummeted-on-concern-gensler-is-expanding-sec-crackdown-.html.

regulations, which effect companies in existence and companies that are preparing to launch products on the blockchain; illegal regulations that fall well beyond the scope of the issuance of powers by congress under the recent ruling in West Virginia v. the EPA; unconstitutional lawsuits against private citizens of the United States, which, when enacted by a regulatory agency, are, on their face, bills of attainder against private citizens and companies of this country; and a concerted media attack on the cryptocurrency industry in a direct and tortious interference with third party contracts through reduction of the faith of the people in a financial product;[2]

2.) The congress of the United States is derelict in its duty to refrain the SEC and CFTC from abusing powers, which may or not be legislative powers unconstitutionally granted to the executive branch, and in doing so is causing irreparable harm to the fledgling cryptocurrency industry in the United States. This dereliction of duty and outright attack on Americans who are working to build a stronger economy for our nation, is a clear and present danger to the national security of the United States as the US dollar is losing market share as the global reserve currency. Any replacement should, as a

---

[2] Hern, Alex (2023). "SEC accuses Coinbase Cryptocurrency Exchange of Breaking US Regulations." The Guardian at https://www.theguardian.com/technology/2023/jun/06/sec-accuses-coinbase-cryptocurrency-exchange-of-breaking-us-regulations. See Also West Virginia v. Environmental Protection Agency (No. 20-1530) at https://www.supremecourt.gov/opinions/21pdf/20-1530_n758.pdf

necessity, be from a company who exists in the United States.

3.) The judiciary of the United States, through error or omission, has neglected to see that federal government oversight of the cryptocurrency industry is illegal under the commerce clause of the United States Constitution. That because of the way that the internet works, the currency does not 'come to' the people using across state lines, rather the citizens who are using cryptocurrency, via the portal, exchanges, wallets, and other means are digitally going to the currency where it is housed. Therefore it never moves and thus does not have a substantial presence in any location other than where it is housed, further, because the use is fully in virtual space, it never moves and the company issuing the currency does not have a substantial relationship with any state, territory, or other entity within the United States, therefore, the regulation, beyond contract, anti-trust, KYC, and anti-terrorism laws is beyond the scope of the federal government;

4.) That, even if the SEC has any authority to issue regulations in an areas clearly not expressed by Congress, cryptocurrencies that do not have a centralized governance feature in their nature do not meet the standards of the Howey test (joint enterprise) and that stablecoins do not meet the profit standard of the Howey test.[3] Further, even if tokens are found to meet these two elements, the very decentralized nature of the cryptocurrency

---

[3] SEC v. W.J. Howey (328 U.S. 293) at https://supreme.justia.com/cases/federal/us/328/293/.

system (centfi excluded) makes the communications a lawful association under the First Amendment of the United States constitution and the SEC, nor the Congress, can make a law prohibiting this association.

5.) Finally, the recent actions of the SEC releasing lawsuits against major cryptocurrency entities during the fledgling stages of bull runs is an unabashed effort to 'barcode' the system.[4] Barcoding is a recognized market manipulation strategy which has been declared a felony in many cases. As the agency responsible for policing this action is the accused in this matter, there should be a special council appointed to investigate the illegal actions of the SEC, a moratorium on any illegal civil cases against cryptocurrency entities at the federal level, and an ad hoc system set up for leaders in the crypto community to be called, when needed, to police major events- such as the FTX scandal, where the punishment by the federal government for stealing billions was house arrest in a mansion and the opportunity to speak before congress.

Sometimes the government of the United States forgets that it is there to protect the rights of the people of the United States, not just to regulate us into servitude. As such, we lay these claims, which are tantamount to sedition, felonious interference with commerce, and unconstitutional bills of attainder against citizens of the United States and citizens of other countries attempting to do business in the

---

[4] Staff (2023). "What does Barcoding Mean in Stocks." Forex at https://www.forex.in.rs/what-does-barcoding-mean-in-stocks/.

United States. These attacks are a violation of the human and constitutional rights of citizens of this great country, and we ask that elected representative in every state, in every town, and in the hallowed halls of Congress look into these problems so that the cryptocurrency market can grow, create jobs and investments for the people of the United States, and allow the United States to take its rightful place as the global economic leader in this young millennium. We must politely but firmly use our right to petition and if we must the right to use the courts to challenge illegal action that takes away our rights.

Special Note

This book was written with AI doing the editing for me. Because I was rushing to finish in just a few days, I wanted to make sure that the text was well edited. Thus, some of the parts may look robotic and not read as smoothly as some of my other books. All the parts are written in a form that is readable by everyone so it will make sense.

Introduction

"If you're not staying on top of your money, you are putting your own financial well being at risk."[5]

As a global community, we have failed to watch over our economic system. We have subbed out our ability to control our own health to the point that we hold very little of value in our own hands. Right now, where you are sitting, how much of your own wealth do you have within reach? I am not talking about the ability to trade stocks on your phone, to use your debit card, or to call your broker- I am talking about the wealth that is on hand if something goes wrong, the government shuts off your accounts, or there is a major emergency.

I think most of us find that we do not have nearly what we would like to have on hand if something like that were to happen; however, we are always constantly in the risk of a major event happening, the government having a liquidity problem (debt ceiling problem), or even an emergency where the comforts of modern life are not available. Dedicated systems that we have outsourced, that require bureaucrats, bank tellers, or other human interaction will not be there if there is a major problem; however, we do have a more modern option in the form of cryptocurrency which operates through an automated system- as long as you have enough electricity to get a signal (and assuming there are signals out there) you have at least some access to your wealth.

---

[5] Susan Orman at https://wellbeing.gmu.edu/famous-quotes-on-financial-stability-and-well-being/,

Now let's face it, if things really do go south and the world is falling apart, electricity is going to be one of the first things to go. As such, we will not have access to our cryptocurrency or our digital currency or the bank accounts we have worked our whole life building up. People are not going to give a damn about gold and silver, or the printed currency that governments are giving out like candy. Food, firearms, and other necessities will be the currency if things get that bad. However, if you are thinking that things are going to get that bad, pick up a prepping book- this book is for people who are ready to deal with the economic cold war that we are facing in the world today.[6] As fiat currencies tank around the world due to inflation, poor leadership, and the very nature of fiat currency, it becomes imperative that we build a better economic system that is beyond the control of government "leaders." When they feel that they own what you own, then you are not a free person- you are simply a slave in a nicer house.

What wealth really comes down to is you being able to control your own future. Do you have the choice if you go to work this week? What about next week? If you must go to work to ensure that you are going to have food on the table, then you are not completely in control of your life. Don't worry, in this world it is called being an adult when you need to work. From the time we get our first job to the time we save up enough to retire; we are at the beck and call of those for whom we work. Regardless of what union stooges tell you, the goal of working is not just to make it

---

[6] Saadi, Sommer & Webb Merryn (2023). "What the New Cold War Means for the Global Economy." Bloomberg at https://www.bloomberg.com/news/articles/2023-03-10/niall-ferguson-on-what-the-new-cold-war-means-for-the-global-economy#xj4y7vzkg.

to the end of the day or to give your kids a better life (which is also very important), it is to save up enough that you do not have to work till you are 70 to beg the government for money to live.

Now, since we live in a world where most developed countries are either republics or democracies, you would think that the governments of these countries would be working to help you get to a point where you were able to do this at about the age of 50, right? No, not really. Governments are built on the idea of control, the old throwback to the feudal era. As a result, people are looked at as cogs in a wheel that sometimes are given the 'right' to vote. Like any good business, the government wants to keep these cogs working within the machine if it can to ensure that it does not have the cost of 'disposing' of the 'out of date' workers (read retirement).

Don't believe me? Well then lets take a look at how the United States spends most of its money. If you listened to the news, then you would think that most of the money for the United States goes to the military. If this is what you think, then you would be looking at our number two budget line item.[7] The most expensive thing to the biggest money spending nation in the world is old people, well to be politically correct it is the department of human services. The government spends more on the elderly than on all other. Between Social Security, Income Security (which also includes poverty not just the elderly), and Healthcare

---

[7] Congressional Budget Office (2023) https://www.cbo.gov/content/budgetary-implications-eliminating-cabinet-department.

we see 48% of the national budget in use.[8] If we agree that about 3% of the 4% spent of veteran affairs is spent on older vets, that means 51% of our budget is spent on older Americans.[9] Now for me, this is not a problem. These people worked, paid taxes, and were what made our country great for 65 years of their lives. For the government, this is a problem because this is the biggest negative budget item on the ledger.

Therefore, we have a problem. The government tends to see the wealth of the older Americans as 'their money' its just that the older people have it till the government can take it in estate tax, then in income tax on the heir, then in sales tax on the stuff bought by the heirs. This means that separating people for their money is a waiting game for the government. One of the biggest reasons that the government is worried about cryptocurrency is that, as a currency, most of it is not taxed until it is converted to U.S. dollars. This means that the tax implication on the estate tax, which are not established yet, are less profitable for the government. Remember, the government wants you to die as soon after you quit working as possible because you go from being a productive member of society to a budget item. If you control your own wealth in the form of stablecoins, however, you are in control of your life and the 'bounty' they give you for social security will not be all that you can live off.

---

[8] https://fiscaldata.treasury.gov/americas-finance-guide/federal-spending/.
[9] Id.

Why Stablecoins are Important.

So, if you are reading this, then you may be wondering why stablecoins are a good investment. Since the crypto market has been beating all the other markets combined for three out of four years (it is a four year cycle) for the last decade, why not invest in some meme coin with 1000% rewards? Well, you could, 90% of meme coins fail, which means you have a better chance putting your life savings in a slot machine than you do putting them in the meme market, but that is your choice. The reason stablecoins are a good investment is because they beat fiat currency when inflation is taken into account. And for investors, especially those looking to retire or to put their kids through college, that is a big deal.

The dollar in your wallet is losing money. With an average inflation rate each year of 3%, a dollar at the beginning of the year is worth $0.97 at the end of the year. Over the last two years, the inflation rate has been nearly 8%, which means that your American Dollar at the beginning of 2022 it is now worth about what a Canadian dollar or Australian Dollar was worth, you lost $0.16. Modern stablecoins, those not based solely on fiat currency have anti-inflationary measures which allow the backing fund to beat the inflation rate. This means that, relative to the basket of goods, your stablecoins keep their value (or grow their value) over the course of time. To understand this, you need to understand the three types of stability.

Negative stability is the most common type of stability that we see in the world today. Some people would say that this is synonymous with depreciation, however, depreciation takes into account the negative stability, which means that it falls a little faster than the going rate. Negative stability means that your dollar is going to be worth a dollar, even if the dollar itself is worth less from one year to the next. The negative stability continually erodes the value of any currency that does not have an anti-inflationary system in place. This is why fiat backed stablecoins are old news (though they are all that the government wants to talk about).

Neutral Stability is the second type of stability that one needs to know about. Neutral stability is what people think of when they think of stability, the value stays the same relative to what people think the value should be. In an ideal world, all fiat currencies should be neutrally stable; however, greed and poor management prevent this from being a reality. As a result, it is very difficult to find a neutrally stable asset, in fact most people will bypass a neutrally stable asset because they are used to seeing negatively stable assets or spend their lives seeking positive neutral assets.

Positive neutral assets are more common in stablecoin after stablecoin 1.3. For an asset to be positive stable, it needs to have an increase in value each year relative to fiat currency, but neutral to the basket of goods. If the value grows beyond the basket of goods, then it is not stable. Also, if it does not beat inflation, it is not positive. So, these types of stablecoins are difficult to manage but when they work properly, they are very successful. All stablecoin 2.0 or greater economies should see positive stability.

The format of the book

Because the topic of stablecoins is so complex, the book itself needs to be simple. Writing on complex topics with a complex structure is often self-defeating; therefore, this book is broken down into three simple parts:

1.) Part I: Foundations of Stability
2.) Part II: Stablecoin Families
3.) Part III: Regulations

These elements are the important elements that people need to understand when dealing with stablecoins. By understanding these elements, they can understand whether the government has their best interests in mind when regulating stablecoins or if they are having the screws put to them.

For any stablecoin discussion, it is important to note that without a proper foundation a coin is not stable. Further, it is important to note that the government, by virtue of their license of economic violence, cannot dictate the terminology of a developing field simply to make it harder for the field to develop. The purpose of the government, in all cases, is to make sure that people's freedoms are protected, and that business can proceed as normal. Any action beyond this or contrary to this is in violation of their duty (unless they are totalitarian, then I guess they do what they want).

Continuing the discussion, we next look at the different families of stablecoin. To make things easy they have been defined in decimals between stablecoin 1.0 (BitUSD) to Stablecoin 2.0 (KIROS). Each family of coins have a set of requirements that indicate what level of development we see for the tokenomics. The more advanced the token, the better it is as a hedge in your investment portfolio. Further, tokens with good tokenomics, like KIROS, make for excellent exchange tokens because they keep their value relative to the basket of goods.

Finally, we take the time to discuss the regulations that are needed for the cryptoverse to become more stable. While the primary focus in this book is the laws needed for better stablecoins, there are many notes needed where the cryptoverse and the stablecoin markets have intertwined interests. There are currently regulations which are very useful for people who are in the digital space, knowing these regulations puts you ahead of the game. There are also laws that we have on our 'wish list' which would just make the world a better place (and give people more access to their wealth. Finally, there is a list of a laws that will just gum up the whole works (which seems to be the direction in which the government is going.) Knowing where and how the regulations work enables us to work together for a better future, a future where people have more freedom.

As we explore the world of stablecoins, keep in mind that the field is constantly changing. Each day something new enters into the market, and each day the government makes adjustments to deal with the changes. This is how economics work. Where we need to be careful is that the government does not try to 'overregulate' the system, when this happens the rights of millions of people suffer. Cryptocurrency is about brining rights to people, all we are asking of the government is to let us do our thing in the cyberverse, there are enough problems in the physical world without the government attacking law abiding companies.

Part I: Foundations of Stability

Introduction to Part I

To be stable, something must have a strong foundation. As a result, stablecoins must be built upon a philosophy that has a firm foundation. Such a simple concept should be simple to follow, correct? Well as with many things that are simple, it can be more difficult to deal with simple matters than it is to deal with complex matters. As a result, we have seen that the government has a difficult time dealing with simple matters, and this includes the philosophy of stablecoin.

Stability is a critical characteristic of stablecoin, setting it apart from other tokens in the cryptocurrency space. Stablecoins aim to maintain a relatively stable value, often pegged to a specific asset or basket of assets. This stability makes stablecoin an attractive option for users who seek to minimize the volatility typically associated with other cryptocurrencies, allowing for greater predictability and usability in day-to-day transactions.

The tokenomics of stablecoin differ from other tokens due to their design and purpose. While many cryptocurrencies are primarily used for speculative investment or as utility tokens within a specific ecosystem, stablecoins are specifically engineered to provide stability and act as a reliable medium of exchange. Tokenomics for stablecoins often involve mechanisms such as collateralization, algorithmic balancing, or backing by real-world assets, ensuring their value remains relatively stable over time.

The importance of asset pools in relation to stablecoin cannot be overstated. The asset pool represents the reserves or underlying assets that back the stablecoin's value. It serves as a safeguard, providing confidence to users that the stablecoin is fully backed and can be redeemed for its underlying assets if needed. A well-managed and transparent asset pool enhances the stability and credibility of the stablecoin, ensuring it maintains its pegged value and mitigating the risk of insolvency.

Different asset type classifications are crucial in the stablecoin ecosystem as they allow for diversification and risk management. Stablecoins can be backed by various asset types such as fiat currencies, cryptocurrencies, commodities, or other financial instruments. By diversifying the asset pool across different asset classes, stablecoins reduce their exposure to a single asset's volatility, enhancing stability and resilience. Furthermore, asset type classifications enable users to choose stablecoins that align with their risk preferences or specific use cases.

Classifying all types of crypto, including stablecoins, is essential for regulatory clarity and investor protection. Clear and consistent classifications enable governments and regulatory bodies to establish appropriate frameworks, ensuring stablecoins comply with relevant laws and regulations. Additionally, classification provides investors with a clear understanding of the nature and risks associated with stablecoins, facilitating informed decision-making, and fostering market transparency.

Investors must recognize that investing in stablecoin can serve as a hedge against market volatility, particularly when entering the market early. Stablecoins offer a relatively stable store of value compared to other cryptocurrencies, making them a potential haven during turbulent market conditions. However, it is important for investors to carefully evaluate the specific stablecoin's design, underlying assets, and the governance framework to assess its long-term viability and effectiveness as a hedge. Early adoption of stablecoin investments may present unique opportunities for capital appreciation, but thorough research and risk assessment are still crucial.

As you explore the evolution of stablecoin, keep these factors in your mind and allow yourself the leeway to understand that stablecoin is more than what the media and the government present it as. It is an evolving field that allows investors to have a hedge against inflation. Inflation is the enemy here, not the government; we just need the government to accept that stopping inflation is something that the public sector can do better than the private sector.

Chapter 1: Stability

"Therefore, everyone who hears these words of mine and puts them into practice is like a wise man who builds his house on the rock. The rain came down, the streams rose, and the winds blew and beat against that house; yet it did not fall, because it had its foundation on the rock. But everyone who hears these words of mine and does not put them into practice is like a foolish man who builds his house on sand. The rain came, the streams rose, and the winds blew and beat against that house, and it fell with a great crash."[10]

Introduction

Stability is one of the most important elements of business. When we look at the world of the cryptoverse, we see that stability is one of the limiting factors preventing mass scale adoption of cryptocurrency across the spectrum. As a result, the cryptocurrency industry is growing more slowly than it could be, should there be real stability in the system. As we move beyond the "Bitcoin Maximalism" which is limiting the expression of cryptocurrency by most governments, expanding to look at how stablecoins bring stability to the market is very important.

> BITCOIN MAXIMALISM IS THE BELIEVE THAT BITCOIN IS THE ONLY CRYPTO

This chapter takes on the issue of whether "Stablecoin 1.0" is really stable. To do this, it is imperative that we look at the foundation on which most stablecoins are built upon,

[10] Matt 7:24-27.

namely fiat currency. Because we know from historical economic data that fiat currency is not truly stable (averaging over 3% inflation for the last five years), then we know that the currency based on the fiat currency cannot be stable. Therefore, by and large, "stablecoin" is a misnomer for anything prior to stablecoin 1.5. Most people see stability as something maintain the same price regardless of the market, which is a unicorn in business reality (reserved for economic classrooms). Fiat currency is stable in its downward trend, which tend to even out at about 3% a year (16% over the last 5 years). However, the inverse of this is also possible. When we look at the basket of goods, we have seen a fairly stable growth of 2% per year for the last five years, which leads to a 10% growth rate over the last 5 years. This positive stability is much more useful for society as the individual's wealth grows with the basket of goods, rather than trailing off with the inflation rate.

One assertion is clear within the digital space, good stablecoins are needed to provide stability to the digital marketplace. Bitcoin is Digital Gold, Ethereum is Digital Silver, and the title of Digital bronze is being battled for by Solana, Avalanche and several other systems. However, these are the backbone of the system, the commodities for which the economy moves money around. In reality, tokens are the currency for the digital universe. With reasonable values, flexible stability, and high levels of use value- Tokens, especially stablecoin tokens, are the future of the cryptoverse for use in exchange. Who wants to calculate that 10,000ths of a Bitcoin needed to buy a Snickers bar, or the gas fees needed to buy a notebook for your child with ETH. Well-developed stablecoin tokens solve that problem, have a minimum burn and transition tax (generally lower than the sales tax), and allow for ease of use in a multitude of situations.

> "Paper Money eventually returns to its intrinsic value—Zero."
>
> Voltaire (1774)

The begin to understand stablecoins, one must understand stability. While many 1st layer stablecoins have done quite well making the negative stability of fiat currency seem like a good thing to build an economic system on, the development of 2nd and 3rd layer stablecoins have allowed us to see that there is a better way. Thus, we set a strong foundation on which we can build a new economic system, not on the failures of the past- but on the promise of a strong new future.

Decline of Fiat Currency

Fiat currency, a form of government-issued money not backed by a physical commodity, has long been the dominant medium of exchange in economies worldwide. However, in recent years, the decline of fiat currency has become a subject of concern, particularly due to its negative effects on stablecoins that are directly pegged to fiat currencies. This chapter section will explore the reasons behind the decline of fiat currency and analyze how it affects stablecoins. Additionally, we will delve into the inherent inflationary nature of fiat currencies and how this further contributes to their decline.

The Decline of Fiat Currency:

Fiat currency faces numerous challenges that have led to its decline in recent times. One primary issue is the loss of trust in centralized financial systems. Historically, governments and central banks have controlled the issuance and management of fiat currencies, often resulting in cases

> "It is no coincidence the century of total war coincided with the century of central banking."
>
> Ron Paul

of hyperinflation, economic instability, and financial crises. Such instances have eroded public confidence, prompting individuals and businesses to seek alternative forms of value storage and exchange.

Furthermore, fiat currencies are vulnerable to political and economic uncertainties. Government policies, geopolitical events, and economic crises can all significantly impact the value and stability of fiat currencies. For example, sudden changes in monetary policy or economic mismanagement can lead to rapid depreciation or devaluation of a currency, causing severe economic repercussions.

Impact on Stablecoins:

Stablecoins, a type of digital currency designed to maintain a stable value, often achieve stability by pegging their value directly to a fiat currency, such as the US dollar or Euro. However, the decline of fiat currency poses challenges to stablecoins and undermines their ability to maintain a stable value.

When the fiat currency to which a stablecoin is pegged experiences a decline in value, the stablecoin's purchasing power is directly affected. This correlation can result in diminished confidence in stablecoins and reduced demand, as users may question their ability to preserve value over time. Moreover, if the decline of the pegged fiat currency is significant or prolonged, stablecoins can face liquidity issues and struggle to maintain stability, potentially causing disruptions in the digital asset ecosystem.

Inflation and Fiat Currencies:

In addition to the challenges mentioned above, fiat currencies are subject to inflation, which further contributes to their decline. Inflation is the sustained increase in the general price level of goods and services in an economy over time. Governments often resort to printing more money to meet their financial obligations and stimulate economic growth, leading to a decrease in the purchasing power of the currency.

Analyzing the provided inflation rates from 2016 to April 2022, it is evident that both global and US inflation rates have been consistently positive, albeit with fluctuations. This means that, on average, prices for goods and services have increased over time. The higher the inflation rate, the greater the erosion of the currency's value. These inflation rates illustrate the persistent loss of purchasing power experienced by holders of fiat currency.

Year	Global Inflation Rate	US Inflation Rate
2016	1.68%	1.21%
2017	1.75%	2.12%
2018	2.38%	2.44%
2019	2.17%	1.81%
2020	1.50%	1.23%
2021	5.74%	7,55%
2022	8.80%	8.38%
2023	6.60%	6.4% (Jan)

The decline of fiat currency and its negative impact on stablecoins that are directly pegged to fiat currencies reveal the inherent vulnerabilities of centralized financial systems.

Loss of trust, political and economic uncertainties, and inflationary pressures all contribute to the diminishing value and stability of fiat currencies. Stablecoins, despite their potential to offer stability in a volatile market, face challenges when their pegged fiat currency experiences a decline in value.[11]

Considering these challenges, alternative financial systems and decentralized digital currencies have gained traction as potential solutions. Cryptocurrencies like Bitcoin and Ethereum, which operate on decentralized networks and are not directly tied to any specific fiat currency, offer potential benefits such as increased transparency, reduced transaction costs, and protection against inflation. However, their volatility and regulatory uncertainties remain obstacles to widespread adoption.

As the decline of fiat currency continues, it is crucial for individuals, businesses, and governments to explore and adapt to emerging financial technologies that provide stability, transparency, and protection against inflation. While the future of fiat currency remains uncertain; it is clear, the need for innovative and resilient financial systems has never been greater.

Basket of Goods and the Relationship with Value

In an era marked by inflationary pressures and fluctuating currency values, individuals and investors seek ways to preserve the purchasing power of their assets. One approach is to utilize a basket of goods as a hedge for

---

[11] U.S. Inflation Calculator (2023). At https://www.usinflationcalculator.com/inflation/current-inflation-rates/.

value. By constructing a diversified portfolio of essential goods and commodities, the goal is to maintain the utility value of the basket, barring any major market shifts. This chapter section explores the concept of using a basket of goods as a hedge against inflation, specifically examining its relationship with the US dollar. The provided relationship between the basket of goods and the US dollar will serve as a reference for analysis.

> "Now imagine one day hyperinflation strikes, and the price of your ribeye increases to one hundred dollars while your daily wage remains ten dollars. What happens to the price of your basket of goods? It cannot rise tenfold because you cannot afford the one-hundred-dollar ribeye. Instead, you make do with the chemical shitstorm that is a soy burger for ten dollars. The CPI, magically, shows zero inflation."
>
> Saifedean Ammous

The Basket of Goods and Inflation Rate:

A basket of goods represents a collection of essential items that individuals consume regularly, such as food, housing, healthcare, transportation, and education. The composition of this basket varies depending on factors such as geographical location, cultural preferences, and economic conditions. The value of the basket of goods can be used as a benchmark to measure changes in the purchasing power of a currency.

Inflation refers to the general increase in prices of goods and services over time. As the value of a currency erodes due to inflation, the cost of purchasing the same basket of goods rises. However, if the basket of goods is carefully constructed to reflect the average consumer's spending patterns, it can act as a hedge against inflation. By investing in the basket of goods, individuals aim to preserve their purchasing power and maintain their standard of living.

Relationship between the Basket of Goods and the US Dollar:

The provided relationship between the basket of goods and the US dollar indicates a consistent increase in the value of the basket over the years. This suggests that the basket of goods has outpaced inflation, providing a potential hedge against the erosion of the US dollar's purchasing power.

In 2016, the value of the basket of goods and the US dollar were at parity, with both having a value of 1.00. Over the subsequent years, as inflation took effect, the value of the basket increased at a relatively higher rate compared to the US dollar. By 2022 (April), the value of the basket had risen to 1.12, while the US dollar's value remained at 1.00. This indicates that the basket of goods has preserved its purchasing power to a greater extent than the US dollar.

Utility Value and Market Shifts:

The utility value of the basket of goods remains relatively stable, barring any major market shifts. The composition of the basket is carefully chosen to represent essential goods and services that individuals require for their daily lives. Therefore, unless there are significant disruptions in the market or unexpected changes in consumer preferences, the utility value of the basket should remain consistent over time.

However, it is important to note that market shifts, such as changes in supply and demand dynamics, technological advancements, or global events, can impact the prices and availability of goods in the basket. These factors may introduce volatility to the utility value of the basket. Therefore, periodic reviews and adjustments to the composition of the basket are necessary to ensure its continued relevance and effectiveness as a hedge against inflation.

Using a basket of goods as a hedge for value provides individuals and investors with a means to protect their assets against inflation. The relationship between the basket of goods and the US dollar, as provided, demonstrates the potential of such an approach. By investing in a diversified portfolio of essential goods and commodities, individuals can strive to maintain their purchasing power and sustain their desired standard of living.

While the utility value of the basket of goods remains relatively stable, market shifts and external factors can introduce volatility. Regular evaluations and adjustments to the basket's composition are essential to ensure its effectiveness as a hedge against inflation. As individuals navigate an inflationary landscape, the use of a basket of goods can serve as a valuable tool in preserving value and mitigating the impact of eroding purchasing power.

Good Marketing and Early Model Stablecoins

Stablecoins emerged as a promising innovation within the cryptocurrency ecosystem, offering the potential for stability and utility in a volatile market. In their early years, stablecoins effectively marketed themselves as financially stable digital assets, designed to maintain a consistent value. However, the reality of value erosion became apparent as time went on, with stablecoins losing value at the same rate as the inflation rate. This chapter section examines the impact of marketing strategies employed by stablecoin projects during their early years, and the consequences faced by holders of stablecoins formed in 2014 in terms of value loss based on the inflation rate between 2014 and 2021.

The Rise of Stablecoins and the Promise of Stability:

> Stablecoins The ground is currently being laid to set the way for a new type of currency –the stablecoin. What is the stablecoin? The stablecoin is an asset that typically features price stability. Cryptocurrency is notoriously unstable, with volatile prices that are often difficult to predict. The advantage of them is that they give the user total control over their holdings.
>
> Ikuya Takashima

Stablecoins entered the cryptocurrency space as a response to the volatility of traditional cryptocurrencies like Bitcoin and Ethereum. These digital assets aimed to offer stability by pegging their value to an external asset, such as a fiat currency like the US dollar or a basket of goods. Stablecoin projects emphasized that their tokens were backed by reserves, held in a trusted bank account or by audited assets, creating an illusion of stability and reliability.

Early Marketing Strategies and Perception of Stability:

To gain traction and attract users, stablecoin projects employed sophisticated marketing strategies. They emphasized the stability and utility of their tokens, highlighting benefits such as low transaction fees, fast settlements, and global accessibility. Moreover, they positioned stablecoins as a secure store of value, an alternative to volatile cryptocurrencies, and a convenient means of transferring funds across borders.

These marketing efforts successfully created a perception of stability and trust, appealing to individuals and businesses seeking a reliable digital asset. Investors were enticed by the notion that stablecoins offered a safe haven amidst the unpredictable cryptocurrency market, as they seemed to hold their value while traditional cryptocurrencies experienced significant price fluctuations.

Value Erosion and the Reality of Inflation:

While stablecoins projected an image of stability, the reality of value erosion became evident over time. Stablecoins that were formed in 2014, for example, would have experienced value loss based on the inflation rate between 2014 and 2021.

To calculate the value loss, we examine the US inflation rate between 2014 and 2021. According to historical data, the US inflation rates were as follows:

2014:	1.62%
2015:	0.12%
2016:	1.28%
2017:	2.13%
2018:	2.44%
2019:	1.81%
2020:	1.23%

2021:	5.74%
Total Value lost due to inflation	16.37%
Value of Basket of goods change ($0.10 to the $1)	Gain of 10% vs. a lost of 16% (26.37 difference)

Using these inflation rates, we can calculate the cumulative value loss of a stablecoin formed in 2014. Applying the inflation rates to the initial value of the stablecoin, we can determine the reduction in purchasing power over the years.

The consequences of this value loss would depend on the specific stablecoin project and its mechanisms for maintaining stability. Some stablecoins may have provisions in place to mitigate value erosion, while others may experience a direct impact.

Challenging the Perception of Stability:

The erosion of value in stablecoins challenges the perception of stability that was created through early marketing efforts. Investors who believed they were holding a stable asset may find themselves facing losses due to inflation. The realization that stablecoins are not immune to the effects of inflation underscores the importance of understanding the underlying mechanisms and risks associated with these digital assets.

> [T]he US dollar is a great example of a fiat stablecoin, as it offers low volatility and so provides a reliable unit of money to invest in both the short term and the long term. However, the US dollar doesn't give the user any form of control, as it is monitored by the Federal Reserve Bank and is dependent on the banking network in the US for commercial use.
>
> Ikuya Takashima

Moving Forward: Transparency and Education:

To ensure the long-term viability and trustworthiness of stablecoins, transparency and education are essential. Stablecoin projects should provide clear information about the mechanisms they employ to maintain stability and disclose any risks associated with holding their tokens. Educating users about the impact of inflation and the potential erosion of value can help them make informed decisions about the suitability of stablecoins for their financial goals.

The early marketing of stablecoins successfully created an illusion of stability and reliability, enticing investors seeking a safe haven within the cryptocurrency market. However, the reality of value erosion became apparent as stablecoins lost value at the same rate as the inflation rate.

The value loss experienced by stablecoins formed in 2014, for instance, demonstrates the consequences faced by holders over time.

The perception of stability projected by stablecoin marketing strategies highlights the importance of transparency and education within the cryptocurrency ecosystem. Investors and users should be aware of the underlying risks associated with stablecoins, including their susceptibility to inflation and the potential erosion of value. By fostering transparency and providing education, the stablecoin industry can work towards building trust and ensuring that users have a comprehensive understanding of the assets they hold.

Better Developed Stablecoins for a Better Developed Cryptoverse

Stablecoins, as a digital asset class, have gained significant attention for their potential to provide stability within the volatile cryptocurrency ecosystem. While traditional stablecoins pegged to a single fiat currency have limitations, a better class of stablecoins, based on a basket of goods or diverse assets, holds the potential to stabilize the system more effectively within a relatively short period. This chapter section explores the advantages of diversified asset-based stablecoins and their ability to enhance stability, mitigate risks, and facilitate rapid system stabilization.

Limitations of Single Fiat Currency Stablecoins:

Stablecoins pegged to a single fiat currency, such as the US dollar, face inherent limitations. These stablecoins are susceptible to the economic and monetary policies of the issuing country, resulting in potential value fluctuations and exposure to inflation risks. Furthermore, reliance on a single fiat currency limits the ability of stablecoins to adapt to a global economy and diversify against regional risks.

Benefits of Diversified Asset-Based Stablecoins:

1. Enhanced Stability: Diversified asset-based stablecoins offer improved stability by pegging their value to a basket of goods or diverse assets. This approach ensures that the stablecoin's value is not solely reliant on one currency or asset, reducing vulnerability to individual market fluctuations. The stability of the basket of goods or diverse assets acts as a buffer against sudden changes in any specific asset's value, allowing for more robust stability overall.
2. Risk Mitigation: By diversifying the underlying assets, diversified asset-based stablecoins reduce the concentration risk associated with single currency stablecoins. Fluctuations in the value of one asset are offset by the stability of other assets within the basket. This diversification reduces the potential impact of a single asset's performance on the stablecoin's overall value, thereby mitigating risk and enhancing resilience.
3. Flexibility and Adaptability: Diversified asset-based stablecoins are better equipped to adapt to global economic shifts and changing market dynamics. The inclusion of a variety of assets allows for a more comprehensive representation of the economy, capturing the relative value of different goods, commodities, and currencies. This flexibility enables stablecoins to respond to regional economic changes, geopolitical events, and inflationary

pressures more effectively, ensuring stable value over time.

4. Rapid System Stabilization: The diversified nature of asset-based stablecoins provides a mechanism for rapid system stabilization. During periods of market volatility or economic crises, stablecoins backed by a basket of goods or diverse assets can absorb shocks more effectively. If one asset within the basket experiences sudden fluctuations, other assets can compensate, minimizing the impact on the overall stablecoin's value. This enables the stablecoin ecosystem to quickly stabilize and restore confidence in the system.
5. Long-Term Inflation Protection: Stablecoins pegged to a basket of goods or diverse assets offer better long-term protection against inflation. As the value of the basket adjusts based on the inflation rates of the underlying assets, the stablecoin can maintain its purchasing power more effectively. By diversifying across different goods, commodities, and currencies, stablecoins can mitigate the risk of being tied to a single inflationary economy and provide a reliable store of value over the long term.

Diversified asset-based stablecoins have the potential to address the limitations of single fiat currency stablecoins and provide a more stable and adaptable system. By pegging their value to a basket of goods or diverse assets, these stablecoins enhance stability, mitigate risks, and facilitate rapid system stabilization. The diversified nature of these stablecoins enables them to absorb shocks, adapt to changing market dynamics, and provide long-term protection against inflation.

As the cryptocurrency ecosystem continues to evolve, the development and adoption of diversified asset-based stablecoins can contribute to a more robust and sustainable digital economy. Their ability to stabilize the system within a relatively short period can enhance market confidence, foster wider adoption, and facilitate the integration of cryptocurrencies into mainstream financial systems.

Dangers of Poor Regulation Diminishing the Value of the Global Economy

The emergence of stablecoins and their potential to revolutionize the financial landscape has gained significant attention. However, the absence of adequate regulation and a limited understanding of the relationship between stablecoins, fiat currency, and inflation pose risks that could cripple the global market and hinder the development of more robust asset-backed stablecoins. This chapter section explores the implications of poor regulation and limited understanding in terms of systemic stability, investor protection, and the future growth of diversified asset-backed stablecoins.

Systemic Stability at Risk:

The absence of comprehensive regulation for stablecoins leaves the global market vulnerable to instability. Poorly regulated stablecoins can amplify systemic risks and create financial contagion, similar to the 2008 financial crisis. Without proper oversight, stablecoins may lack the mechanisms to absorb shocks or maintain a stable value, leading to sudden value collapses or unpredictable market behavior. A lack of systemic stability could have far-reaching consequences, undermining market confidence and impeding economic growth.

> When people tell you the stock market is more stable than crypto over the long term, remind them the stock market collapsed 777.68 points in one day in 2008. It also lost $14 billion in one day in 1929.

Investor Protection Challenges:

Inadequate regulation exposes investors to various risks associated with stablecoins. Investors may not fully comprehend the intricacies of stablecoin mechanics, including the relationship between stablecoins and inflation. This lack of understanding can lead to misplaced trust and investments in unstable or poorly managed stablecoin projects. Additionally, the absence of clear regulatory frameworks may result in inadequate investor protection measures, leaving investors exposed to fraud, market manipulation, or financial losses.

Inflation Misalignment and Economic Disruptions:

A fundamental understanding of the relationship between stablecoins and inflation is crucial for their successful integration into the global market. Stablecoins, especially those pegged to fiat currencies, rely on stable value preservation over time. However, if the underlying fiat currency experiences significant inflation, the value of stablecoins can become misaligned, jeopardizing their intended stability. Insufficient understanding of inflation dynamics and its potential impact on stablecoins may lead to economic disruptions, price distortions, and reduced confidence in stablecoin ecosystems.

Impediments to Robust Asset-Backed Stablecoin Development:

Poor regulation and limited understanding can hinder the development of robust asset-backed stablecoins.[12] Diverse asset-backed stablecoins hold the potential to provide stability, mitigate risks, and offer long-term value preservation. However, without proper regulatory frameworks and a comprehensive understanding of their underlying assets, these stablecoins may face barriers to entry, limited market adoption, and challenges in establishing trust among users and investors. Insufficient understanding of asset valuation methodologies, liquidity management, and risk mitigation strategies can impede the growth of diversified stablecoins as reliable alternatives to traditional fiat currencies.

Harmonized Regulatory Frameworks:

To address the risks associated with poor regulation and limited understanding, harmonized regulatory frameworks

---

[12] Meichler, Mattis (2023). "Coinbase Requests for Clarity on Crypto Rules 'Should be Denied', says SEC." Decrypt at https://decrypt.co/140298/coinbase-request-clarity-crypto-rules-should-be-denied-says-sec. This shows that the SEC is more worried about illegal civil cases against citizens than regulatory clarity.

need to be established. These frameworks should encompass areas such as capital requirements, transparency, investor protection, risk management, and governance standards. Collaborative efforts between governments, financial authorities, and industry participants are essential to ensure effective oversight, foster market stability, and encourage innovation in diversified asset-backed stablecoins.

Education and Awareness:

Promoting education and awareness among market participants, policymakers, and the public is vital in mitigating the risks associated with stablecoins and inflation. Enhancing financial literacy about stablecoin mechanisms, inflation dynamics, and the potential implications for value preservation can empower individuals and investors to make informed decisions. Additionally, regulatory bodies should actively engage in knowledge-sharing initiatives, disseminate guidance, and promote transparency to ensure a more informed and resilient market environment.

> "The government wants to fool its people by doubling the minimum wage but at the same time printing more than enough money into circulation. We can't solve problems by creating more problem for generations to come which is why Bitcoin and some other decentralized finance applications powered by blockchain technology looks to be our only hope of creating a better future."
>
> Olawale Daniel

Poor regulation and limited understanding regarding stablecoins, fiat currency, and inflation pose significant risks to the global market and the development of robust asset-backed stablecoins. To mitigate these risks, it is crucial to establish comprehensive regulatory frameworks, enhance investor protection measures, and foster a deeper understanding of stablecoin mechanics and inflation dynamics. By doing so, we can promote market stability, encourage innovation, and pave the way for the growth of diversified asset-backed stablecoins that provide long-term value preservation and contribute to a more resilient global financial ecosystem.

Conclusion

The decline of fiat currency, characterized by a loss of trust, political and economic uncertainties, and inflationary pressures, has raised concerns about its impact on stablecoins. Stablecoins, which are pegged to fiat currencies, face challenges when their underlying currencies experience a decline in value. This can result in diminished confidence, reduced demand, liquidity issues, and disruptions in the digital asset ecosystem.

To address the limitations of fiat currency and enhance stability in the cryptocurrency ecosystem, alternative financial systems and decentralized digital currencies have gained traction. Cryptocurrencies like Bitcoin and Ethereum, which operate on decentralized networks and are not directly tied to any specific fiat currency, offer potential benefits such as increased transparency, reduced transaction costs, and protection against inflation. However, their volatility and regulatory uncertainties remain obstacles to widespread adoption.

In an era marked by inflationary pressures, individuals and investors seek ways to preserve the purchasing power of their assets. Utilizing a basket of goods as a hedge against inflation can help maintain the utility value of the basket and mitigate the erosion of a currency's purchasing power. The relationship between the basket of goods and the US dollar indicates that the basket has outpaced inflation, providing a potential hedge against the erosion of the US dollar's value.

Early marketing strategies employed by stablecoin projects created an illusion of stability and reliability, attracting investors seeking a safe haven within the cryptocurrency market. However, the reality of value erosion became apparent as stablecoins lost value at the same rate as the inflation rate. This emphasizes the importance of transparency and education within the stablecoin industry to ensure users understand the risks and mechanisms for maintaining stability.

To enhance stability and mitigate risks in the cryptocurrency ecosystem, diversified asset-based stablecoins offer advantages over single fiat currency stablecoins. By pegging their value to a basket of goods or diverse assets, these stablecoins enhance stability, mitigate concentration risks, adapt to global economic shifts, and facilitate rapid system stabilization during market volatility or economic crises.

In conclusion, the decline of fiat currency and its negative effects on stablecoins highlight the vulnerabilities of centralized financial systems. As individuals, businesses, and governments navigate this changing landscape, exploring and adapting to emerging financial technologies that provide stability, transparency, and protection against inflation is crucial. The use of diversified asset-based stablecoins and the adoption of innovative and resilient financial systems can contribute to a better-developed and more stable cryptoverse.

Chapter 2: Tokenomics of Stablecoins

"Three eras of currencies: commodity based, government based, and math based." [13]

And

"If crypto succeeds, it's not because it empowers better people. It's because it empowers better institutions."[14]

Introduction

While tokenomics is always an important factor when looking at a cryptocurrency, in the case of stablecoin the analysis of tokenomics must go significantly deeper. Because stablecoins rely on not only their own stability, but the stability of all entities within their ecosystem, it is imperative that there is a stable structure underlying the mechanism of the stablecoin. As such, one must look to the nations and systems on which the stablecoin is built to see if underlying "fault lines" may cause the stablecoin to fracture when pressures outside of the cryptoverse start to form.

Location is one of the most important elements of where a stablecoin is formed. Because the politics and international relations of a given country have such a strong relationship with stablecoins, one should always look at where the coin is incorporated. Coins incorporated in the United States, the UK, the EU, or Japan have a higher stability rating than coins that are formed in unstable countries or countries with unknown regulatory regime. Likewise, it is important

---

[13] Chris Dixon (2019) at https://www.cryptoaltruism.org/blog/15-quotes-about-the-potential-of-blockchain-and-crypto.
[14] Vitalik Buterin at https://www.cryptoaltruism.org/blog/15-quotes-about-the-potential-of-blockchain-and-crypto.

to look at where the company managing the stablecoin is formed, sometimes it occurs that the coin and the company are formed in different nations, which on its own can have its own problems.

Digital location is also very important for stablecoins. The chain on which a coin is formed can be as important and the nation in which the coin is formed. Stable, fast, and low-cost chains are much more effective than problematic, slow, and high gas fee chains. If you need the equivalent of 12% of your transaction in gas fees to move your cryptocurrency, the stablecoin is not stable! Therefore, it is imperative to ensure that the chain, and the country in which the chain is formed, is very stable so that the underpinnings of a stablecoin are strong.

Beyond location, it is important to look at the corporate nature of the company holding the stablecoin. As we note in Chapter 6, one of the most common problems we see with stablecoins in the market today is a coin "wagging the dog" of the company. In these cases, the company is formed as an afterthought to the token. When this happens, the corporate structure of the company tends to be less function in relation to the needs of the token, rather than a snug fit which allows the two entities to work together yet independently of each other. Therefore, it becomes very important that the structure of the company takes into account proper disbursement, proper reserves and proper anti-inflationary measures.

All of these measures are important in addition to the normative tokenomics items.[15] Because the nature of

---

[15] For more details on tokenomics, please review Smithmyer et al (2022). "Dragons of the Digital Age" Elite Exclusivity: Detroit OR

stablecoins is more complex, the added levels of protection allow for a consumer and institution to have more faith in the operations of the token. Building a better future means that there needs to be a set of "best practices" for new tokens. The sooner we build these in the developed world, the sooner we can weed out 'rub pulls' which hurt millions of people.

Which Country is the Stablecoin Formed in?

In the fast-evolving world of cryptocurrency, stablecoins have emerged as a crucial tool for providing stability and facilitating transactions within the digital ecosystem. While the technology behind stablecoins is decentralized, the nation in which a stablecoin is formed plays a pivotal role in its success. This essay explores the importance of the nation's stability, favorable cryptocurrency laws, robust power grid, and strong property rights protection for the development and sustainability of a stablecoin.

> "The root problem with conventional currency is all the trust that's required to make it work. The central bank must be trusted not to debase the currency, but the history of fiat currencies is full of breaches of that trust. Banks must be trusted to hold our money and transfer it electronically, but they lend it out in waves of credit bubbles with barely a fraction in reserve. We have to trust them with our privacy, trust them not to let identity thieves drain our accounts.
>
> Satoshi Nakamoto

Smithmyer, Christopher (2023). "Digital Exchequer" Elite Exclusivity: Detroit.

Stability: A Foundation for Trust and Confidence

A stablecoin relies on the trust and confidence of its users, and the stability of the nation in which it is formed is fundamental to foster this trust. Political stability, economic strength, and social harmony contribute to an environment where stablecoins can thrive. Stability provides assurance to market participants, investors, and users that the value of the stablecoin will be preserved over time, reinforcing its utility as a medium of exchange, unit of account, and store of value.

Favorable Cryptocurrency Laws: Nurturing Innovation and Regulation

A nation's legal framework and regulatory environment greatly impact the development and adoption of stablecoins. Progressive and comprehensive cryptocurrency laws are essential to foster innovation, attract investment, and promote responsible use of stablecoins. Clear regulations that provide legal certainty and protect investors inspire confidence and encourage the growth of stablecoin projects. Additionally, a robust legal framework ensures that illicit activities, such as money laundering and terrorist financing, are effectively addressed, safeguarding the reputation of the stablecoin and its issuing entity.

Robust Power Grid: Ensuring Uninterrupted Operations

The reliable supply of electricity is critical for the seamless operation of digital infrastructure, including stablecoin networks. A stable power grid, capable of meeting the energy demands of stablecoin transactions and associated blockchain processes, is vital. Interruptions or fluctuations in power supply can disrupt the stablecoin's network, leading to transaction delays, reduced user confidence, and potential financial losses. A nation with a well-developed power grid and advanced energy infrastructure minimizes the risk of such disruptions, ensuring the stability and efficiency of the stablecoin ecosystem.

Strong Property Rights Protection: Upholding Ownership and Security

Respecting property rights and providing a robust legal framework for ownership protection is essential for stablecoin users and investors. A nation that safeguards property rights through transparent and fair judicial processes builds trust in the stablecoin ecosystem. Protection against arbitrary property seizures without proper judicial action is crucial, as it ensures that stablecoin users' assets are secure, fostering long-term confidence and participation in the stablecoin market. Investors and users seek stablecoin projects domiciled in nations with a proven track record of upholding property rights and providing effective legal remedies in case of disputes.

The nation in which a stablecoin is formed holds significant importance in shaping its success and sustainability. Stability, supported by a stable political and economic environment, builds trust and confidence among users. Favorable cryptocurrency laws provide regulatory clarity, fostering innovation while curbing illicit activities. A robust power grid ensures uninterrupted operations, maintaining the efficiency of stablecoin networks. Strong property rights protection upholds ownership and security, enhancing trust and enabling long-term participation. By recognizing the critical role of the nation, stakeholders can foster an environment conducive to the growth and resilience of stablecoins, further advancing the digital economy.

What Country is the Company that is Holding the Stablecoin Formed in (if separate from Stablecoin)?

> "We have elected to put our money and faith in a mathematical framework that is free of politics and human error."
>
> Tyler Winkelvoss

The nation in which a company holding a stablecoin is formed plays a critical role in shaping the success and credibility of the stablecoin project. Beyond the stability and laws of the nation, factors such as international relations, tax policy, transparency, cryptocurrency regulations, and the fairness of the judicial system hold substantial importance. This essay explores the significance of these elements in relation to the domicile of the stablecoin company, emphasizing the need for good relations, favorable tax policies, transparency, robust cryptocurrency protections, and a fair judicial system.

Good Relations with Mainstream Nations: Strengthening Trust and Collaboration

Establishing and maintaining good relations with mainstream nations is essential for a stablecoin company. Positive diplomatic ties and cooperation enhance the company's credibility and foster trust among global stakeholders. Collaboration with other nations can facilitate regulatory harmonization, international partnerships, and the adoption of stablecoins on a global scale. By operating within a country that maintains strong relationships with mainstream nations, the stablecoin company can navigate regulatory challenges, promote cross-border transactions, and enhance its reputation as a reliable and trustworthy entity.

Favorable Tax Policy and Transparent Records: Ensuring Compliance and Trust

> "You Don't Pay Taxes, they take taxes."
>
> Chris Rock

A stablecoin company's domicile should have a favorable tax policy that promotes a fair and sustainable business environment. A transparent and predictable tax regime encourages compliance and fosters trust among investors and users. Clear tax regulations enable the stablecoin company to plan its financial activities effectively, while transparent records and reporting demonstrate accountability and help prevent illicit financial practices. By operating within a nation with sound tax policies and a commitment to transparency, the stablecoin company can establish a solid reputation, attract investment, and build long-term partnerships.

Strong Cryptocurrency Protections: Safeguarding Investors and Users

The level of cryptocurrency protection provided by a stablecoin company's domicile is paramount for the security and confidence of investors and users. A nation that has established comprehensive and robust regulations for cryptocurrencies demonstrates a commitment to consumer protection, security, and fraud prevention. It should have measures in place to combat money laundering, terrorist financing, and other illicit activities associated with cryptocurrencies. By choosing a jurisdiction with strong cryptocurrency protections, the stablecoin company can provide a safe and secure platform, attracting a wider user base and fostering market growth.

Fair Judicial System: Ensuring Justice and Equality

The fairness and impartiality of the judicial system in the stablecoin company's domicile are crucial considerations. A transparent and equitable legal framework, providing equal protection and recourse to all individuals, irrespective of nationality or citizenship, reinforces trust in the stablecoin project. The judicial system should ensure that property rights are respected, contracts are enforceable, and disputes are resolved in a fair and efficient manner. A jurisdiction with a fair judicial system promotes stability, reduces legal risks, and ensures that the stablecoin company can operate with confidence, attracting investors and users from diverse backgrounds.

The nation in which a stablecoin company is domiciled holds significant importance in shaping the success and reputation of the stablecoin project. Good relations with mainstream nations foster trust and collaboration, while a favorable tax policy and transparent records ensure compliance and accountability. Strong cryptocurrency protections safeguard investors and users, while a fair judicial system provides justice and equality for all. By carefully considering these factors and selecting a domicile that encompasses these attributes, stablecoin companies can establish themselves as reputable, secure, and globally recognized entities, contributing to the growth and acceptance of stablecoins in the digital economy.

What Chain is the Stablecoin Formed on?

The choice of the underlying blockchain infrastructure for a stablecoin is crucial in determining its efficiency, scalability, and compatibility with the broader ecosystem. This essay delves into the importance of selecting the right chain for a stablecoin, differentiating between a token with its own standalone chain and a token working on an established chain. It examines the need for a high level of transactions per second (TPS) rate, low gas fees, compatibility with exchanges, and the ability to build child chains or plasmas for value maintenance and storing assets in NFT form.

Standalone Chain vs. Established Chain: Considerations for Stability and Ecosystem Integration

Choosing between a standalone chain or an established chain requires careful evaluation of stability, security, and ecosystem integration. A standalone chain provides complete control and flexibility over the stablecoin's infrastructure, allowing for tailored features and governance mechanisms. However, it requires significant development efforts and may face challenges in achieving widespread adoption and network effects. On the other hand, utilizing an established chain, such as Ethereum or Binance Smart Chain, offers the advantages of an existing ecosystem, robust infrastructure, and a ready user base. It enables seamless integration with other decentralized applications (dApps), wallets, and exchanges, enhancing interoperability and usability.

High TPS Rate: Enabling Scalability and Mass Adoption

A high TPS rate is essential for a stablecoin chain to handle a large volume of transactions efficiently, ensuring scalability and enabling mass adoption. As the demand for stablecoins increases, particularly in decentralized finance (DeFi) applications and cross-border transactions, the blockchain infrastructure must be capable of processing transactions quickly and seamlessly. A high TPS rate ensures that the stablecoin can handle increased transaction volumes, minimizing network congestion, and maintaining a smooth user experience. This scalability is crucial for the stablecoin's utility and its ability to compete with traditional financial systems.

Low Gas Fees: Enhancing Affordability and Accessibility

The cost of gas fees associated with transactions on the blockchain can significantly impact the usability and attractiveness of a stablecoin. High gas fees limit the affordability and accessibility of transactions, particularly for small-value transfers or frequent microtransactions. Users may be deterred from using the stablecoin if the fees outweigh the benefits. Therefore, it is crucial to select a chain that offers low gas fees, allowing for cost-effective and efficient transactions. Low gas fees make stablecoins more accessible to a broader user base, including individuals and businesses, promoting their adoption and usage across various sectors.

Exchange Compatibility: Facilitating Liquidity and Market Integration

The ability of a stablecoin to work seamlessly with exchanges is critical for its liquidity, market integration, and wider acceptance. Compatibility with major centralized and decentralized exchanges allows users to easily trade, purchase, and redeem the stablecoin, enhancing its liquidity and utility. Additionally, integration with popular exchanges helps stabilize the stablecoin's value by providing price discovery and ensuring arbitrage opportunities. By ensuring compatibility with exchanges, the stablecoin gains exposure to a broader market, expanding its use cases and attracting more participants.

Enhanced Functionality: Child Chains, Plasmas, and NFTs for Value Maintenance

To maintain value and provide additional functionalities, a stablecoin chain should have the ability to implement child chains, plasmas, and support non-fungible tokens (NFTs) for storing assets such as corporate papers or other holdings. Child chains allow for the creation of sidechains that inherit the security of the main chain while providing scalability and faster transaction processing. This ensures that the stablecoin can accommodate increased usage without sacrificing security or efficiency.

Similarly, plasmas, as layer-two solutions, can be implemented to improve the scalability and speed of the stablecoin chain, mitigating congestion and reducing transaction costs. These enhancements contribute to the stability and usability of the stablecoin, making it more attractive for a wide range of applications.

Additionally, the ability to tokenize assets such as corporate papers or other holdings as NFTs on the stablecoin chain offers numerous advantages. NFTs provide unique identification, immutability, and transparency, enabling secure storage and transfer of digital assets. By leveraging NFTs, stablecoin chains can offer enhanced security, traceability, and accessibility for critical documents and assets, providing an innovative solution for value storage and management.

The choice of the chain for a stablecoin plays a pivotal role in its functionality, scalability, and integration with the broader ecosystem. Considerations such as opting for a standalone chain or an established chain, ensuring a high TPS rate, low gas fees, compatibility with exchanges, and implementing enhanced functionalities like child chains, plasmas, and NFTs are crucial for the success and adoption of a stablecoin. By carefully assessing these factors, stablecoin projects can select a chain that meets their specific requirements, providing a robust and efficient infrastructure to support their goals in the rapidly evolving digital economy.

What Country is the Native Chain of the Stablecoin Formed In?

The nation in which the native chain for a stablecoin is established holds significant importance in determining the efficiency and effectiveness of the stablecoin ecosystem. This essay delves into the crucial role of a stable nation in supporting a stablecoin's native chain, considering factors such as international relations, infrastructure, the legal system, business interoperability, and regulations. A stable nation provides a conducive environment for the development and adoption of stablecoins, ensuring trust, regulatory clarity, and global interoperability.

International Relations: Building Trust and Collaboration

A stable nation's positive international relations are crucial for the success of a stablecoin operating on its native chain. Cooperative relationships with other nations foster trust, facilitate cross-border transactions, and encourage global acceptance of the stablecoin. Collaboration between stablecoin issuers, regulatory bodies, and financial institutions across countries enables seamless integration of stablecoins into the global financial system. Strong international relations also help in navigating potential challenges, such as regulatory compliance, cross-border remittances, and regulatory harmonization, ensuring a stablecoin's viability on a global scale.

Infrastructure: Robust Technology and Power Grid

A stablecoin's native chain requires a stable nation with robust technological infrastructure and a reliable power grid. A stable and well-developed technological infrastructure is essential for maintaining the stability, security, and scalability of the native chain. It enables efficient transaction processing, high throughput, and low latency, ensuring a seamless user experience. Additionally, a reliable power grid ensures uninterrupted operation and availability of the stablecoin's network, minimizing downtime and transaction disruptions. A stable nation's investment in technology and infrastructure reinforces the trust and confidence of users, businesses, and investors in the stablecoin ecosystem.

Legal System: Clarity, Protection, and Enforcement

A stable nation with a well-defined legal system provides the necessary clarity, protection, and enforcement mechanisms for a stablecoin operating on its native chain. Clear legal frameworks and regulations surrounding cryptocurrencies and stablecoins foster trust, attract investment, and provide a predictable business environment. Legal protections for investors and users safeguard their interests, ensuring transparent and fair transactions. Moreover, a robust legal system facilitates dispute resolution and provides recourse in case of fraud, hacking, or other illicit activities. By establishing a stablecoin within a nation with a reliable legal system, stakeholders can have confidence in the stability, security, and enforceability of their transactions, enhancing the stablecoin's credibility and adoption.

Business Interoperability: Facilitating Integration and Adoption

A stable nation with a supportive business environment and interoperability facilitates seamless integration of stablecoins within existing financial systems. Collaboration between stablecoin issuers, banks, financial institutions, and businesses promotes interoperability, allowing for efficient conversion, settlement, and transfer of stablecoins across different payment networks. A stablecoin operating on a native chain within a stable nation can establish partnerships and interoperability protocols with existing financial infrastructure, including payment processors, remittance services, and merchant networks. This integration enables businesses and individuals to transact with stablecoins easily, enhancing the overall utility and adoption of the stablecoin.

Regulations: Ensuring Compliance and Mitigating Risks

Clear and well-defined regulations tailored to the stablecoin ecosystem are essential for maintaining stability and mitigating risks. A stable nation with effective regulatory frameworks provides guidance on issues such as anti-money laundering (AML), know-your-customer (KYC) requirements, and consumer protection. These regulations help ensure compliance, reduce fraud, and mitigate risks associated with stablecoin operations. Additionally, regulatory oversight enhances the stability and transparency of stablecoin issuance, ensuring that stablecoins are adequately backed and maintaining price stability. A stablecoin operating on a native chain within a nation with favorable regulations can attract both domestic and international participants, fostering a secure and compliant stablecoin ecosystem.

The nation in which the native chain for a stablecoin is established plays a critical role in its efficiency and effectiveness. A stable nation with positive international relations, robust infrastructure, a reliable legal system, supportive business interoperability, and clear regulations provides a conducive environment for the development and adoption of stablecoins. These factors instill trust, enable seamless integration, protect stakeholders' interests, and ensure compliance, contributing to the stability, usability, and global acceptance of stablecoins. As the stablecoin ecosystem continues to evolve, the choice of a stable nation becomes increasingly crucial for building a sustainable and successful stablecoin project.

What is the Corporate Structure of the Company Holding the Stablecoin?

> "Every company has two organizational structures; the formal one is written on the charts; the other is the everyday relationship of the men and women in the organization."
>
> Harold Green

The corporate structure of the company backing a stablecoin plays a crucial role in ensuring the stability, functionality, and long-term viability of the stablecoin ecosystem. This essay explores the significance of the corporate structure, focusing on the importance of the company's prior formation before launching the token on-chain. Additionally, we discuss how a company with a full ecosystem, comprising multiple subsidiaries, can effectively align with the decentralized nature of the cryptoverse, fostering innovation, scalability, and adaptability.

Prior Formation: Aligning the Token with Company Goals

The formation of the company backing a stablecoin before its token goes on-chain is vital for aligning the token's features, functionality, and purpose with the strategic goals of the company. Establishing the company first allows for careful planning, strategy development, and coordination between the token's design and the company's overall vision. By integrating the stablecoin into the company's existing infrastructure, it becomes easier to ensure that the stablecoin meets the specific needs of the company, its stakeholders, and the target market.

I. Strategic Alignment: Prior formation enables the company to define its objectives, business model, and target audience, ensuring the stablecoin's design and features are in line with the company's overall strategy.

II. Regulatory Compliance: By forming the company prior to launching the token, there is an opportunity to navigate the complex regulatory landscape, obtain necessary licenses, and comply with relevant regulations, reducing legal risks and enhancing the stablecoin's credibility.

III. Resource Allocation: Establishing the company beforehand allows for proper resource allocation, such as talent acquisition, technology infrastructure, and operational support, to facilitate the successful deployment and management of the stablecoin.

II. Full Ecosystem: Leveraging Decentralization

A company with a full ecosystem, comprising multiple subsidiaries, can effectively embrace the decentralized nature of the cryptoverse, unlocking numerous benefits in terms of innovation, scalability, and adaptability.

I. Innovation: A diversified ecosystem promotes innovation by fostering collaboration between different subsidiaries, each specializing in various aspects of the stablecoin ecosystem. This allows for the exploration of new technologies, features, and use cases, driving continuous improvement and ensuring the stablecoin remains competitive in a rapidly evolving landscape.

II. Scalability: A decentralized corporate structure enables the scaling of operations, as different subsidiaries can focus on specific areas, such as custody, liquidity provision, auditing, or regulatory compliance. Each subsidiary can develop expertise in their respective domain, enabling efficient scaling without compromising the core functionalities of the stablecoin.

III. Adaptability: The decentralized nature of a full ecosystem facilitates adaptability to changing market conditions, regulatory requirements, and user preferences. Subsidiaries can autonomously respond to specific challenges, ensuring agility and quick decision-making, which is vital in the fast-paced crypto industry.

Inter-subsidiary Synergy: Maximizing Efficiency

The corporate structure of a company backing a stablecoin with multiple subsidiaries allows for synergies and collaboration among different entities within the ecosystem, enhancing overall efficiency and effectiveness.

I. Specialization: Each subsidiary can specialize in a specific aspect of the stablecoin ecosystem, such as token issuance, wallet development, compliance services, or exchange operations. This specialization allows for streamlined processes, optimized workflows, and improved quality control.

II. Risk Mitigation: By dividing responsibilities among subsidiaries, risk can be effectively mitigated. For example, custody services can be separated from token issuance or exchange operations, reducing the risk of a single point of failure, and enhancing security.

III. Global Reach: With subsidiaries located in different jurisdictions, a company can tap into various markets, comply with local regulations, and cater to a diverse range of users. This global presence improves the stability and accessibility of the stablecoin, attracting a broader user base and facilitating international adoption.

The corporate structure of the company backing a stablecoin is of utmost importance in ensuring the stability, functionality, and long-term success of the stablecoin ecosystem. Prior formation allows for strategic alignment, regulatory compliance, and resource allocation, ensuring the stablecoin meets the company's goals and regulatory requirements. Moreover, a full ecosystem with multiple subsidiaries provides numerous benefits, including innovation, scalability, and adaptability, while inter-subsidiary synergy maximizes efficiency and risk mitigation. By carefully designing and integrating a robust corporate structure, stablecoin projects can foster trust, drive innovation, and thrive in the decentralized and ever-evolving cryptoverse.

What is the Disbursement Structure of the Stablecoin?

The disbursement structure of a stablecoin plays a critical role in ensuring its stability, resilience, and long-term viability. This essay highlights the importance of carefully considering the allocation of funds within a stablecoin ecosystem. Specifically, we will explore the significance of distributing no more than 17% to the development team, maintaining at least 30% in fractional reserves, allocating a portion of the token for company development, and implementing liquidity thresholds tied to the value of the coin. Furthermore, we will provide an analogy highlighting the comparative reserve grades of stablecoins and traditional banking systems in the United States.

Balanced Allocation: Ensuring Long-Term Stability

The disbursement structure of a stablecoin should prioritize balanced allocations to different stakeholders, ensuring long-term stability and fostering trust within the ecosystem.

I. Development Team: Limiting the allocation to the development team to a maximum of 17% prevents concentration of power and ensures that incentives remain aligned with the stablecoin's success. This safeguards against potential conflicts of interest and encourages sustainable development.

II. Fractional Reserves: Maintaining at least 30% in fractional reserves strengthens the stability of the stablecoin by backing its value with assets of equal or greater worth. Adequate reserves provide a buffer against market volatility, ensuring the stablecoin remains pegged to its intended value.

III. Company Development: Allocating a portion of the stablecoin to company development enables ongoing innovation, ecosystem growth, and infrastructure development. This allocation ensures that the company can adapt to changing market conditions, improve user experience, and sustain long-term viability.

Liquidity Thresholds: Ensuring Adequate Reserves

Implementing liquidity thresholds tied to the value of the stablecoin is crucial for maintaining its peg and safeguarding against potential risks or instability.

I. Reserves and Stability: A well-defined liquidity threshold framework ensures that the stablecoin maintains adequate reserves relative to its value. The grading system, such as the one mentioned (S grade 100-110%, A grade 90-100% or 110%+, B grade 60-90%, C grade 30-60%, D grade 10%-30%, F grade less than 10%), provides transparency and allows users to assess the stability of the stablecoin.

II. Market Confidence: By adhering to defined liquidity thresholds, stablecoin issuers demonstrate their commitment to maintaining stability and bolstering market confidence. Users can have greater trust in stablecoins with higher reserve grades, knowing that they are well-protected against potential value fluctuations.

III. Comparative Analysis: It is interesting to note that under the mentioned terms, banks in the

> United States would receive an F grade for reserves. This highlights the importance of rigorous reserve requirements for stablecoins, as they aim to provide a reliable and secure alternative to traditional financial systems.

The disbursement structure of a stablecoin is essential for its stability, functionality, and long-term success. Balanced allocation to the development team, fractional reserves, and company development ensures sustainable growth and innovation. Moreover, implementing liquidity thresholds tied to the value of the coin safeguards against potential risks, maintains market confidence, and provides transparency. By carefully considering and implementing these disbursement strategies, stablecoin projects can foster trust, attract users, and establish themselves as reliable and robust financial instruments in the decentralized landscape.

What is the Reserve Structure of the Stablecoin?

Proper reserves are a critical component of stablecoins, ensuring liquidity and ecosystem maintenance. In this essay, we will emphasize the significance of robust and transparent reserve mechanisms for stablecoins. Specifically, we will discuss the importance of fractional reserves, explore the potential risks associated with the lack of proper reserves, and highlight the benefits of implementing strong reserve strategies.

> Unite Liberally with just frugality, always reserve something for the hand of charity; and never let your door be closed to the voice of suffering humanity.
>
> Patrick Henry

Fractional Reserves: Ensuring Stability and Trust

Fractional reserves refer to holding a portion of the stablecoin's value in reserve assets, backing its value and ensuring stability. This practice is fundamental to the proper functioning of stablecoins and fosters trust within the ecosystem.

- Value Stability: Fractional reserves act as a safeguard against price volatility, ensuring that stablecoins maintain a consistent value and remain pegged to the underlying asset or currency they represent. Reserves provide a buffer against market fluctuations, allowing stablecoins to fulfill their intended purpose of facilitating secure and predictable transactions.
- User Confidence: Robust reserves inspire confidence among users and investors, as they can trust that the stablecoin issuer has the means to honor redemptions and maintain liquidity. Transparency in reserve management further enhances user confidence by allowing stakeholders to verify that the reserves are adequately maintained.
- Risk Mitigation: Fractional reserves help mitigate counterparty and systemic risks. By diversifying reserves across different assets, stablecoin issuers can reduce the impact of a single asset's failure or market disruption. This approach enhances the stability of the stablecoin ecosystem and protects against potential losses.

Risks of Insufficient Reserves: Lessons from the Past

The lack of proper reserves has proven to be a significant failing in the cryptocurrency space. Understanding the risks associated with inadequate reserves is crucial for building a resilient stablecoin ecosystem.

- Liquidity Crisis: Insufficient reserves can lead to liquidity crises, where stablecoins struggle to honor redemptions or maintain their pegged value. This can result in a loss of trust, a mass exodus of users, and a collapse of the stablecoin's ecosystem.
- Market Manipulation: Without proper reserves, stablecoins become susceptible to market manipulation and speculative attacks. Large-scale redemptions or coordinated trading activities can exploit the lack of reserves and destabilize the stablecoin's value.
- Regulatory Scrutiny: Insufficient reserves may attract regulatory scrutiny and raise concerns about potential fraudulent or deceptive practices. Regulators are increasingly focusing on stablecoins to ensure that they meet the necessary financial standards and protect consumer interests.

Benefits of Strong Reserve Strategies: Stability and Ecosystem Growth

Implementing robust reserve strategies brings numerous benefits to stablecoins, contributing to their stability, growth, and long-term success.

- Enhanced Stability: Strong reserves provide a solid foundation for stablecoins, allowing them to weather market fluctuations and maintain their peg. This stability attracts users, businesses, and investors, fostering growth and wider adoption.
- Increased Liquidity: Proper reserves ensure the availability of liquidity within the stablecoin

ecosystem, facilitating seamless transactions and enabling users to convert stablecoins into other assets when needed. This liquidity is crucial for the stablecoin's utility and usability in various financial applications.

- Ecosystem Expansion: With adequate reserves, stablecoin issuers can allocate resources toward ecosystem development, such as building partnerships, integrating with exchanges, and funding innovation. A robust ecosystem attracts developers, entrepreneurs, and investors, driving further growth and diversification of use cases.

Proper reserves are integral to the stability, functionality, and long-term success of stablecoins. Fractional reserves ensure value stability, inspire user confidence, and mitigate risks. The lack of adequate reserves can lead to liquidity crises, market manipulation, and regulatory scrutiny. Conversely, implementing strong reserve strategies enhances stability, provides liquidity, and enables ecosystem growth. By prioritizing robust and transparent reserve mechanisms, stablecoin projects can build trust, attract users, and establish themselves as reliable and resilient financial instruments in the evolving landscape of cryptocurrencies.

What is the Anti-Inflationary Mechanism of the Stablecoin?

A smart burn system, where a portion of tokens is burned on each transaction and reintroduced to the system through the purchase of new assets, plays a crucial role in maintaining the value and supply management of stablecoins. In this essay, we will highlight the significance of a smart burn system and how it enables continuous value development while limiting the need for excessive token issuance from the reserve. Additionally, we will discuss the allocation of a small portion of the burn for the maintenance of the corporate structure.

Sustaining Value through Token Burn

A smart burn system involves temporarily systematically removing a portion of tokens from circulation on each transaction. This process has several important benefits for stablecoins, particularly in terms of sustaining value and managing supply.

Token burn mechanisms help manage the overall supply of stablecoins. As tokens are burned, the circulating supply decreases, creating scarcity and potentially increasing the value of the remaining tokens. This controlled supply mechanism contributes to price stability and prevents inflationary pressures.

With a reduction in supply due to the burn, the relative scarcity of the remaining tokens may lead to increased demand and subsequently drive up their value. This mechanism incentivizes token holders to participate in the stablecoin ecosystem and contributes to the perception of long-term value appreciation.

By implementing a smart burn system, stablecoin projects can allocate a portion of the burn for ecosystem development. This includes initiatives such as funding new asset purchases for the asset pool, promoting innovation, and supporting the growth of the stablecoin ecosystem.

Once the tokens have been removed from circulation, they are entered into a 'tipple' wallet, which will then re-circulate the coins, using them to purchase assets for the asset pool. This model allows for the pool to always be growing. As the pool grows, the stability rating of the token can increase from 'C' where most tokens start, to 'S' grade. Once the ecosystem starts to reach 'S' Grade, the system can be expanded to maintain the grade.

Limiting Token Issuance from Reserves

A key advantage of a smart burn system is that it reduces the reliance on issuing new tokens from the reserve. This approach ensures that stablecoins maintain their value and avoid excessive dilution, while also allowing for the organic development of value within the system.

By limiting token issuance, stablecoins can preserve the value of existing tokens. Excessive token issuance can dilute the value of each individual token and erode confidence in the stablecoin's stability and purchasing power. A smart burn system helps mitigate this risk by managing supply and promoting value preservation.

Instead of constantly issuing new tokens, a smart burn system encourages the efficient use of reserves. The funds allocated for purchasing new assets in the asset pool contribute to the overall value of the stablecoin, ensuring that the reserve remains robust and can support the stability of the stablecoin ecosystem.

A stablecoin that demonstrates a sustainable tokenomics model through a smart burn system is more likely to gain trust and attract users and investors. The transparent and controlled supply management conveys a commitment to maintaining value and highlights the stability and long-term viability of the stablecoin project.

Allocation for Corporate Structure Maintenance

In addition to value development and supply management, a small portion of the token burn can be allocated for the maintenance of the corporate structure. This ensures the smooth operation and growth of the stablecoin project.

I. Corporate Structure Sustainability: Allocating a small percentage of the burn for corporate structure maintenance allows stablecoin projects to allocate resources for activities such as legal compliance, audits, and administrative costs. These funds support the sustainability of the corporate structure and ensure ongoing adherence to regulatory standards.

II. Governance and Transparency: By utilizing a portion of the burn for corporate maintenance, stablecoin projects can enhance governance and transparency. These resources enable the implementation of robust governance frameworks, including independent audits and regular reporting, fostering trust among stakeholders.

III. Decentralized Nature: The allocation of a small portion of the burn for corporate structure maintenance aligns with the decentralized ethos of the cryptoverse. By maintaining a decentralized and sustainable corporate structure, stablecoin projects can adapt to the evolving regulatory landscape, promote decentralization, and improve the overall resilience of the stablecoin ecosystem.

A smart burn system plays a vital role in sustaining value, managing supply, and ensuring the long-term viability of stablecoins. By implementing this mechanism, stablecoin projects can control supply, maintain value, and foster trust among users and investors. Furthermore, allocating a small portion of the token burn for the maintenance of the corporate structure supports the stability and governance of the stablecoin project. Through these measures, stablecoins can establish themselves as reliable, transparent, and resilient financial instruments within the cryptoverse.

Conclusion

In conclusion, the implementation of enhanced tokenomics brings numerous benefits to stablecoin ecosystems. By focusing on key aspects such as reserves, liquidity, supply management, and value development, stablecoin projects can achieve greater stability, sustainability, and trust among users and investors.

Firstly, the establishment of proper reserves is paramount. Fractal reserves provide a solid foundation for stablecoins, ensuring that they have sufficient assets to back their value. By maintaining adequate reserves, stablecoin projects can instill confidence in users, mitigating the risk of potential insolvency and preserving the stability of the coin.

Secondly, the liquidity of stablecoins is crucial for their functionality and acceptance within the broader financial ecosystem. Ensuring a healthy level of liquidity allows stablecoins to serve their intended purpose as efficient mediums of exchange, enabling seamless transactions and facilitating wider adoption. By actively managing liquidity thresholds and adhering to transparent reporting standards, stablecoin projects can provide stability and liquidity to the market.

The implementation of a smart burn system brings additional advantages to stablecoin ecosystems. By systematically burning a portion of tokens on each transaction and reintroducing them through the purchase of new assets, stablecoins can sustain value and manage supply effectively. This mechanism contributes to price stability, increases token value, and reduces reliance on excessive token issuance from reserves. Furthermore, allocating a small portion of the burn for corporate structure maintenance ensures the longevity and sustainability of the stablecoin project.

Moreover, the interoperability of stablecoins within the larger blockchain ecosystem enhances their usability and accessibility. By operating on established chains or building their own native chains, stablecoin projects can leverage the existing infrastructure, benefit from high transaction processing speeds, and enjoy lower gas fees. This interoperability also allows stablecoins to work seamlessly with exchanges, build child chains or plasmas, and create NFTs to represent corporate documents or other holdings, thereby expanding the scope of their applications.

Lastly, the stability of the nation in which the stablecoin operates and the corporate structure behind it are essential considerations. A stable nation provides a supportive regulatory environment, legal protections for cryptocurrencies, robust infrastructure, and positive international relations. These factors contribute to the efficient operation of stablecoins and foster trust among users and investors.

A decentralized and diversified corporate structure further aligns with the decentralized nature of the cryptoverse. By establishing multiple subsidiaries and fostering a decentralized ecosystem, stablecoin projects can adapt to changing regulatory landscapes, enhance governance, improve transparency, and increase resilience. This decentralized structure enhances the stability and longevity of the stablecoin project, promoting the principles of decentralization and trust in the cryptoverse.

In conclusion, the importance of enhanced tokenomics cannot be understated in the realm of stablecoins. By incorporating proper reserves, ensuring liquidity, implementing smart burn systems, leveraging interoperability, and establishing stable nations and decentralized corporate structures, stablecoin projects can establish themselves as reliable, sustainable, and valuable assets within the cryptocurrency market. Through these measures, stablecoins can pave the way for a more inclusive and efficient financial ecosystem in the digital age.

Chapter 3: Asset Pools

"Take care of yourself, you are your most valuable asset."[16]

Introduction

Assets are one of the greatest equalizers in the world. Financially, a company with strong assets can compete on nearly equal footing regardless of the amount of time that it has been around. Even when companies come up against nations, their assets become representative of whether they can win in court, move, or simply pay the fees for doing business in the country. Intrinsically, the assets of a corporate are their identity- thus they become their own best asset.

When we look at asset pools in stablecoins, we notice that there is a clear difference between asset pools and backing pools. An asset pool is a permanent pool of assets that is used to create a modicum of stability in an unstable world, the basket of goods. A backing pool is simply a reserve of assets that can be tapped by the company to buy back tokens when the user wants to sell off and there is no one in the market to buy the tokens. Asset pools are more refined, companies with true asset pools do not swap directly with the consumer when there is a run on the market, rather they buy from the market after liquidating a portion of their asset. This is a distinction between early model stablecoins and later model stablecoins.

Another important area of stablecoin asset pools is ensuring that the assets and the holders of those assets are diversified. Diversification can create an added level of

---

[16] Matshona Dhliwayo at https://graciousquotes.com/asset/.

stability for the system, even if one of the assets starts to falter the rest can 'pick up the slack.' As a result, there should always be several "funds" holding the assets of a stablecoins (even if they are owned by the same holding company). Within those funds there should also be a diversification of the assets held, the more diverse the assets, the more stable the fund can be.

Rounding out the asset pool discussion are the topics of launch strategies, sales ratio, and burn ratio. The launch strategy cannot be for 100% backing in the initial launch, if this strategy is used then the coin will not have the capital reserves to run the company (as seen with the Terra attack). The sales ratio should be somewhere between 20-30% of the incoming funds going towards running the company. This creates a 'less than 100%' asset to fund ratio, but for years banks were operating at 20%, then 10%, then 2%, now they have no requirement for their valuation of their funds. Stablecoins, when using a proper system, will always be over 70% of the present rate by maturity.

Building out the system is very important when developing the structure of a stablecoin. People expect their funds to be backed by something. Stablecoin are nearly unique in the cryptocurrency realm as they have all four valuation models: Transactional, Transitional, Market and Asset to build their pricing. This allows them to have a higher level of confidence in the community as opposed to other cryptocurrency types. Further, because they are not speculatory (they are a hedge not a profit engine) this eliminates them from classification as a commodity or security by competent officials.

Purpose of an Asset Pool (investment)

In the ever-evolving world of cryptocurrencies, stablecoins have emerged as a vital component, providing stability, liquidity, and a bridge between traditional finance and the digital realm. At the core of stablecoin functionality lies the concept of an asset pool, which serves as a fundamental pillar for maintaining the stability and value of these digital assets. In this article, we will explore the importance of an asset pool in relation to stablecoins, highlighting its role in mitigating volatility, ensuring liquidity, and supporting the overall ecosystem.

Stability and Mitigating Volatility:

> "Volatility is the price of admission. The prize inside are superior long-term returns. You have to pay the price to get the return."
>
> Morgan Housel

One of the primary objectives of stablecoins is to provide stability in an otherwise volatile cryptocurrency market. Unlike other cryptocurrencies that experience substantial price fluctuations, stablecoins aim to maintain a relatively constant value, typically pegged to a fiat currency such as the US dollar. The asset pool plays a critical role in achieving this stability.

By backing the stablecoin with a pool of assets, such as cash, treasury bills, or other stable financial instruments, the value of the stablecoin is anchored to the value of these assets. This ensures that the stablecoin's price remains relatively stable and less susceptible to market volatility. The asset pool acts as a reserve that can be used to stabilize the stablecoin's value, especially during times of market stress or increased demand.

Liquidity and Market Confidence:

Liquidity is a vital factor in any financial system, and stablecoins are no exception. Users and investors rely on stablecoins to facilitate seamless transactions, serve as a medium of exchange, and store value. An effective asset pool contributes significantly to maintaining liquidity within the stablecoin ecosystem.

The asset pool allows stablecoin projects to ensure sufficient liquidity by backing the outstanding supply of stablecoins with readily available assets. This ensures that users can easily convert stablecoins into their underlying assets or fiat currencies without significant price impact. The presence of a well-managed asset pool enhances market confidence, as it provides reassurance that the stablecoin issuer has the necessary reserves to meet the demand for redemptions and maintain the stability of the coin.

Trust and Transparency:

Transparency is a vital aspect of stablecoin projects, as it fosters trust and confidence among users and investors. The asset pool plays a crucial role in enhancing transparency within the stablecoin ecosystem.

Stablecoin projects that maintain a transparent and auditable asset pool provide stakeholders with visibility into the composition and value of the assets backing the stablecoins. This transparency instills trust by allowing users to verify that the stablecoin is adequately collateralized and the issuer's claims are substantiated. Regular audits and public reporting of the asset pool's holdings contribute to the credibility and accountability of stablecoin projects, further strengthening trust within the community.

Ecosystem Expansion and Value Creation:

An asset pool can also serve as a catalyst for ecosystem expansion and value creation within the stablecoin project. By managing the assets strategically, stablecoin issuers can generate returns and create additional value for the ecosystem.

For instance, the assets in the pool can be invested in interest-bearing instruments or utilized in decentralized finance (DeFi) protocols to generate yield. The income generated from these investments can be used to cover operational expenses, fund development, and contribute to the long-term sustainability of the stablecoin project. Additionally, by participating in the broader DeFi ecosystem, stablecoin projects can offer users opportunities for yield farming, lending, and other financial services, thereby enhancing the overall value proposition of the stablecoin.

Risk Management and Contingency Planning:

A well-managed asset pool provides stablecoin projects with the means to mitigate risks and effectively manage unforeseen circumstances. By diversifying the asset pool across different types of assets, jurisdictions, and counterparties, stablecoin issuers can reduce exposure to specific risks and increase resilience.

Furthermore, the asset pool serves as a contingency reserve, enabling stablecoin projects to address unexpected challenges or maintain the stablecoin's value during periods of market stress. For example, if the stablecoin experiences a sudden surge in demand or faces potential disruptions in the market, the asset pool can be utilized to stabilize the stablecoin's value and restore market confidence.

> "The biggest risk is not taking any risk… in world that is changing really quickly, the only strategy that is guaranteed to fail is not taking risks."
>
> Mark Zuckerberg.

In conclusion, the importance of an asset pool in stablecoin ecosystems cannot be overstated. The asset pool serves as a foundation for stability, liquidity, and trust within the stablecoin project. It mitigates volatility, ensures liquidity for seamless transactions, and supports the overall value proposition of stablecoins. Moreover, a well-managed asset pool enhances transparency, fosters market confidence, and enables ecosystem expansion and value creation. By effectively managing the asset pool, stablecoin projects can navigate risks, maintain stability, and contribute to the growth of the cryptocurrency ecosystem as a whole.

Purpose of a Backing Pool (Speculation)

In the realm of stablecoins, two terms often arise: asset pool and backing pool. While these terms may appear similar, they represent distinct concepts that have significant implications for the stability, utility, and purpose of stablecoins. In this article, we will explore the difference between an asset pool and a backing pool, emphasizing their roles as stability factors for investment and valuation tools for speculation, respectively. We will also highlight the importance of stablecoins being primarily used for investment rather than speculation, despite their potential for reliable growth during their maturing years.

Asset Pool: Stability Factor for Investment

An asset pool, in the context of stablecoins, refers to the reserve of assets that backs the value of the stablecoin. It serves as a stability factor for investment, ensuring that the stablecoin maintains a relatively constant value over time. The asset pool comprises various assets such as cash, government securities, or other stable financial instruments.

The primary purpose of an asset pool is to provide stability and mitigate volatility within the stablecoin ecosystem. By backing the stablecoin with real-world assets, stablecoin issuers offer investors a reliable and less volatile alternative to traditional cryptocurrencies. This stability makes stablecoins attractive for various use cases, including remittances, payments, and as a store of value.

Investors who hold stablecoins backed by an asset pool can be confident in their investment's stability and predictable value. This feature makes stablecoins an attractive option for risk-averse investors seeking to preserve their capital and avoid the wild price fluctuations associated with speculative cryptocurrencies.

Backing Pool: Valuation Tool for Speculation

On the other hand, a backing pool refers to a pool of funds or collateral that is used to back the value of a cryptocurrency without necessarily being composed of stable or low-volatility assets. Unlike an asset pool, a backing pool is primarily a valuation tool for speculation, rather than a stability factor for investment.

Backing pools are commonly found in speculative cryptocurrencies that rely on the value of their native tokens or other volatile assets to determine their worth. The assets in a backing pool can include cryptocurrencies, tokens, or even non-financial assets. The value of the cryptocurrency backed by a backing pool is subject to the fluctuations and volatility of these underlying assets.

It is important to note that stablecoins, by design, aim to provide stability and reliable value, distinguishing them from speculative cryptocurrencies. While speculation can be an inherent part of any financial market, stablecoins are intended to be primarily used as investment vehicles rather than speculative instruments.

Importance of Stablecoins for Investment

Stablecoins play a crucial role in the cryptocurrency ecosystem by providing a stable and reliable option for investors. They offer a bridge between traditional finance and the digital realm, combining the benefits of blockchain technology with the stability of fiat currencies.

The importance of stablecoins for investment lies in their ability to provide a secure and predictable investment vehicle. Unlike speculative cryptocurrencies, stablecoins offer stability, low volatility, and a reliable store of value. This makes them suitable for various investment strategies, such as portfolio diversification, hedging against market fluctuations, and facilitating efficient cross-border transactions.

Moreover, stablecoins are often used as a medium of exchange within decentralized finance (DeFi) protocols. They enable users to access decentralized lending, borrowing, and yield farming opportunities while minimizing exposure to the volatile nature of other cryptocurrencies. This further reinforces the importance of stablecoins for investment, as they provide the necessary stability and liquidity for participating in DeFi ecosystems.

Reliable Growth and Maturity

While stablecoins are primarily intended for hedging investment, it is worth noting that they can still offer reliable growth during their maturing years. As stablecoin projects gain traction and establish themselves within the cryptocurrency market, they attract more users, liquidity, and partnerships. This growth contributes to the stability and market value of the stablecoin, enhancing its attractiveness as an investment asset.

As stablecoins mature, they can also introduce additional features and functionalities that expand their use cases beyond basic value storage. For example, some stablecoin projects have integrated smart contract capabilities, enabling programmable transactions and the creation of decentralized applications (dApps). This evolution further reinforces stablecoins' position as investment instruments that offer both stability and utility within the broader cryptocurrency ecosystem.

The distinction between an asset pool and a backing pool is essential in understanding the nature and purpose of stablecoins. While asset pools serve as stability factors for investment, backing pools primarily function as valuation tools for speculation. Stablecoins, designed to provide stability, low volatility, and reliable value, play a crucial role in the investment landscape, offering a secure alternative to traditional cryptocurrencies. By leveraging asset pools and focusing on investment rather than speculation, stablecoins contribute to the growth and maturity of the cryptocurrency ecosystem while providing investors with a trusted and predictable investment option.

Diversification of holdings

In the world of stablecoins, diversification of asset pools plays a critical role in ensuring their stability, resilience, and long-term viability. A diversified asset pool refers to a balanced composition of various assets that back the value of a stablecoin, reducing the risks associated with any single asset. In this article, we will explore why asset pool diversification is crucial for stablecoins, examining its benefits, risk mitigation capabilities, and the broader implications for investors and the stability of the cryptocurrency market.

Risk Mitigation and Stability:

One of the primary reasons for asset pool diversification in stablecoins is risk mitigation. By diversifying the assets within the pool, stablecoin issuers minimize the exposure to any single asset's performance. This diversification reduces the vulnerability of stablecoins to price volatility, default risks, or sudden shocks in specific asset markets.

> Mutual funds have historically offered safety and diversification. And they spare you the responsibility of picking individual stocks.
>
> Ron Chernow

A well-diversified asset pool includes a mix of cash, government securities, low-risk bonds, and other stable financial instruments. This blend provides stability, liquidity, and the potential for consistent returns. The combination of different asset classes helps smooth out the impact of market fluctuations, ensuring the stablecoin's value remains relatively constant over time.

Resilience in Market Turbulence:

Asset pool diversification is particularly crucial during times of market turbulence. In periods of economic uncertainty, specific assets may experience significant volatility or even potential defaults. By diversifying the asset pool, stablecoin issuers can mitigate the impact of such events on the stablecoin's value.

For example, during a financial crisis, if a stablecoin's asset pool is heavily concentrated in a single asset class that experiences severe losses, it can jeopardize the stablecoin's stability and viability. On the other hand, a well-diversified asset pool can absorb the impact of individual asset failures, providing stability and ensuring the stablecoin continues to maintain its pegged value.

Enhanced Liquidity and Market Depth:

Diversification of asset pools also enhances the liquidity and market depth of stablecoins. By including multiple assets with varying liquidity profiles, stablecoin issuers ensure a robust market for buying and selling the stablecoin. This liquidity is essential for stablecoin users, as it enables seamless transactions and supports the stablecoin's utility as a medium of exchange.

Furthermore, a diverse asset pool contributes to market depth, which refers to the availability of buy and sell orders at various price levels. A deep market allows stablecoin holders to enter and exit positions with minimal slippage and ensures a fair and efficient price discovery process. This fosters trust and confidence in stablecoins, attracting more users and strengthening their overall market presence.

Risk Diversification for Investors:

Asset pool diversification also benefits stablecoin investors. By diversifying the underlying assets, stablecoins reduce the concentration risk associated with investing in a single asset or asset class. This risk diversification appeals to risk-averse investors who seek stable and predictable returns.

Diversification enables investors to allocate their funds across a range of assets, spreading the risk and reducing their exposure to any one asset's performance. This strategy helps protect capital, smooth out investment returns, and provides investors with a balanced and less volatile investment option compared to traditional cryptocurrencies.

Stability for the Cryptocurrency Market:

The importance of asset pool diversification extends beyond individual stablecoins. It contributes to the overall stability and credibility of the cryptocurrency market as a whole. Stablecoins with diversified asset pools serve as anchors within the market, providing stability during periods of volatility and acting as reliable reference points for price valuation.

Moreover, a diverse and stable stablecoin market helps facilitate smoother and more efficient trading across various cryptocurrencies and decentralized finance (DeFi) platforms. Stablecoins serve as an intermediary between volatile cryptocurrencies and traditional fiat currencies, enabling seamless transactions, liquidity provision, and risk management within the broader cryptocurrency ecosystem.

Asset pool diversification is of paramount importance for stablecoins, ensuring their stability, resilience, and attractiveness to investors. Diversifying the asset pool mitigates risks, enhances stability, and provides a predictable investment option within the volatile cryptocurrency market. It enhances liquidity, market depth, and overall market stability, fostering trust and confidence in stablecoins. As the cryptocurrency ecosystem continues to evolve, asset pool diversification remains a key factor in establishing stablecoins as a reliable and viable financial instrument for various use cases.

Multi-fund holding models.

In the realm of investment and asset management, the holding model adopted plays a crucial role in determining the level of diversification and resilience of the portfolio. This article delves into the significance of multi-fund holding models compared to mono-fund holding models. By exploring the benefits and advantages of multi-fund approaches, we will illustrate how the second level of diversification achieved through this strategy enhances insulation from market factors and contributes to long-term success.

Enhanced Diversification:

Diversification is a fundamental principle in investment management that aims to reduce risk by spreading investments across various assets and asset classes. While mono-fund holding models focus on diversifying within a single asset class or investment type, multi-fund holding models take diversification to the next level.

In a multi-fund approach, investors allocate their assets across different funds that cover a broad range of asset classes, such as stocks, bonds, real estate, commodities, and alternative investments. This secondary layer of diversification allows for exposure to multiple market segments, geographical regions, and economic sectors. By encompassing a more extensive array of assets, multi-fund holding models provide greater protection against market volatility, economic downturns, and sector-specific risks.

Risk Mitigation and Market Resilience:

Multi-fund holding models offer inherent risk mitigation benefits and enhance market resilience. By distributing investments across different funds and asset classes, investors reduce their exposure to any single fund's performance or the performance of a particular market segment. If one fund or asset class experiences a downturn, the impact on the overall portfolio is minimized due to the presence of other funds that may perform well during the same period.

This insulation from market factors contributes to portfolio stability and reduces the risk of substantial losses. It allows investors to navigate market cycles with more confidence, as the performance of a single fund or asset class has less significant implications for the overall portfolio's value. As a result, multi-fund holding models offer a more resilient investment strategy that can weather market fluctuations and preserve capital.

Capitalizing on Market Opportunities:

One of the significant advantages of multi-fund holding models is the ability to capitalize on market opportunities across various asset classes. Different asset classes tend to perform differently under varying economic conditions and market cycles. By diversifying into multiple funds, investors position themselves to take advantage of the potential upside in different markets.

For instance, during periods of economic expansion, equities may outperform other asset classes, while fixed-income investments might provide stability during economic downturns. Multi-fund holding models enable investors to allocate their assets strategically, increasing the likelihood of participating in favorable market conditions while mitigating the risks associated with a concentrated investment approach.

Flexibility and Adaptability:

Multi-fund holding models offer investors a higher degree of flexibility and adaptability. They allow for dynamic asset allocation and the ability to rebalance the portfolio to reflect changing market conditions, investment goals, and risk tolerance. This flexibility ensures that the portfolio remains aligned with the investor's objectives and adjusts to evolving market dynamics.

Additionally, multi-fund holding models provide the opportunity to tap into emerging trends and capitalize on new investment opportunities. As the investment landscape evolves, new asset classes or sectors may gain prominence, and by diversifying across multiple funds, investors can allocate capital to these emerging areas, potentially reaping significant returns.

Long-Term Performance and Risk-Adjusted Returns:

Research has consistently shown that multi-fund holding models tend to deliver superior long-term performance and risk-adjusted returns compared to mono-fund approaches. The combination of enhanced diversification, risk mitigation, and market opportunity capture contributes to a more balanced and stable portfolio.

By spreading investments across a broad spectrum of asset classes, investors reduce the potential for large drawdowns and enhance the potential for consistent returns over the long run. The ability to manage risk effectively and participate in different market segments positions multi-fund holding models for long-term success and outperformance relative to mono-fund approaches.

In the world of investment management, the choice between multi-fund and mono fund holding models plays a significant role in achieving diversification and market resilience. Multi-fund holding models offer distinct advantages by providing an additional layer of diversification, enhancing risk mitigation, and capitalizing on market opportunities. The ability to adapt to changing market dynamics, capitalize on emerging trends, and deliver superior long-term performance makes multi-fund holding models a preferred choice for investors seeking to achieve sustainable growth and weather market fluctuations. By embracing the benefits of multi-fund diversification, investors position themselves for success in the ever-evolving landscape of investment and asset management.

Derivative Launch Strategies to Build Worth

In the realm of stablecoins, the use of derivative strategies for initial distribution can offer unique advantages, including the potential for profit margins for early buyers and investors. By employing these strategies in conjunction with a basket of goods, stablecoin issuers can create opportunities for both stability and profitability. This article explores the benefits of utilizing derivative strategies for the initial distribution of stablecoins and highlights the potential for early adopters and investors to capitalize on these opportunities.

Initial Distribution and Profit Margin:

Derivative strategies can be effectively utilized during the initial distribution phase of stablecoins to provide early buyers with a profit margin. This approach involves creating derivative products tied to the stablecoin's underlying assets or a related market index. These derivatives enable investors to gain exposure to the stablecoin's value fluctuations without directly owning the stablecoin itself.

By allowing early buyers to purchase these derivatives at a discounted price during the initial distribution, stablecoin issuers can incentivize early adoption and generate liquidity in the market. The profit margin arises from the potential price appreciation of the stablecoin, as its value stabilizes and gains acceptance in the market. This approach not only attracts investors but also facilitates the smooth introduction of stablecoins to the broader market.

Basket of Goods and Profit Opportunities:

When stablecoins are backed by a basket of goods or a diversified portfolio of assets, derivative strategies can offer significant profit opportunities for early adopters and investors. The basket of goods serves as a benchmark or underlying asset for the stablecoin's value. By using derivative instruments tied to the basket of goods, investors can profit from price discrepancies between the stablecoin and the assets in the basket.

For example, if the stablecoin is designed to track the value of a specific commodity basket, investors can purchase derivatives that benefit from price movements in the commodities comprising the basket. As the stablecoin gains acceptance and the underlying assets appreciate, early adopters and investors can experience substantial profits through their derivative positions.

Hedging and Risk Management:

While derivative strategies provide profit opportunities, they also serve a vital role in hedging and risk management for stablecoin issuers. Derivatives can be used as hedging instruments to mitigate potential risks associated with the stablecoin's value fluctuations. By entering into derivative contracts, stablecoin issuers can offset the potential downside risks by transferring them to other parties in the market.

> "[Cryptocurrency] is a bank in cyberspace, run by incorruptible software offering a global, affordable, simple and secure savings account to billions of people that don't have the option or desire to run their own hedge fund."
>
> Michael J. Saylor

This hedging mechanism ensures that the stablecoin's value remains relatively stable, providing confidence and reliability to users and investors. Moreover, derivative strategies enable stablecoin issuers to manage market volatility and fluctuations, ensuring a more robust and resilient stablecoin ecosystem.

Early Adoption Incentives and Market Liquidity:

Derivative strategies not only benefit early buyers and investors but also contribute to market liquidity and stability. By offering derivative products tied to stablecoins, issuers create incentives for investors to participate in the market, enhancing liquidity and trading volumes. The availability of derivatives allows investors to engage in various trading strategies, including hedging, arbitrage, and speculation.

Increased liquidity and trading activity in the market contribute to price stability and reduce the risk of market manipulation. This, in turn, fosters confidence in the stablecoin and encourages wider adoption by users and businesses. The combination of derivative strategies and market liquidity supports the growth and sustainability of the stablecoin ecosystem.

Regulatory Considerations and Investor Protection:

While derivative strategies present lucrative opportunities, it is essential to address regulatory considerations and ensure investor protection. Stablecoin issuers must comply with applicable regulations governing derivative products and provide transparency to investors regarding the risks associated with these strategies.

Regulatory oversight helps maintain market integrity and protects investors from potential market abuses. By adhering to regulatory requirements and implementing robust risk management measures, stablecoin issuers can foster trust and confidence among investors, further promoting the widespread adoption of stablecoins.

> "Overwinding happens when hedge funds destroy companies by attempting to leverage derivatives against otherwise productive long-term assets."
>
> Douglas Rushkoff

Derivative strategies offer a powerful tool for the initial distribution of stablecoins, providing profit margins for early buyers and investors. By leveraging these strategies in conjunction with a basket of goods or diversified asset portfolios, stablecoin issuers create opportunities for both stability and profitability. The use of derivatives incentivizes early adoption, enhances market liquidity, and enables risk management for stablecoin issuers. However, it is crucial to navigate regulatory considerations and prioritize investor protection to ensure the long-term success and sustainability of stablecoin ecosystems. As the cryptocurrency landscape continues to evolve, the judicious use of derivative strategies can pave the way for the widespread adoption of stablecoins and their integration into various sectors of the global economy.

Sale Ratio (70/30)

When it comes to the allocation of funds raised from the sale of stablecoins, striking the right balance between the asset pool and corporate operations is crucial for the stability and growth of the stablecoin ecosystem. A 70/30 split, with 70% of the funds allocated to the asset pool and 30% to corporate operations, has emerged as an ideal distribution ratio. This article explores the reasons why this allocation is considered optimal, ensuring the asset pool's adequacy while supporting the necessary operations for the stablecoin's success.

Asset Pool Stability and Growth:

The asset pool serves as a critical component of a stablecoin's value proposition, providing the necessary collateral to back the stablecoin's value. Allocating 70% of the funds raised to the asset pool ensures its stability and growth. By maintaining a significant portion of funds within the asset pool, stablecoin issuers can instill confidence in users and investors, as the pool's size reflects the coin's backing capacity.

Furthermore, a robust asset pool contributes to price stability and resilience, as it enables efficient market-making and redemption processes. It ensures that the stablecoin can be exchanged for its underlying assets at a predictable and reliable value. This stability is vital for users seeking a reliable medium of exchange or store of value.

Adequate Liquidity and Market Functionality:

The 70/30 split allows for sufficient liquidity within the asset pool, enabling smooth operations and market functionality. Allocating the majority of funds to the asset pool ensures that there are ample reserves available to facilitate exchanges and meet redemption demands. This liquidity is essential for maintaining a healthy trading environment and supporting market participants' confidence in the stablecoin.

Moreover, adequate liquidity within the asset pool reduces the risk of price volatility and improves price discovery mechanisms. It enables stablecoin issuers to fulfill their commitment of providing a stable and pegged value to users, promoting trust and adoption in the market.

Funding Corporate Operations and Development:

While the asset pool is crucial for stability, allocating a portion of funds to corporate operations is equally important. The 30% allocation allows stablecoin issuers to cover operational expenses, invest in research and development, and build a robust ecosystem around the stablecoin.

Corporate operations encompass various activities, including marketing, compliance, legal support, technology infrastructure, and talent acquisition. Allocating funds to these areas ensures the smooth functioning of the stablecoin ecosystem and supports its growth in terms of adoption, partnerships, and technological advancements.

Flexibility and Adaptability:

The 70/30 split provides a level of flexibility and adaptability for stablecoin issuers. It allows them to respond to market dynamics and changing regulatory environments effectively. By maintaining a significant portion of funds in the asset pool, issuers can adjust their operations to align with market demands without compromising the stability of the stablecoin.

In times of increased market volatility or heightened redemption requests, the asset pool's size can be leveraged to stabilize the stablecoin's value. Meanwhile, the allocated funds for corporate operations provide the necessary resources for stablecoin issuers to adapt to evolving market conditions and meet regulatory obligations effectively.

The 70/30 split, with 70% allocated to the asset pool and 30% to corporate operations, strikes a balance between ensuring the stability and growth of the stablecoin ecosystem. By maintaining a robust asset pool, stablecoin issuers can instill confidence in users and provide a reliable value proposition. Simultaneously, allocating funds to corporate operations supports operational efficiency, ecosystem development, and adaptability to market dynamics. This distribution ratio allows stablecoin issuers to meet the demands of both asset-backed stability and operational requirements, ensuring a successful and sustainable stablecoin venture.

Smart burn ratio (98/2)

Implementing a burn mechanism, particularly a smart burn, within the framework of a stablecoin can contribute to its stability, growth, and value appreciation. In this discussion, we will delve into why a burn rate of approximately 2% is considered optimal and how allocating 2% of that burn rate to the company's operations facilitates proper functioning and helps maintain a controlled burn rate.

Smart Burn and Stability:

A smart burn mechanism involves the temporary systematic reduction of token supply by burning a percentage of tokens with each transaction to a tipple wallet. This approach ensures a controlled reduction in supply over time, ultimately increasing scarcity and potentially enhancing the token's value.

> "Hedge Funds are not especially liquid. Many are 'gated'- meaning there are only small windows when you can withdraw your money."
>
> Barry Ritholtz

A burn rate of around 2% strikes a balance between controlling token inflation and maintaining price stability. It allows for a gradual decrease in supply without causing abrupt shocks to the market. This gradual reduction in supply can create a positive psychological effect, stimulating demand and potentially supporting long-term price appreciation.

Sustainable Company Operations:

Allocating a portion of the 2% burn rate to the company's operations, specifically 2%, enables the stablecoin company to function effectively while keeping the burn rate at a manageable level. It allows the company to cover necessary expenses such as research and development, marketing, compliance, and talent acquisition.

Sustaining company operations is vital for the stablecoin's overall success. It ensures ongoing technological advancements, ecosystem development, and the ability to adapt to market demands and regulatory changes. By allocating a portion of the burn rate to the company, stability, and long-term sustainability of the stablecoin project can be achieved.

Controlling the Burn Rate:

Maintaining a controlled burn rate is crucial for stablecoin projects to avoid excessive token supply reduction, which can lead to illiquidity or price volatility. By allocating 2% of the burn rate to the company's operations, the stablecoin project can maintain a balance between reducing token supply and ensuring sufficient liquidity and market stability. Additionally, each time coins are cycled they increase the value of the asset pool, increasing the relative value of the coin to 'outrun' inflation.

The company's portion of the burn rate allows for flexibility in managing operations and mitigates potential risks associated with an overly aggressive burn rate. It provides stability by ensuring that the company has the necessary resources to support the stablecoin's development and growth, while still allowing for a controlled reduction in token supply.

Long-Term Sustainability:

Maintaining a sustainable burn rate is essential for the long-term viability of the stablecoin project. An excessively high burn rate could deplete the token supply too rapidly, impacting market stability and hindering the stablecoin's ability to function effectively as a medium of exchange or store of value.

By implementing a 2% burn rate and allocating 2% of that rate to the company's operations, stablecoin projects can strike a balance between stability, growth, and sustainable operations. This approach ensures that the company has the means to fulfill its responsibilities while simultaneously supporting the token's value proposition through controlled supply reduction.

A 2% burn rate, with 2% (0.04%) allocated to the company's operations, offers a balanced approach to token burning within the stablecoin ecosystem. It allows for controlled token supply reduction, fostering scarcity and potential value appreciation. Simultaneously, allocating a portion of the burn rate to the company's operations ensures sustainable and effective project management, supporting ongoing development, market stability, and long-term success. Striking the right balance between the burn rate and company allocation is crucial for creating a stable and prosperous stablecoin venture.

Conclusion

Asset pools play a vital role in the stability, growth, and resilience of stablecoins. They serve as a fundamental pillar for maintaining the peg to a specific asset or basket of goods, ensuring price stability and trust among users. Throughout this chapter, we have explored the importance of asset pools, their characteristics, and the benefits they offer to the stablecoin ecosystem.

Asset pools provide stability by backing stablecoins with tangible assets, such as fiat currencies, commodities, or other cryptocurrencies. This backing instills confidence in users, as they know that the value of their stablecoins is supported by real-world assets, reducing the risk of price volatility. Moreover, asset pools enable stablecoins to fulfill their primary purpose as a reliable medium of exchange and store of value.

Diversification within asset pools is a critical factor that enhances stability and mitigates risk. By holding a diverse range of assets, stablecoin projects can minimize the impact of market fluctuations on the value of their tokens. Diversification across different asset classes, industries, and geographies ensures that the stability of the stablecoin is not dependent on the performance of a single asset, reducing vulnerability to external shocks.

Furthermore, the distinction between asset pools and backing pools is important to understand. Asset pools focus on stability and long-term value preservation, as they act as a stability factor for investment. On the other hand, backing pools primarily serve as a valuation tool for speculative purposes. Stablecoins should be primarily used for investment and transactional purposes, rather than as speculative assets, to maintain their stability and fulfill their intended purpose.

The use of a multi-fund holding model, where stablecoin projects hold multiple asset pools, provides an additional layer of diversification and insulation from market factors. This approach allows for a more robust and resilient stablecoin ecosystem, as the risks associated with individual assets are spread across multiple pools. By adopting a multi-fund holding model, stablecoin projects can effectively navigate market fluctuations and protect the value and stability of their tokens.

Incorporating derivative strategies into the initial distribution of stablecoins can be an effective way to incentivize early adopters and investors. Derivatives provide opportunities for profit through exposure to a basket of goods, allowing participants to benefit from the price movements of various assets. This approach can create additional value for stablecoin holders and encourage wider adoption of the stablecoin within the market.

A balanced allocation of funds raised from stablecoin sales is crucial for maintaining the stability and value of the asset pool. The suggested 70/30 split, with 70% of funds going to the asset pool and 30% allocated to corporate operations, ensures that the asset pool remains at an appropriate level relative to the total value of the stablecoins in circulation. This distribution supports the stability of the stablecoin, as the asset pool retains a significant portion of the total value, reinforcing confidence in the pegged value.

Lastly, implementing a smart burn mechanism with a 2% burn rate, where 2% is allocated to the company's operations, helps maintain stability and sustainable operations. The controlled reduction in token supply through the burn mechanism enhances scarcity and potential value appreciation. Simultaneously, allocating a portion of the burn rate to the company's operations ensures that the stablecoin project can function effectively and adapt to market demands while maintaining a balanced burn rate.

In summary, asset pools are a critical component of stablecoin ecosystems, providing stability, value preservation, and trust among users. Diversification, multi-fund holding models, proper fund allocation, and smart burn mechanisms are essential elements that contribute to the long-term success and sustainability of stablecoin projects. By adhering to these principles, stablecoins can fulfill their role as reliable and efficient digital assets within the cryptocurrency landscape.

Chapter 4: Asset Classification

"Investors who followed a broadly diversified asset allocation that included a total market fund and small value index fund would have fared very well during the past decade."[17]

Introduction

Within the realm of finance, people want to know what is backing any given asset. This means that each asset needs to be backed by something, which in the case of a stablecoin is other assets. The different classes of assets that are backing stablecoins can determine the stability of the stablecoins. When it comes down to it, there are two key areas of assets- highly regulated assets and limited regulation assets. Each of these can be a valuable took for a stablecoin investor, but knowing each type of asset can be a challenge.

> "A dramatic spending increase to fund the SEC and the CFTC, as envisioned by the authors of the Dodd-Frank legislation, would further the mindset that our nation's problems can be solved with more spending, not more efficiency."
>
> Scott Garrett

The highly regulated classifications of assets are securities and commodities. In the United States the SEC regulates the securities and the CFTC regulates commodities. Despite the current management of the SEC and CFTC, it is important that these types of assets are properly regulated to ensure that the consumer and the institutional

[17] Richard A Ferri at https://www.goodreads.com/work/quotes/357067-all-about-asset-allocation.

investor are protect. The more direct the data on these types of assets, the better it is for the system.

Less regulated asset are assets where the government tends to be more lenient on the asset because its value is more dependent on the market. This categorization includes currencies, antiques/collectables, and property. While there is substantial regulation around these different areas, there is not the regulatory agency framework which brings the inefficiency that securities and commodities have, thus they are considered less regulated.

All assets have value in the asset pool of a stablecoin, the balance that must be reached is the level of liquidity needed to manage the fund. In times of strong market growth, less liquid methods of asset storage can be useful in 'riding the wave' at a stable rate. In a bear market, it can be useful to have more of the asset fund in very liquid elements, so they can be moved at a moment's notice. Balancing this is a game best played by humans being advised by AI systems, which can increase the profitability of the fund and the stability of the token.

Securities

Securities are essential financial instruments that play a significant role in the global economy. They represent various types of investments, each with its unique characteristics and regulatory requirements. In this comprehensive exploration of securities, we will discuss their definition, classification, and distinct features. Specifically, we will focus on stocks, bonds, derivatives, commercial paper, convertible bonds, bond funds, and private equity. Additionally, we will address how stablecoins differ from traditional securities, particularly regarding the Howey Test and the absence of an expectation of profit.

Securities and the Howey Test

> We're Going to try and prove to the market that you can do a legal coin offering. If the SEC doesn't crack down, this party will be amazing- the biggest party in town for a long time. If they do crack down, a lot of people are going to feel a lot of pain."
>
> Naval Ravikant

Definition of Securities: Securities are financial instruments that represent ownership or debt obligations issued by corporations, governments, or other organizations. They are bought and sold in financial markets, allowing investors to participate in the profits, interest, or capital appreciation generated by the issuer.

The Howey Test: The Howey Test is a legal assessment developed by the U.S. Supreme Court to determine whether a transaction qualifies as an investment contract, thus falling under the definition of a security. It consists of four key elements:

- Investment of Money
- Expectation of Profits
- Common Enterprise
- Efforts of Others

Stablecoins, as digital assets designed to maintain a stable value, do not typically meet the criteria outlined in the Howey Test for securities. Stablecoins are primarily intended to facilitate efficient transactions and serve as a medium of exchange, rather than providing investors with an expectation of profit.

Stocks

Definition of Stocks: Stocks, or shares, represent ownership in a company. When individuals purchase stocks, they become shareholders and gain voting rights, dividend entitlements, and the potential to participate in capital appreciation. Characteristics of Stocks:

- Equity Ownership
- Dividends
- Capital Appreciation

Bonds

Definition of Bonds: Bonds are debt instruments issued by governments, municipalities, or corporations to raise capital. Investors who purchase bonds become creditors and lend money to the issuer for a specified period, receiving periodic interest payments and the return of principal upon maturity. Characteristics of Bonds:

- Fixed Income
- Maturity Date
- Credit Ratings

Derivatives

Derivatives are financial contracts whose value is derived from an underlying asset, such as stocks, bonds, commodities, or currencies. They enable investors to speculate on price movements, hedge risks, or gain exposure to specific assets without owning them directly. Characteristics of Derivatives:

- Leverage
- Risk Management

- Diverse Instruments
- Section: Commercial Paper

Commercial Paper

Commercial paper refers to short-term debt instruments issued by corporations to finance their immediate operational needs. It is typically unsecured and has maturities ranging from a few days to one year.

- Characteristics of Commercial Paper:
- Short-Term Debt
- Issued by Corporations
- Low Risk

Convertible Bonds

Convertible bonds are hybrid securities that combine elements of debt and equity. They allow bondholders to convert their bonds into a predetermined number of the issuer's common stock shares within a specified period. Characteristics of Convertible Bonds:

- Dual Nature
- Conversion Ratio
- Upside Potential

Bond Funds

Definition of Bond Funds: Bond funds are investment vehicles that pool investors' capital to purchase a diversified portfolio of bonds. They offer investors exposure to a broad range of fixed-income securities while providing professional management and liquidity. Characteristics of Bond Funds:

- Diversification
- Professional Management

- Liquidity

Private Equity

Definition of Private Equity: Private equity refers to investments made in privately held companies that are not publicly traded. Investors provide capital in exchange for equity ownership and actively participate in the management and growth of the company. Characteristics of Private Equity:

- Long-Term Investments
- Active Involvement
- Illiquidity

Securities encompass a wide range of financial instruments that enable individuals and organizations to invest, diversify their portfolios, and participate in the growth and profits of various entities. While stablecoins differ from traditional securities due to the absence of an expectation of profit, they offer stability, usability, and efficiency as digital assets. Understanding the distinctions between stocks, bonds, derivatives, commercial paper, convertible bonds, bond funds, and private equity is crucial for investors seeking to navigate the financial landscape effectively. By grasping the characteristics and classifications of securities, investors can make informed decisions and build robust portfolios aligned with their investment goals and risk tolerance.

Commodities

Commodities play a vital role in global trade and investment, representing essential raw materials and goods. In this comprehensive exploration of commodities, we will discuss their definition, classification, and distinct features. Specifically, we will focus on the legal test for determining what qualifies as a commodity, the differentiation between hard and soft commodities, and the main types of commodities, including metals, energy, minerals, livestock, agriculture, coins with their own chains, and wood.

What is a Commodity?

A commodity refers to a primary raw material or good that is interchangeable with other similar goods and is typically used in the production of other goods or as a tradable asset. Commodities are subject to supply and demand dynamics, with their prices often influenced by global market forces.

> "Capital is a money, capital is commodities. By virtue of it being value, it has acquired the occult ability to add value to itself. If it brings forth living offspring, or, at least, lays the golden egg."
>
> Karl Marx

The legal definition of a commodity can vary, and different jurisdictions may adopt specific tests to determine whether a particular asset qualifies as a commodity. One widely used legal test is the "Commodity Exchange Act (CEA)" in the United States, which defines a commodity as "a physical commodity or a financial instrument that is the subject of a futures contract."

Hard Commodities vs. Soft Commodities

Differentiating Hard and Soft Commodities: Commodities can be categorized as either hard or soft commodities based on their characteristics. The distinction lies in the tangibility and durability of the underlying asset.

Hard Commodities: Hard commodities are natural resources extracted from the earth, typically in solid or physical form. These commodities include metals, energy resources, minerals, and wood. They are often used as inputs in manufacturing, construction, and infrastructure projects.

Soft Commodities: Soft commodities, on the other hand, encompass agricultural products and livestock. These commodities are generally perishable and require specific storage conditions. Soft commodities include crops, livestock, and other agricultural goods.

Types of Commodities

- Metals: Metals are one of the key categories of commodities, known for their industrial applications and value as investment assets. They include precious metals such as gold, silver, platinum, and palladium, as well as base metals like copper, aluminum, and nickel. Metals serve as essential inputs for various industries, including construction, electronics, and automotive.
- Energy: Energy commodities encompass resources used to generate power and fuel various sectors of the economy. Crude oil, natural gas, and coal are prominent examples of energy commodities. These resources are critical for transportation, electricity generation, and industrial processes.
- Minerals: Minerals comprise a diverse range of commodities extracted from the earth's crust. They include iron ore, copper ore, bauxite, uranium, and rare earth metals. Minerals are essential raw materials for manufacturing, construction, and the production of consumer goods.

- Livestock/Meat: Livestock commodities involve the trading of animals for meat production, such as cattle, hogs, and poultry. The prices of livestock commodities are influenced by factors like feed costs, demand for meat products, and supply chain dynamics.
- Agriculture: Agricultural commodities encompass various crops and agricultural products. They include wheat, corn, soybeans, rice, coffee, cocoa, sugar, cotton, and more. Agricultural commodities are influenced by factors like weather conditions, global demand, and government policies.
- Coins with Their Own Chains: Coins with their own chains refer to cryptocurrencies that operate on their own blockchain networks. These commodities include popular cryptocurrencies like Bitcoin, Ethereum, and Litecoin. They function as both digital currencies and investment assets, offering decentralization, security, and potential price appreciation.
- Wood: Wood, as a commodity, includes timber and other forest products. It serves as a raw material for construction, furniture manufacturing, and various wood-based industries. Factors affecting the wood market include sustainable forestry practices, supply chain logistics, and demand for timber products.

Commodities are essential components of global trade and investment, providing raw materials, energy resources, and agricultural products. Understanding the legal definitions, distinctions between hard and soft commodities, and the various types of commodities is crucial for investors, traders, and businesses operating in commodity markets. By comprehending the unique characteristics, supply-demand dynamics, and market influences of metals, energy, minerals, livestock, agriculture, coins with their own chains, and wood, stakeholders can make informed decisions and navigate these markets effectively.

Currencies

> "A Weak currency is the sign of a weak economy, and a weak economy leads to a weak nation."
>
> Ross Perot

Currencies serve as a fundamental element in asset pools, providing liquidity and stability to financial systems. In this discussion, we will explore the usefulness of different types of currencies in asset pools. While fiat currencies, such as traditional paper money, may face inflationary factors, we will explore how smart burn systems can help overcome these challenges and ensure ample backing for asset pools. Additionally, we will examine the concepts of commodity money, representative money, secure digital money (cryptocurrency), semi-secure digital money (central bank digital currencies - CBDCs), and secondary digital cash (in-game cash) to shed light on their roles within the broader currency landscape.

Fiat Money

Fiat Money Defined: Fiat money refers to currency that is not backed by a physical commodity like gold or silver but derives its value from the trust and confidence placed in the issuing government. It is the most common form of currency used globally.

One of the main challenges with fiat money is the potential for inflation due to factors such as increased money supply and changes in economic conditions. Inflation erodes the purchasing power of fiat currencies over time, leading to concerns about their long-term stability. To address the inflationary concerns associated with fiat money, asset pools can implement a smart burn system. This system involves burning a portion of the tokens or coins on each transaction, reducing the total supply in circulation. By reducing the supply, the value of the remaining tokens can be preserved or even increased, offering confidence in the asset pool's backing.

Commodity Money and Representative Money

Commodity money refers to currencies that have intrinsic value based on a physical commodity. Historically, commodities such as gold, silver, or even livestock have been used as mediums of exchange. Their value is derived from their scarcity, durability, and desirability. Representative money is a form of currency that represents a claim on a physical commodity. For example, in the past, paper currency could be exchanged for a specific amount of gold or silver. The value of representative money is tied to the underlying commodity it represents. Commodity money and representative money provide stability and intrinsic value, as they are backed by tangible assets. However, their use in asset pools may be limited due to logistical challenges and the need for physical storage. While these forms of currency may not be directly integrated into asset pools, they can serve as historical examples of currencies based on tangible assets.

Secure Digital Money (Cryptocurrency)

Secure digital money refers to cryptocurrencies built on secure and decentralized blockchain networks. These digital currencies, such as Bitcoin and Ethereum, offer transparency, immutability, and strong cryptographic security. Similar to fiat money, secure digital money can also face inflationary pressures due to the potential creation of new tokens or coins. However, smart burn systems can be implemented to mitigate inflation risks and maintain the value of the cryptocurrency. By periodically burning a portion of the tokens, the supply can be controlled, promoting scarcity, and potentially increasing the value of the remaining tokens.

Semi-Secure Digital Money (Central Bank Digital Currencies-CBDCs)

Semi-secure digital money refers to central bank digital currencies (CBDCs) issued and regulated by central banks. CBDCs aim to provide secure digital alternatives to physical cash while maintaining the backing and stability associated with traditional fiat currencies. CBDCs are designed to operate similarly to traditional fiat currencies, and their inflationary factors are tied to the monetary policies of the respective central banks. While smart burn systems may not be directly applicable to CBDCs, they can still benefit from the transparency and efficiency offered by blockchain technology.

Secondary Digital Cash (In-Game Cash)

Secondary digital cash refers to digital currencies used within virtual environments or video games. These currencies hold value within the game's ecosystem and can be used for in-game transactions. Inflation within secondary digital cash systems is typically managed by game developers who control the issuance and availability of in-game currencies. While smart burn systems may not be directly relevant in this context, the stability and value of these currencies depend on the game's design and economy.

> "The design of central bank digital currencies is a global effort of collaboration rather than competition."
>
> Agustin Carstens

Currencies play a vital role in asset pools, providing liquidity and stability. While fiat currencies may face inflationary factors, smart burn systems offer a potential solution by reducing token supply and maintaining ample backing. Additionally, understanding the different types of currencies, such as commodity money, representative money, secure digital money (cryptocurrency), semi-secure digital money (CBDCs), and secondary digital cash, helps us appreciate their unique characteristics and applications within asset pools. By leveraging innovative technologies and adopting appropriate mechanisms, asset pools can enhance confidence and ensure the stability and value of the currencies they utilize.

Antiquities/Collectables

Antiques and collectibles possess inherent value and uniqueness, making them an intriguing asset class for backing pools. In this discussion, we will explore how various types of collectibles, including artwork, historical items, fad items, vehicles, gemstones/jewelry, digital collectibles, literature (books), and property, can be utilized within backing pools. We will delve into their potential for value appreciation, consider their liquidity challenges, and assess their overall contribution to the stability and growth of asset pools.

Artwork (including Music)

Artwork, including paintings, sculptures, and other forms of visual expression, represents a tangible and appreciating asset. The value of artwork often increases at a faster rate than traditional market rates, driven by factors such as rarity, historical significance, artist reputation, and demand from collectors and investors. While artwork can yield substantial returns, it generally suffers from low liquidity. The sale of valuable artwork often requires specialized expertise, extensive research, and connections within the art market. Consequently, converting artwork into cash may not be immediate or straightforward, making it more suitable for long-term asset holding rather than frequent liquidity needs.

Historical Items

Historical items, such as artifacts, manuscripts, and memorabilia, carry historical and cultural significance. These items can appreciate in value due to their rarity, connection to historical events or figures, and the interest of collectors, museums, and institutions. Similar to artwork, historical items may encounter liquidity challenges. Their specialized nature often necessitates engagement with experts and niche markets to find suitable buyers. However, the value appreciation potential and the historical importance associated with these items make them compelling additions to backing pools.

Fad Items

Fad items encompass collectibles that experience significant popularity and demand during a particular period. Examples include limited edition toys, trading cards, or trendy gadgets. These items can exhibit rapid value appreciation during the height of their popularity. Fad items are subject to market trends and may experience a decline in demand once the initial hype subsides. As a result, liquidity can be volatile, as the value of these items heavily relies on the prevailing fad. Careful consideration should be given to their inclusion in asset pools, balancing the potential for short-term gains with long-term sustainability.

Vehicles

Classic cars, motorcycles, and other vintage vehicles hold a unique appeal for collectors and enthusiasts. These vehicles can appreciate in value over time, driven by factors such as rarity, condition, historical significance, and desirability among automotive enthusiasts. The liquidity of vehicles can vary based on factors such as the demand for specific models, condition, and the availability of interested buyers. While selling a high-value vehicle may require more effort and time, the potential for substantial returns and the enjoyment of owning these assets make them attractive for inclusion in backing pools.

Gemstones/Jewelry

Precious gemstones and high-quality jewelry are valued for their beauty, rarity, and long-term investment potential. These assets can appreciate in value due to factors such as the scarcity of gemstones, craftsmanship, and market demand. Gemstones and jewelry generally have lower liquidity due to the specialized nature of the market. However, their intrinsic value and enduring appeal make them suitable for long-term asset holdings within backing pools.

Digital Collectibles

The rise of blockchain technology has introduced a new form of collectibles known as digital assets or non-fungible tokens (NFTs). These digital collectibles include virtual artworks, music, virtual land, and other unique digital items. NFTs enable verifiable ownership and scarcity in the digital realm. Digital collectibles offer unique advantages, such as instant transferability and transparent ownership records. However, their liquidity can vary depending on the popularity and demand for specific digital assets. Marketplaces and platforms dedicated to NFTs provide avenues for buying, selling, and trading digital collectibles.

Literature (Books)

Rare and collectible books, especially first editions or limited editions, can hold significant value. Factors such as author reputation, historical importance, condition, and demand from collectors contribute to the appreciation potential of literary works. The liquidity of rare books relies on finding interested collectors, specialized booksellers, or auction houses. The time and effort required to locate suitable buyers may be greater than for other assets. Nonetheless, the enduring appeal and intellectual value of literature make it an intriguing addition to backing pools.

Property

Property, including real estate holdings, offers stability and long-term growth potential. Real estate can appreciate over time, generate rental income, and provide diversification within an asset pool. Properties with historical significance or unique characteristics can add additional value. While property can provide a reliable source of income and capital appreciation, liquidity may be limited due to the time and effort required for buying or selling real estate. However, property can serve as a stable and valuable asset within a diversified backing pool.

Antiques and collectibles encompass a broad range of assets that can contribute to the stability and growth of backing pools. While they may face challenges related to liquidity, their potential for value appreciation, unique characteristics, and appeal to collectors make them valuable additions to asset portfolios. By carefully selecting and managing these assets, asset pool managers can enhance diversification, attract investors seeking alternative investments, and create a well-rounded and resilient portfolio. The inclusion of antiques and collectibles adds depth and uniqueness to the overall investment strategy, creating opportunities for reliable growth and long-term value preservation.

Conclusion

In this chapter, we have explored various asset types that can be utilized within backing pools to enhance stability, diversification, and potential value appreciation. We have delved into different categories of assets, including traditional securities such as stocks, bonds, derivatives, and commercial paper, as well as alternative assets like collectibles, commodities, and cryptocurrencies. Each asset type brings its unique characteristics, opportunities, and challenges to the table.

Securities, such as stocks and bonds, offer the advantage of being well-established investment instruments with clear regulatory frameworks. They provide investors with ownership rights and income potential, making them a cornerstone of many asset portfolios. Derivatives and commercial paper offer additional tools for risk management and short-term liquidity needs, further enhancing the versatility of asset pools.

Commodities, including metals, energy, minerals, livestock/meat, agriculture, coins with their own chains, and wood, provide tangible assets with intrinsic value. They can act as hedges against inflation, geopolitical uncertainties, and market volatility. By incorporating commodities into backing pools, investors can diversify their exposure and potentially benefit from long-term price appreciation.

Collectibles and antiquities add a unique dimension to asset pools, offering the potential for value appreciation beyond traditional market rates. Artwork, historical items, fad items, vehicles, gemstones/jewelry, digital collectibles, literature (books), and property all bring their own allure and opportunities. While liquidity challenges may exist for some of these assets, their inclusion can bring uniqueness, cultural value, and the potential for significant returns over time.

Currencies play a crucial role in the asset pool ecosystem. Fiat money serves as a medium of exchange and a benchmark for value, despite its inflationary nature. Secure digital currencies, such as cryptocurrencies, provide decentralized and secure means of transferring value, while central bank digital currencies (CBDCs) offer a hybrid approach with potential benefits in terms of efficiency and financial inclusion. Additionally, secondary digital cash in the form of in-game currencies demonstrates the evolution of digital economies and the value of virtual assets.

Lastly, we explored the use of properties as assets within backing pools. Real estate offers stability, income generation, and the potential for long-term appreciation. Properties with historical significance or unique features can provide additional value and diversification.

In conclusion, a well-designed and diversified asset pool considers a combination of different asset types. By harnessing the strengths and managing the challenges of each asset class, investors can build resilient portfolios that mitigate risks and capture opportunities. The careful selection, management, and monitoring of asset types within backing pools are essential to maintain stability, provide value preservation, and meet the expectations of investors seeking reliable growth. As the investment landscape evolves, asset pools must adapt and leverage the diverse range of assets available to achieve optimal results.

Chapter 5: Coin Classifications

"There is a constant interaction and overlap between the physical, virtual, augmented, and mixed worlds."[18]

Introduction

Understanding what you are trading is one of the most important elements of being successful in cryptocurrency. There are armies of speculators out there that say they are 'investors' because they have money in cryptocurrency. Investors are people who have taken the time to research the asset in which they are putting their money. Speculators on the other hand are people who think just because something is a fad that they will make money on it. While investors have a reasonable track record (30-50% success rate), speculators usually fall in the 1 in 100 is successful range.

One of the first things that an investor needs to understand is the difference between coins and tokens. Because most stablecoins are actually tokens, there is a lot of confusion in the area. It is important, however, that the investor know the difference because tokens tend to move in very different patterns than native chain coins do. While there is limited relationship between the token and the chain it is on, they can move entirely independent of the chain, therefore, the distinction must be noted.

This chapter then goes on to discuss the different types of tokens that one may run into while they are trading. Each type of token has a unique ability, and many of them have different financial implications based on the current

---

[18] Simone Puorto at https://www.goodreads.com/quotes/tag/cryptocurrency

classification in the regulatory system. Therefore, it is important to have a 'cheat sheet' of the different types of tokens and how the function in the cryptoverse before entering into the realm of hedging your investments with stablecoins.

Coins v. Tokens

In the rapidly evolving world of cryptocurrencies, the terms "coins" and "tokens" are often used interchangeably. However, there are distinct differences between the two that are crucial to understand. In this discussion, we will explore the fundamental characteristics, use cases, and variations of coins and tokens within the cryptoverse.

> "If the cryptocurrency market overall or a digital asset is solving a problem, it's going to drive some value."
>
> Brad Garlinghouse

Coins:

Coins, also known as cryptocurrencies or digital currencies, are native assets that operate on their own independent blockchain networks. They are typically designed to serve as a medium of exchange and store of value. Bitcoin (BTC), the first and most well-known cryptocurrency, is a prime example of a coin.

Coins possess several key characteristics. First, they have their own blockchain infrastructure, which distinguishes them from tokens that are built on existing blockchain platforms. This independent nature allows coins to have their unique consensus algorithms, mining processes, and governance structures.

Furthermore, coins often prioritize decentralization and aim to provide a transparent and secure financial system. Their underlying blockchain technology ensures immutability, traceability, and resistance to censorship. Coins can be transferred directly between users without the need for intermediaries, such as banks or financial institutions.

Coins typically have a limited supply, often predetermined by their protocol. This scarcity contributes to their potential value appreciation over time. Additionally, coins are commonly used as a means of payment, investment, and a store of wealth.

Tokens:

Tokens, on the other hand, are digital assets that rely on existing blockchain networks, such as Ethereum, Binance Smart Chain, or Solana. They are created and hosted on these platforms and leverage their infrastructure and functionalities.

Tokens represent various digital assets, including utility tokens, security tokens, governance tokens, non-fungible tokens (NFTs), and more. Their functionalities and purposes can vary significantly depending on their specific design and the underlying blockchain platform.

Utility tokens are among the most common types of tokens. They are typically used to access and utilize specific products, services, or platforms within a decentralized ecosystem. For example, in a decentralized application (DApp) ecosystem, utility tokens may grant users access to certain features, pay for transaction fees, or participate in decentralized governance.

Security tokens, on the other hand, are tokens that represent ownership in an underlying asset, such as shares in a company or real estate. These tokens are subject to securities regulations and offer investors ownership rights, dividends, or profit-sharing opportunities.

Governance tokens provide holders with voting rights and influence over the decision-making processes within a decentralized organization or platform. These tokens allow community members to participate in shaping the future direction and policies of the project.

Non-fungible tokens (NFTs) are unique tokens that represent ownership of a specific digital or physical asset. NFTs have gained significant attention in recent years, particularly in the art and collectibles space, as they enable the ownership and transfer of digital artwork, virtual real estate, game items, and more.

The Distinction:

The key distinction between coins and tokens lies in their underlying infrastructure and purpose. Coins operate on their own independent blockchains, while tokens are built on existing blockchain networks. Coins serve as native currencies or stores of value, whereas tokens can have various utilities, including access to services, ownership representation, governance, or unique asset representation.

Coins often prioritize decentralization, security, and privacy, whereas tokens inherit the characteristics and capabilities of the blockchain platforms on which they are built. Tokens offer greater flexibility in terms of functionality and programmability, as they can leverage the features and smart contract capabilities of the underlying blockchain.

Furthermore, coins often have their own mining mechanisms or consensus algorithms to secure their networks, while tokens do not typically require mining and rely on the security provided by the underlying blockchain platform.

In conclusion, the distinction between coins and tokens is vital in understanding the diverse landscape of the cryptoverse. Coins, as native assets, operate on their own independent blockchains, providing decentralized financial systems and serving as mediums of exchange and stores of value. Tokens, on the other hand, rely on existing blockchain networks, offering a wide range of utilities such as access to services, ownership representation, governance participation, or unique asset representation.

While coins prioritize decentralization, security, and fungibility, tokens leverage the functionalities and programmability of their underlying blockchain platforms, allowing for greater flexibility and customization. Both coins and tokens play essential roles in the evolving cryptocurrency ecosystem, driving innovation, facilitating transactions, and enabling new forms of digital ownership and value exchange.

Stablecoins (Tokens)

Stablecoins have emerged as a significant category within the cryptocurrency ecosystem, offering stability in an otherwise volatile market. These digital assets are designed to maintain a stable value by pegging their price to a specific underlying asset, often a fiat currency like the US dollar. In this discussion, we will explore the concept of stablecoins, their functioning principles, and highlight some of the most common stablecoins referred to as "Stablecoin 1.0."

What are Stablecoins?

Stablecoins are cryptocurrencies that aim to minimize price volatility by pegging their value to an underlying asset. This underlying asset can be a fiat currency, a commodity, or a basket of assets. By pegging the stablecoin's value, users can have a reliable means of exchange, store of value, and unit of account within the cryptocurrency ecosystem. AS you will see in Part II, stablecoins go well beyond the simple coins in stablecoin 1.0.

Stablecoin 1.0:

Stablecoin 1.0 refers to the first generation of stablecoins that emerged in the market. These stablecoins primarily rely on fiat currency reserves to maintain their stability. Let's explore some of the most common stablecoins in this category:

- Tether (USDT): Tether is one of the earliest and most widely adopted stablecoins. It is pegged to the US dollar on a 1:1 basis, meaning that each USDT token is backed by one US dollar held in reserve. Tether has gained popularity as a convenient medium for trading and transferring value within the crypto ecosystem.
- USD Coin (USDC): Developed by Circle and Coinbase, USD Coin is an Ethereum-based stablecoin that is also pegged to the US dollar. It operates under a regulated framework and undergoes regular audits to ensure full backing by fiat reserves. USDC has gained traction among cryptocurrency traders and decentralized finance (DeFi) applications.
- TrueUSD (TUSD): TrueUSD is another stablecoin that aims to maintain a 1:1 peg against the US

dollar. It operates on the Ethereum blockchain and undergoes regular attestations to verify its reserves. TrueUSD provides transparency and reliability for users seeking stability in their transactions and holdings.

- Dai: Unlike the aforementioned stablecoins that rely on centralized entities to maintain stability, Dai operates on the decentralized Ethereum blockchain. It is collateralized by digital assets held in smart contracts and is governed by MakerDAO, a decentralized autonomous organization (DAO). The value of Dai is maintained through a system of overcollateralization and incentivized incentives to maintain its peg to the US dollar.

Functioning of Stablecoin 1.0:

Stablecoin 1.0 typically functions through a reserve-based model, where the stablecoin issuer holds a reserve of the underlying asset that backs the stablecoin's value. The reserve is audited regularly to ensure that it matches the circulating supply of the stablecoin. The stablecoin can be issued or redeemed at a 1:1 ratio against the underlying asset.

When a user purchases stablecoins, the issuer holds an equivalent amount of the underlying asset as collateral. This collateralization mechanism provides confidence to users that the stablecoin can be redeemed for the underlying asset at any time. It creates a link between the stability of the stablecoin and the value of the underlying asset.

The stability of stablecoin 1.0 is maintained by managing the reserve and adjusting the supply of stablecoins based on market demand. If the stablecoin's value deviates from its peg, market mechanisms are employed to restore stability. For example, if the stablecoin's value drops below the peg, the issuer may use the reserve to buy back stablecoins and reduce the circulating supply, thereby increasing the value.

Stablecoin 1.0, represented by popular examples like Tether, USD Coin, TrueUSD, and Dai, has paved the way for stability in the cryptocurrency market. These stablecoins rely on fiat currency reserves or collateralized assets to maintain a pegged value. They provide users with stability, convenience, and a reliable means of exchange within the cryptocurrency ecosystem. As the cryptocurrency industry continues to evolve, stablecoins will play a crucial role in enabling various use cases and fostering wider adoption of cryptocurrencies.

Utility Tokens

In the cryptocurrency world, utility tokens have gained significant attention and utility within various blockchain-based platforms and projects. Unlike cryptocurrencies that primarily serve as a medium of exchange or store of value, utility tokens are designed to provide access to specific products, services, or functionalities within a decentralized ecosystem. In this discussion, we will explore the concept of utility tokens, their purpose, and their role in the crypto landscape.

What is a Utility Token?

A utility token, also known as an application token or user token, represents a digital asset that has a specific purpose within a decentralized application (DApp), platform, or ecosystem. These tokens are typically created and issued through initial coin offerings (ICOs) or token sales to raise funds for project development.

Unlike cryptocurrencies like Bitcoin or Ethereum, which serve as general-purpose digital currencies, utility tokens have a narrower scope and are tied to the functionality or services offered by the issuing project. They act as a gateway to access and utilize certain features or benefits within the platform or network.

Purpose and Functionality:

Utility tokens serve a variety of purposes within decentralized ecosystems. Let's explore some common use cases and functionalities of utility tokens:

- Access and Usage: Utility tokens grant holders access to specific products, services, or functionalities within a platform or network. For example, within a decentralized file storage platform, utility tokens may be required to upload, download, or store files.
- Payment and Incentives: Utility tokens can be used as a means of payment for goods or services within the ecosystem. They enable efficient and cost-effective transactions between participants. Additionally, utility tokens may also be used to incentivize users for their participation, contributions, or engagement within the platform.
- Governance and Voting Rights: In some cases, utility tokens provide holders with voting rights and decision-making power within the decentralized ecosystem. Token holders can participate in governance mechanisms, such as voting on protocol upgrades, policy changes, or project proposals.

- Staking and Rewards: Utility tokens often have staking mechanisms where token holders lock their tokens for a specific period to support network security, consensus protocols, or decentralized validation processes. In return, they receive rewards or additional utility tokens as incentives.
- Discounted Services or Benefits: Some projects offer discounted services or exclusive benefits to token holders. These perks can include reduced transaction fees, premium features, early access to new functionalities, or priority customer support.
- Liquidity and Trading: Utility tokens can be traded on cryptocurrency exchanges, providing liquidity, and enabling users to buy, sell, or trade them against other cryptocurrencies or fiat currencies.

Regulatory Considerations:

It is important to note that the regulatory treatment of utility tokens may vary across jurisdictions. While some utility tokens may not fall under the category of securities, others may be subject to securities regulations depending on their characteristics and the legal framework of the respective jurisdiction.

Utility tokens play a vital role in the blockchain ecosystem, providing access, functionality, and incentives within decentralized platforms and networks. These tokens enable developers and projects to create innovative business models, foster user engagement, and incentivize active participation.

> "Good Regulation should be conducive to the business and to customer protection."
>
> Jamie Dimon

However, it is crucial for investors and users to conduct thorough research and due diligence before engaging with utility tokens to ensure they understand the specific functionalities, risks, and regulatory considerations associated with each token.

Global Exchange Tokens

Global exchange tokens, also known as exchange-based tokens or exchange utility tokens, are a type of digital asset that are primarily used within cryptocurrency exchanges. These tokens serve specific functions within the exchange ecosystem, providing users with various benefits, privileges, or incentives for their participation. In this discussion, we will explore the concept of global exchange tokens and their role in the crypto industry.

Global exchange tokens are designed to facilitate seamless and efficient trading activities on cryptocurrency exchanges. They are typically native to a specific exchange platform and can be used to access and unlock various features, services, or advantages within that particular exchange ecosystem. These tokens often have utility beyond mere speculative investments and are closely tied to the functionality and operations of the exchange.

Let's delve into some key aspects and use cases of global exchange tokens:

- Trading Fee Discounts: One of the primary benefits of global exchange tokens is the potential to receive discounted trading fees. Exchange platforms often provide incentives to users who hold a certain amount of their native tokens, allowing them to enjoy reduced fees when executing trades. This discount mechanism encourages users to acquire and hold these tokens, leading to increased liquidity and trading volume on the exchange.

- Token Utility and Access: Global exchange tokens can grant users access to specific services or features within the exchange platform. This can include priority access to new token listings, participation in token sales or initial exchange offerings (IEOs), advanced trading functionalities, or exclusive access to certain trading pairs. By holding these tokens, users can enjoy enhanced capabilities and benefits that enhance their overall trading experience.
- Staking and Governance: Some global exchange tokens incorporate staking mechanisms, allowing users to lock their tokens within the exchange ecosystem. In return, stakers may receive additional rewards, such as a share of trading fees generated by the platform. Additionally, certain exchange tokens may provide holders with voting rights or decision-making power in the governance of the exchange. Token holders can participate in voting processes for listing new tokens, making platform changes, or determining strategic directions for the exchange.
- Token Burns and Scarcity: To increase the value and scarcity of global exchange tokens, some exchanges implement token burn mechanisms. Token burns involve removing a portion of the token supply from circulation, typically by sending them to a burn address from which they cannot be accessed. By reducing the token supply, exchange platforms aim to create scarcity, potentially leading

to increased demand and value for the remaining tokens.

- Airdrops and Rewards: Global exchange tokens may also be distributed to users through airdrop campaigns or rewards programs. Airdrops can be used as a promotional strategy to attract new users to the exchange platform or to reward existing users for their loyalty and engagement. These airdrops can come in the form of free tokens or additional benefits tied to token holdings.

It is important to note that the specific functionalities and benefits of global exchange tokens can vary between different exchanges. Each exchange may have its own unique token economy and utility model, tailored to meet the specific needs and objectives of their platform and user base.

> "Regulations need to catch up with innovation."
>
> Henry Paulson

Global exchange tokens play a crucial role in the cryptocurrency exchange ecosystem by providing users with enhanced functionalities, incentives, and benefits. These tokens not only facilitate efficient trading activities but also promote user engagement, liquidity, and the overall growth of the exchange platform. By leveraging global exchange tokens, users can access discounted trading fees, participate in governance processes, enjoy exclusive features, and potentially earn additional rewards. As the crypto industry continues to evolve, global exchange tokens will likely remain an integral part of the exchange experience, shaping the landscape of digital asset trading.

Niche Tokens

Niche tokens, also known as specialized tokens, are a specific category of digital assets in the cryptocurrency ecosystem that cater to specific industries, communities, or use cases. These tokens are designed to serve a particular niche market, addressing specific needs, and providing unique functionalities. In this discussion, we will explore the concept of niche tokens and how they function in the cryptoverse.

Niche tokens differ from mainstream cryptocurrencies like Bitcoin or Ethereum, which aim to be widely adopted and used as general-purpose digital currencies or platforms. Niche tokens, on the other hand, are built with a narrower focus, targeting specific industries, communities, or applications. They offer specialized features, services, or solutions tailored to meet the requirements of their target audience.

Here are some key aspects and characteristics of niche tokens:

- Industry-Specific Solutions: Niche tokens often emerge to address specific challenges or inefficiencies within particular industries or sectors. For example, there are tokens dedicated to decentralized finance (DeFi), real estate, supply chain management, gaming, healthcare, renewable energy, and more. These tokens aim to provide industry-specific solutions by leveraging blockchain technology and its unique features such as transparency, immutability, and decentralization.
- Community Engagement and Participation: Niche tokens tend to foster strong communities and user engagement. They attract individuals who are

passionate about the specific industry or use case the token represents. These communities often actively contribute to the development, adoption, and promotion of the token and its associated ecosystem. Community members may participate in governance, propose improvements, or collaborate on projects related to the niche token.

- Tokenomics and Utility: Niche tokens typically have their own tokenomics, which define the supply, distribution, and utility of the tokens. The tokens may serve as a medium of exchange within a specific ecosystem, enabling transactions or access to specialized services. They may also incorporate additional features such as staking, voting rights, or rewards to incentivize token holders and promote engagement.
- Partnerships and Integration: Niche tokens often seek partnerships and integrations with relevant entities in their respective industries. Collaborations with businesses, organizations, or platforms can help drive adoption and create synergies between the token and existing industry infrastructure. These partnerships may involve integrating the token into existing applications or systems, allowing users to access specialized services or unlock specific functionalities.
- Ecosystem Development: Niche tokens are often accompanied by the development of a comprehensive ecosystem comprising decentralized applications (dApps), platforms, or protocols. The ecosystem aims to provide a robust infrastructure and toolset that supports the use case or industry targeted by the token. It may include marketplaces,

data analytics tools, governance mechanisms, and other components to enhance the token's utility and usability.

It's important to note that niche tokens may face challenges in terms of adoption and liquidity compared to more established cryptocurrencies. Their success often depends on factors such as market demand, industry acceptance, regulatory considerations, and the ability to deliver practical solutions.

Niche tokens play a vital role in the cryptoverse by catering to specific industries, communities, or use cases. These tokens offer specialized solutions, services, or features that address specific needs within their respective niches. By leveraging blockchain technology, niche tokens aim to enhance efficiency, transparency, and innovation in various industries. They foster community engagement, encourage industry-specific collaborations, and contribute to the development of focused ecosystems. While niche tokens may face unique challenges, their targeted approach and specialized functionalities provide opportunities for innovation and advancement within specific domains of the cryptoverse.

Memecoins

Memecoins have gained significant attention and popularity in the cryptocurrency world. Born from internet culture and driven by memes, these digital assets have captured the imagination of investors and traders alike. In this discussion, we will explore the concept of memecoins and how they function in the cryptoverse.

Memecoins are a specific category of cryptocurrencies that rely heavily on viral memes and internet culture for their branding and appeal. These tokens often feature well-known internet memes, characters, or references as part of their visual identity and marketing strategy. They aim to leverage the power of social media and internet communities to create hype and generate interest.

Here are some key aspects and characteristics of memecoins:

- Viral Memes and Internet Culture: Memecoins embrace popular internet memes and cultural references to capture attention and engage with a broad audience. These tokens often have catchy names, logos, and branding that resonate with online communities. By associating themselves with familiar and humorous content, memecoins create a sense of relatability and entertainment value.
- Community-Driven Engagement: Memecoins thrive on the active participation of their community. They encourage users to share memes, engage in online discussions, and spread the word about the token through social media platforms, forums, and dedicated communities. Memecoin communities often have a strong sense of camaraderie, with members sharing a common interest in internet culture and the potential for financial gains.
- Speculation and Trading: Memecoins are often driven by speculative trading activity, where investors aim to profit from short-term price fluctuations. Due to their volatile nature and hype-driven market sentiment, memecoins can experience significant price swings in relatively short periods. Investors may seek to capitalize on these price

movements, often leading to frenzied buying and selling activity.

- Limited Fundamental Value: Unlike traditional cryptocurrencies or tokens that may have underlying utility or value propositions, memecoins typically lack fundamental value beyond their internet meme appeal. Their value is primarily derived from market demand, hype, and speculation rather than intrinsic factors. This makes memecoins highly speculative investments, with prices largely influenced by sentiment and market trends.
- Risks and Caveats: It's important to note that memecoins carry inherent risks. Due to their speculative nature, memecoins can experience extreme price volatility and are prone to pump-and-dump schemes or market manipulation. Additionally, the reliance on internet memes and viral trends can lead to short-lived popularity and potential decline in value once the initial hype subsides.
- Social Impact and Donations: Some memecoins have embraced the concept of "meme philanthropy" by incorporating charitable donations or social impact initiatives into their ecosystem. These tokens may allocate a portion of their transaction fees or token supply to support specific causes or charities, leveraging the viral nature of memes to raise awareness and funds.

It's important to approach memecoins with caution and conduct thorough research before considering investment. While some memecoins may experience significant gains, the speculative nature of these tokens means that they also carry a high risk of losses. Investors should be mindful of market conditions, exercise due diligence, and diversify their portfolio to mitigate risks.

Memecoins represent a unique category of cryptocurrencies that harness the power of viral memes and internet culture to create a distinct brand and generate hype. These tokens rely on social media engagement, community-driven marketing, and speculative trading to drive their value. Memecoins offer a way for individuals to participate in the cryptocurrency market with a touch of humor and entertainment. However, it's important to approach memecoins with caution, understanding the risks associated with their speculative nature.

NFTs

Non-Fungible Tokens (NFTs) have gained significant attention and popularity in recent years, revolutionizing the world of digital assets and ownership. In this discussion, we will explore the concept of NFTs, their unique characteristics, and the potential future implications they hold, including their role in contracts and self-fulfilling agreements.

NFTs are cryptographic tokens that represent ownership or proof of authenticity of a unique digital asset or piece of content. Unlike cryptocurrencies such as Bitcoin or Ethereum, which are fungible and can be exchanged on a one-to-one basis, NFTs are non-fungible, meaning each token has its own distinct value and cannot be exchanged on a like-for-like basis.

- Unique Digital Assets: NFTs can represent a wide range of digital assets, including but not limited to artwork, music, videos, collectibles, virtual real estate, and more. What sets NFTs apart is their ability to establish provable ownership and uniqueness in the digital realm, allowing creators and collectors to buy, sell, and trade digital assets with confidence.

- Token Standards: The most common standard for NFTs is the ERC-721 standard on the Ethereum blockchain. It provides a framework for creating, buying, selling, and transferring ownership of NFTs. Other token standards, such as ERC-1155 and Flow's Fungible Token Standard (FT), also support the creation and management of NFTs.
- Proof of Ownership and Authenticity: NFTs leverage blockchain technology to establish proof of ownership and authenticity. The decentralized nature of blockchain ensures transparency and immutability, making it virtually impossible to alter or counterfeit ownership records. This feature is particularly valuable for artists and creators, as it enables them to prove the originality and ownership of their digital creations.
- Marketplaces and Secondary Sales: NFTs are typically bought, sold, and traded on online marketplaces specifically designed for digital assets. These marketplaces provide a platform for creators to showcase and sell their NFTs, while collectors and enthusiasts can browse, purchase, and engage in secondary sales. Some popular NFT marketplaces include OpenSea, Rarible, and SuperRare.
- Potential for Contracts: One fascinating aspect of NFTs is their potential to serve as a vehicle for contracts and self-fulfilling agreements. Smart contracts, which are self-executing agreements with predefined terms and conditions written in code, can

be attached to NFTs. This opens up possibilities for creating dynamic and automated contracts that can execute actions based on predefined triggers or conditions.

- Self-Fulfilling Contracts: In the future, NFTs could play a significant role in self-fulfilling contracts, where the terms and conditions of the agreement are embedded within the token itself. By attaching smart contracts to NFTs, the conditions can be programmed to automatically trigger certain actions when predefined criteria are met. This removes the need for intermediaries or third parties to oversee and enforce contractual obligations.
- Potential Applications: The potential applications of self-fulfilling contracts using NFTs are vast. For example, in the world of intellectual property, NFTs can automate royalty payments to artists or content creators every time their work is resold in the secondary market. Real estate transactions could also benefit from self-fulfilling contracts, where ownership transfer and property transfers occur automatically upon fulfilling predefined conditions.
- Challenges and Considerations: While the concept of self-fulfilling contracts through NFTs is exciting, there are challenges and considerations to address. One significant challenge is the need for reliable and accurate data inputs to trigger the execution of smart contracts. Additionally, legal frameworks and regulations surrounding contractual agreements need to adapt to the rapidly evolving landscape of blockchain technology.

As the NFT space continues to evolve, it is likely that we will see more innovation and exploration of NFTs in the context of contracts and self-fulfilling agreements. Developers, legal experts, and industry stakeholders will need to collaborate to establish best practices, standardize frameworks, and ensure the legal enforceability of smart contracts embedded in NFTs.

> "NFTs, crypto collecting like Pokémon for adults."
>
> Anonymous

NFTs have revolutionized the concept of digital ownership, providing a unique way to authenticate and trade digital assets. With the potential for self-fulfilling contracts, NFTs could transform the way we engage in contractual agreements, automating processes and ensuring the secure execution of predefined conditions. While there are challenges to overcome and legal considerations to address, the future of NFTs in contract fulfillment holds immense promise for various industries and opens up new possibilities for creators, collectors, and participants in the digital economy.

SFTs

Semi-fungible tokens (SFTs) are a unique type of digital asset that combines characteristics of both fungible and non-fungible tokens. While traditional fungible tokens are identical and interchangeable, and non-fungible tokens represent unique assets, SFTs introduce a level of customization and uniqueness within a specified category or group of assets. In this discussion, we will explore the concept of SFTs and highlight their usefulness as governance tokens.

Understanding Semi-Fungible Tokens (SFTs): SFTs are digital tokens that possess certain standardized properties within a specific category or group but allow for customization or differentiation within those parameters. Unlike non-fungible tokens (NFTs), which represent unique assets, SFTs maintain some level of interchangeability, allowing for specific distinctions within a broader set.

- Customization and Categorization: SFTs are often used to represent assets that share common attributes but possess unique variations within those parameters. For example, in a gaming context, SFTs could represent different variations of in-game items or characters, each with its own attributes, strengths, or abilities. These tokens can be categorized and customized to provide a diverse range of options while still maintaining a level of interchangeability.
- Usefulness as Governance Tokens: One of the key applications of SFTs is their use as governance tokens within decentralized systems and platforms. Governance tokens allow token holders to participate in decision-making processes, such as voting on protocol upgrades, proposing changes, or determining project directions. SFTs with specific customization features can be used to represent different levels of participation or voting power within the governance framework.
- Customized Voting and Participation: By leveraging SFTs as governance tokens, projects and platforms can introduce different levels of participation and influence based on the unique properties of each token. For example, SFTs could be designed to represent varying levels of staked or locked assets, reputation scores, or specific roles within the community. This customization allows for more

nuanced decision-making and ensures that voting power aligns with the contributions or expertise of token holders.

- Enhanced Governance Flexibility: SFTs provide greater flexibility and adaptability in governance systems compared to traditional fungible tokens. With SFTs, governance structures can be designed to consider diverse factors, such as token holders' expertise, duration of participation, or specific community roles. This allows for more inclusive and representative decision-making processes, fostering a sense of ownership and accountability within the community.
- Nurturing Community Engagement: By using SFTs as governance tokens, projects and platforms can enhance community engagement and incentivize active participation. The ability to customize and differentiate token properties creates a sense of uniqueness and value for token holders. This can encourage community members to contribute their expertise, actively participate in discussions, and vote on proposals, ultimately strengthening the overall governance framework.
- Challenges and Considerations: While SFTs offer exciting opportunities for governance and community participation, there are challenges to consider. Designing effective governance models using SFTs requires careful thought and consideration to ensure fairness, avoid centralization, and prevent manipulation. Additionally, clear communication and transparency are crucial to ensuring that token holders understand

the implications and value associated with different customization features.

Semi-fungible tokens (SFTs) introduce a unique dimension to the world of digital assets by combining fungibility and customization. As governance tokens, SFTs allow for customized voting and participation, enabling projects and platforms to foster inclusive decision-making processes and community engagement. By leveraging the flexibility and uniqueness of SFTs, governance frameworks can be tailored to align with token holders' contributions and expertise, nurturing a vibrant and empowered community. However, careful consideration must be given to designing robust and fair governance models using SFTs, and transparent.

Conclusion

In conclusion, the world of cryptocurrencies encompasses a diverse range of digital assets, each with its own unique characteristics and functionalities. Throughout this section, we have explored different coin classifications and their respective roles in the cryptoverse.

We began by examining stablecoins, which provide stability and peg their value to an external asset. Stablecoins serve as a crucial bridge between the volatile crypto market and traditional fiat currencies, facilitating seamless transactions and offering stability to users.

Moving on, we delved into utility tokens, which are designed to grant access to a specific product or service within a blockchain ecosystem. These tokens play a fundamental role in incentivizing user participation, promoting network growth, and enabling efficient operations within decentralized platforms.

Next, we explored global exchange tokens, which serve as native tokens within cryptocurrency exchanges. These tokens often play a vital role in platform governance, providing users with various benefits such as reduced fees, voting rights, and access to exclusive features.

We then examined niche tokens, which cater to specific industries or communities. These tokens are designed to address specific needs, solve industry-specific challenges, or cater to unique user bases. Niche tokens foster innovation and specialization within their respective sectors, driving adoption and growth.

Additionally, we explored memecoins, which have gained significant popularity for their humorous and viral nature. While memecoins often carry a degree of speculation and entertainment value, they highlight the creative and social aspects of the crypto community.

Furthermore, we delved into non-fungible tokens (NFTs), which represent unique assets on the blockchain, such as digital art, collectibles, and virtual real estate. NFTs have revolutionized ownership and provenance in the digital realm, providing creators and collectors with new avenues for monetization and asset representation.

Lastly, we discussed semi-fungible tokens (SFTs), which combine elements of fungible and non-fungible tokens. SFTs offer customization and differentiation within a specific category, making them valuable for governance purposes, where specific attributes and roles can be assigned to token holders.

By understanding the different classifications of coins, we gain a holistic view of the various functionalities and purposes that digital assets serve. These classifications reflect the evolving nature of the crypto landscape, where innovation and diversity drive the development of new token types to cater to different user needs and market demands.

As the crypto industry continues to evolve, new coin classifications may emerge, presenting exciting opportunities and challenges. It is crucial for individuals and businesses to stay informed and adapt to the changing landscape, leveraging the potential of these digital assets while being mindful of risks and regulatory considerations.

In conclusion, the classification of coins provides a framework for understanding the unique features and roles they play within the cryptoverse. From stablecoins to NFTs, each coin classification brings its own set of benefits and use cases, contributing to the overall growth and maturation of the cryptocurrency ecosystem.

Chapter 6: Problems and Solutions with Stablecoins

"Stablecoins The ground is currently being laid to set the way for a new type of currency –the stablecoin. What is the stablecoin? The stablecoin is an asset that typically features price stability. Cryptocurrency is notoriously unstable, with volatile prices that are often difficult to predict. The advantage of them is that they give the user total control over their holdings. On the other hand, the US dollar is a great example of a fiat stablecoin, as it offers low volatility and so provides a reliable unit of money to invest in both the short term and the long term. However, the US dollar doesn't give the user any form of control, as it is monitored by the Federal Reserve Bank and is dependent on the banking network in the US for commercial use. To get a combination of the two –full user control and reduced volatility –is an exciting prospect. Maker is a company that is currently working on a project to make this happen by creating a currency known as the Dai, which is set to become a stablecoin that combines user control with price stability. Social Networks"[19]

Introduction

When it comes to stablecoins, there are some problems that the world is just now figuring out how to face. While stablecoins are more useful than fiat currency, they have the difficulty of being so 'new' that many people do not understand how they work. As a result, some people are less likely to trust them because there are several problems in the early structure of some of the original stablecoins

---

[19] Ikuya Takashima at https://www.goodreads.com/quotes/8841294-stablecoins-the-ground-is-currently-being-laid-to-set-the.

(the Terra attack for example). As the system becomes more stable, the problems of the stablecoins will be ironed out. In this chapter we look at some of the problems.

Wagging the dog is one of the most dangerous problems that we are seeing in the cryptocurrency market today. While building the token first then building the company may be devastating to a normal coin, there a huge problem with building a stablecoin first then building the company around it. Because stablecoins are so unique, the features for building out the system must be considered before the coins should even be put on chain, or else a company may end up with a coin where everyone gets rich but the owners.

Tokenomics issues of stablecoins can also be a problem. Valuation is one of the most common problems that is seen within the stablecoin space. Because creators want to keep things very simple, most early stablecoins were based on fiat currency. As we have noted at several points, fiat currency is not stable. Therefore, quadratic valuation is more efficient than mono-asset valuation. Next, most early adopting stablecoins use a 1:1 system or an algorithmic system to maintain ecosystem valuation. This can be problematic when facing inflation, smart burns seem to be more efficient here. Further, many stablecoins are just set up for the initial sale and partnership agreements, there is not post-monetization growth strategy, which can lead to 'desperate' companies making partnership agreements with dangerous funds. If you do not plan for your money in the system, someone else will plan how to get your money.

Education and speculation are also big problems for the stablecoin market. With regulators not understanding how stablecoins work, speculators on the market influence them on how to create regulations. This can be devastating for people who are actually using the stablecoin as a hedge against market problems. The community must, therefore, come together to build a consensus of how the system should work so that regulators have good information to ensure that they are able to work within the system.

Wagging the dog

In the rapidly evolving world of cryptocurrencies, the creation and management of tokens require careful planning and consideration. One critical aspect often overlooked is the establishment of a robust corporate structure that aligns with the token's intended purpose. Neglecting this crucial element can lead to instability and hinder the token's long-term success. In this section, we will explore the challenges that arise when tokens are created without a well-defined corporate structure and highlight the benefits of designing a corporate ecosystem around the token's economics.

> "Innovation is the lifeblood of an organization. Knowing how to lead and work with creative people requires knowledge and action that often goes against typical organizational structure. Protect unusual people from bureaucracy and legalism typical of organizations."
>
> Max de Pree

The Pitfalls of Token Creation without Corporate Structure:

When a token is launched without a solid corporate structure in mind, it can result in various challenges and limitations that impede its stability and growth. Here are some key issues that can arise:

1. Lack of Purpose and Direction: Without a clear corporate structure, the token may lack a defined purpose, leaving investors and users uncertain about its utility and value proposition. This ambiguity can lead to skepticism, hampering adoption and limiting the token's potential.
2. Inconsistent Tokenomics: Tokenomics, including factors such as supply, distribution, and use cases, are vital for establishing stability and value. Without a corporate structure to support and guide the tokenomics, there is a risk of inconsistencies or haphazard decisions that can disrupt the token's equilibrium.
3. Governance Challenges: Tokens often require governance mechanisms to make critical decisions, such as protocol upgrades, consensus changes, or fund allocation. In the absence of a well-defined corporate structure, effective governance becomes difficult, potentially leading to conflicts and inefficiencies in decision-making.
4. Regulatory Compliance: The lack of a corporate structure can make it challenging to navigate the ever-evolving regulatory landscape. Compliance with regulations is crucial to ensure the token's legitimacy and protect the interests of investors. Without proper corporate governance and compliance measures, the token project may face legal uncertainties and regulatory challenges.

The Solution: Designing a Corporate Ecosystem for Tokenomics:

To address the issues associated with token creation without a corporate structure, it is imperative to establish a comprehensive ecosystem that aligns with the token's economics. This involves designing a corporate structure with multiple subsidiaries that support and enhance the token's functionality and stability. Here's why this approach is crucial:

1. Integrated Tokenomics: A well-designed corporate ecosystem ensures that all aspects of tokenomics, such as token issuance, reserve management, governance, and ecosystem development, are aligned and coordinated. This integration enhances the stability and reliability of the token, fostering trust among users and investors.
2. Strategic Decision-Making: A corporate structure provides a framework for making strategic decisions that benefit the token and its ecosystem. By establishing subsidiary entities, each with a specific focus, the project can allocate resources efficiently, foster innovation, and adapt to changing market conditions.
3. Legal and Regulatory Compliance: A robust corporate structure enables the token project to navigate legal and regulatory requirements effectively. By incorporating compliance measures into the corporate ecosystem, the project can mitigate regulatory risks, ensuring transparency, and building trust with regulators and users alike.
4. Ecosystem Expansion: With a well-designed corporate structure, the token project can explore opportunities for ecosystem expansion and strategic

partnerships. Subsidiaries within the corporate ecosystem can focus on specific areas such as research and development, marketing, or ecosystem integration, allowing for greater scalability and growth.

In the cryptoverse, token creation without a proper corporate structure can lead to significant challenges and undermine the stability and potential of the token. To avoid these pitfalls, it is crucial to design a corporate ecosystem that supports and enhances the token's functionality, governance, and overall stability. By establishing subsidiary entities and integrating tokenomics within a robust corporate structure, token projects can ensure long-term success, regulatory compliance, and the ability to adapt to evolving market dynamics. Ultimately, a well-designed corporate structure is the key to wagging the dog effectively, providing stability and a solid foundation for tokens to thrive in the cryptoverse.

Valuation understanding

Cryptocurrencies have revolutionized the financial landscape, offering new opportunities and challenges. However, a significant obstacle to wider adoption and understanding is the common misconception that cryptocurrencies are not backed by anything and lack inherent value. In this section, we will delve into the issue of valuation in the crypto space, address the prevailing misconception, and explore the Quadratic Valuation System as a functional solution.

The Problem of Misunderstanding Valuation:

One of the most common criticisms of cryptocurrencies is the perception that they are not backed by anything tangible, leading some to question their value and sustainability. This misconception often arises from a lack of understanding of the underlying mechanisms and dynamics of crypto markets. It is essential to address this issue to foster a more accurate perception of cryptocurrency valuation.

The Quadratic Valuation System:

To overcome the misunderstanding of crypto valuation, the Quadratic Valuation System provides a comprehensive framework that considers multiple factors contributing to a cryptocurrency's value. Let's explore the four key components of this valuation system:

1. Market Value: Market value refers to the price at which people are willing to buy and sell a cryptocurrency. This component recognizes the role of market dynamics, demand, and investor sentiment in determining the value of a digital asset. The perception of scarcity and utility of the cryptocurrency plays a significant role in its market value.
2. Transactional Value: Transactional value reflects the value added by the increased adoption and use of the cryptocurrency. As more individuals, companies, and nations embrace a particular token, its transactional value increases. This value is derived from the efficiency, speed, and cost-effectiveness of transactions facilitated by the cryptocurrency's underlying blockchain technology.
3. Transitional Value: Transitional value represents the value derived from the data passing over the blockchain network. Cryptocurrencies that enable secure and transparent data transfers can generate

significant value as the volume and importance of data being transmitted increases. This value arises from the potential for improved data integrity, privacy, and security offered by blockchain technology.

4. Asset Value: Asset value refers to the value of the asset pool backing the cryptocurrency. This component is especially relevant for stablecoins, which are designed to maintain a stable value relative to a specific asset or basket of assets. The asset value is determined by factors such as the composition and liquidity of the asset pool, ensuring that the cryptocurrency has ample backing to support its value.

Functional Solution: Addressing Misconceptions:

> "Strive not to be a success, but rather to be of value."
>
> Albert Einstein

The Quadratic Valuation System provides a functional solution to address the misconception that cryptocurrencies lack inherent value. By considering multiple dimensions of value, this framework offers a more holistic view of crypto valuation. Educating the public, investors, and regulators about this comprehensive valuation system is crucial for overcoming the misunderstanding surrounding cryptocurrency value.

Additionally, transparency, accurate reporting, and regulatory compliance within the crypto ecosystem are vital for building trust and addressing concerns about valuation. Implementing industry-wide standards and best practices, including disclosure requirements and auditing processes, can help establish a reliable foundation for cryptocurrency valuation.

Misunderstanding the valuation of cryptocurrencies, often stemming from the misconception that they are not backed by anything, poses a significant challenge to wider adoption and acceptance. The Quadratic Valuation System, incorporating market value, transactional value, transitional value, and asset value, provides a comprehensive framework for understanding the true value of cryptocurrencies. By embracing this system and promoting transparency and regulatory compliance, the crypto industry can overcome misconceptions, foster trust, and pave the way for greater adoption and acceptance in the financial landscape.

Ecosystem Control

Ecosystem control in stablecoins can become a challenge when companies release the tokens and attempt to shift the responsibility to the community without proper decentralization. This approach can hinder the stability and long-term success of the stablecoin. To address this issue, a gradual decentralization process is necessary. This includes releasing tokens to the public, introducing a governance token (SFT) for community participation, and implementing a smart burn function. By adopting these measures, companies can ensure a more equitable distribution of power and decision-making, maintain compliance with regulatory frameworks, and enhance the stability and trustworthiness of the stablecoin ecosystem.

Post-Monetization Growth

One prevalent issue is the lack of a post-monetization strategy after the initial release of the coins. Many stablecoin projects rely solely on the hope that the community will continue using the token without a clear plan for sustaining value and ensuring long-term viability. This approach can lead to uncertainty and instability within the ecosystem.

One effective solution to address this challenge is the implementation of a smart burn system. The smart burn mechanism plays a crucial role in maintaining the stability and profitability of the stablecoin. By designating a portion of the tokens to be burned on each transaction, the smart burn system reduces the circulating supply, which in turn can drive up the value of the remaining tokens. This creates a positive feedback loop that promotes scarcity and potential price appreciation.

The smart burn system offers several advantages for the post-monetization phase of a stablecoin. Firstly, it provides a means of generating profit and maintaining the value of the token over time. As transactions occur and tokens are burned, the reduction in supply can contribute to increased scarcity and demand, potentially driving up the token's price. This can incentivize users to hold and transact with the stablecoin, as they stand to benefit from its value appreciation.

Moreover, the smart burn system offers a sustainable revenue source for maintaining the stability of the stablecoin ecosystem. A portion of the burned tokens can be allocated to funding ongoing development, maintenance, and operational expenses of the project. This ensures that there are sufficient resources to support the infrastructure, security measures, and improvements necessary for a robust and reliable stablecoin system.

> "Even if you are on the right track, you will get run over if you just sit there."
>
> Will Rodgers

Additionally, the smart burn system helps mitigate potential inflationary pressures that may arise with the continuous issuance of new tokens. Instead of relying solely on the creation of additional tokens to meet demand, the smart burn mechanism reduces supply while simultaneously increasing the value of the remaining tokens. This counteracts the risk of devaluation and contributes to a more stable and sustainable ecosystem.

The smart burn system also aligns with the principles of transparency and decentralization. Since the burn process occurs on-chain, the transactions and token burn events are visible and auditable by anyone. This fosters trust and accountability within the stablecoin community, as stakeholders can verify the proper implementation of the burn mechanism and ensure its fairness.

Furthermore, the smart burn system incentivizes long-term token holders by creating an opportunity for them to benefit from the growth and success of the stablecoin. As the value of the remaining tokens appreciates due to the reduction in supply, holders can potentially realize capital gains. This encourages loyalty and participation in the stablecoin ecosystem, as users have a vested interest in its continued success.

In summary, the implementation of a smart burn system offers a promising solution to address the lack of post-monetization strategies in many stablecoin projects. By burning tokens on each transaction, the smart burn mechanism can generate profit, maintain token value, fund ongoing development, and mitigate inflationary pressures. It also promotes transparency, decentralization, and incentivizes long-term engagement from the stablecoin community. As such, the smart burn system represents a valuable tool for ensuring the stability and long-term viability of stablecoins in the ever-evolving landscape of the cryptocurrency market.

Education in Cryptocurrency

Lack of education in the field of cryptocurrency is indeed a significant problem, especially when it comes to stablecoins. It is crucial to address this issue as it can lead to misconceptions, misinformation, and unwarranted criticism from individuals, including legislators who may not fully understand the intricacies of stablecoin technology. One prominent example is the case of Elizabeth Warren, a senator known for her critical stance on stablecoins and cryptocurrencies in general.[20] However, the solution lies in providing accessible and comprehensive education on stablecoins, including within the public school

---

[20] Warmbrodt, Zachary (2023). "Elizabeth Warren is building an anti-crypto army. Some conservatives are onboard." Politico at https://www.politico.com/news/2023/02/14/elizabeth-warren-anti-crypto-ftx-00082624. This just shows the disconnect between the government and the people who are working on the ground.

system, to bridge the knowledge gap and foster informed discussions.

One of the primary challenges in the cryptocurrency space, including stablecoins, is the lack of awareness and understanding among legislators and policymakers. Many lawmakers approach the subject with skepticism and, at times, hostility due to misconceptions and a limited understanding of the underlying technology. This can result in proposals and regulations that may not align with the true potential and benefits of stablecoins. Education is vital to dispel myths and ensure that lawmakers can make informed decisions that promote innovation, consumer protection, and economic growth.

> "Growth, in some curious way, I suspect, demands being always in motion just a little bit, one way or another."
>
> Norman Mailer

To address this issue, it is essential to establish educational initiatives that focus on providing accessible and accurate information about stablecoins. One approach is to develop free or low-cost community education programs that aim to educate the general public, including lawmakers, about the fundamentals, use cases, and benefits of stablecoins. These programs can include seminars, workshops, and online courses that cover topics such as blockchain technology, decentralized finance (DeFi), and the specific characteristics and mechanics of stablecoins.

Integrating cryptocurrency education into the public school system is another effective way to address the knowledge gap. By including stablecoins and related topics in the curriculum, students can gain early exposure to the concepts, technology, and potential impact of cryptocurrencies on the global economy. This can be implemented at various educational levels, from high school to higher education institutions. Such initiatives can be designed to provide a balanced view, covering both the opportunities and risks associated with stablecoins, while encouraging critical thinking and fostering a future workforce that is knowledgeable about digital currencies.

Additionally, collaborations between educational institutions, industry experts, and blockchain companies can play a crucial role in developing educational materials and resources. These partnerships can help create comprehensive and up-to-date educational content that covers various aspects of stablecoins, including their design, operation, regulatory considerations, and potential use cases. By leveraging the expertise of industry professionals, these educational materials can provide real-world insights and practical examples to enhance understanding.

Furthermore, engaging with lawmakers directly and providing them with accurate and evidence-based information is crucial. Organizing seminars, conferences, or roundtable discussions specifically tailored for legislators can help bridge the knowledge gap and facilitate meaningful conversations. These forums can provide an opportunity for industry experts to share their insights, clarify misconceptions, and address concerns related to stablecoins. By fostering open dialogue and creating a platform for collaboration, legislators can gain a deeper understanding of stablecoins, leading to more informed decision-making and the development of balanced and effective regulatory frameworks.

In addition to formal education, promoting awareness through public campaigns and media channels is essential. Educating the general public about stablecoins and cryptocurrencies can help dispel myths, reduce stigmas, and create a more informed and receptive environment. This can be achieved through media collaborations, public service announcements, and social media campaigns that focus on highlighting the benefits, use cases, and potential risks associated with stablecoins. Increasing public awareness can lead to a more informed public discourse, which, in turn, can influence legislators and policymakers to approach stablecoins with a greater understanding and appreciation.

The lack of education in the field of cryptocurrency, particularly concerning stablecoins, poses a significant challenge. Misconceptions and a limited understanding among legislators can lead to misguided policies and regulations. However, by implementing comprehensive educational initiatives, including free or low-cost community programs and integration into the public school system, we can bridge the knowledge gap and promote a more informed and nuanced understanding of stablecoins. Collaboration between educational institutions, industry experts, and policymakers, along with public awareness campaigns, can contribute to creating a more favorable environment for stablecoins and cryptocurrencies, where innovation and consumer protection can coexist.

Speculation v. Investment

One of the key problems in the stablecoin space revolves around the distinction between speculation and investment. Speculation refers to the process of engaging in financial activities without thorough research or analysis, often driven by the desire to follow a market fad or make quick profits. On the other hand, investment involves conducting diligent research, analyzing the fundamentals, and making informed decisions based on long-term value and growth potential. This problem is particularly relevant in the context of stablecoins, where misconceptions and unrealistic expectations can arise due to a lack of understanding about their purpose and function.

> "Good and bad ideas both come from the same fountain of speculation and experiment."
>
> Shaun Tan

It is not uncommon to see individuals who complain about stablecoins, particularly those involved in speculation, expressing disappointment or frustration because they did not make the significant profits they anticipated. However, it is crucial to recognize that stablecoins are designed primarily as a hedge against inflation and as a means of preserving value, rather than as a tool for generating substantial profits. Their primary objective is to maintain a stable value relative to a specific reference asset, typically a fiat currency like the US dollar or a basket of assets.

Stablecoins provide stability and serve as a reliable store of value in volatile cryptocurrency markets. They are particularly useful for traders who seek to temporarily park their funds and avoid the price fluctuations associated with other cryptocurrencies. Moreover, stablecoins facilitate easier and faster transactions within the crypto ecosystem, allowing users to swiftly move in and out of positions without worrying about the price volatility that plagues other digital assets.

However, the misconception arises when individuals approach stablecoins with a speculative mindset, expecting them to generate significant profits similar to high-risk assets like cryptocurrencies or stocks. Stablecoins, by design, aim to maintain a stable value, with minimal price fluctuations. Their purpose is not to offer substantial returns but rather to preserve purchasing power and protect against the erosion of value caused by inflation. Thus, evaluating the success of stablecoins based on speculative criteria is a fundamental misunderstanding of their intended function.

To address this problem, it is crucial to educate market participants about the purpose and characteristics of stablecoins. Providing clear and accurate information about their value proposition, such as stability, ease of use, and efficient transactions, can help set realistic expectations and prevent individuals from approaching stablecoins purely from a speculative standpoint.

Another aspect to consider is the importance of conducting thorough research and due diligence before engaging with stablecoins or any other financial instrument. Investment requires a careful analysis of the underlying asset, the stability mechanisms employed, the management team, and the overall ecosystem. This approach allows investors to make informed decisions based on the long-term value and growth potential of stablecoins.

Furthermore, raising awareness about the role of stablecoins as a hedge against inflation can help individuals understand their significance in a broader economic context. Stablecoins provide an alternative to traditional fiat currencies that are subject to inflationary pressures. By preserving value and offering a stable means of exchange, stablecoins can play a vital role in a diversified investment portfolio and serve as a reliable store of wealth.

The distinction between speculation and investment is a crucial problem in the stablecoin space. Misconceptions and unrealistic expectations often arise when individuals approach stablecoins with a speculative mindset, seeking substantial profits rather than recognizing their primary function as a hedge against inflation. Educating market participants about the purpose, characteristics, and benefits of stablecoins, as well as promoting the importance of thorough research and due diligence, can help address this problem. By fostering a better understanding of stablecoins as a stable store of value and facilitating efficient transactions, we can promote their responsible use and maximize their potential benefits within the cryptocurrency ecosystem.

Conclusion

The stablecoin space faces several key problems that need to be addressed in order to ensure the stability, functionality, and widespread adoption of stablecoins. Through this chapter, we have explored some of these challenges and discussed potential solutions that can help overcome them.

One of the major problems in the stablecoin space is the lack of understanding and education about stablecoins, both among the general public and policymakers. This leads to misconceptions and unrealistic expectations, particularly in terms of the profit-generating potential of stablecoins. By providing free or low-cost community education on stablecoins, including in the public school system, we can promote a better understanding of their purpose, benefits, and limitations. This education can help individuals differentiate between speculation and investment, setting realistic expectations and fostering responsible use of stablecoins.

Another significant problem is the lack of a post-monetization strategy for stablecoins. Many stablecoins are released without a clear plan for maintaining stability and ensuring ongoing usability. Gradually decentralizing control through the release of tokens, introducing governance mechanisms through the issuance of governance tokens, and implementing smart burn systems can help ensure the long-term sustainability and stability of stablecoins. These strategies allow for community involvement and participation in the decision-making process while maintaining stability and value preservation.

The issue of valuation also poses a challenge in the stablecoin space. The concept of valuation is often misunderstood, leading to statements like "it is not backed by anything" or overly simplistic assessments of value. Implementing a quadratic valuation system that considers market value, transactional value, transitional value, and asset value can provide a more comprehensive and accurate representation of the value of stablecoins. This approach takes into account the dynamics of the market, the growth potential through increased usage, the value of data passing over the blockchain, and the backing assets in the pool. By promoting a better understanding of the valuation mechanisms, we can enhance confidence in stablecoins and facilitate more accurate assessments of their worth.

Ecosystem control is another important problem to address in the stablecoin space. Releasing stablecoins and then simply attempting to "turn the problem over to the community" without a well-defined corporate structure or governance model can lead to instability and lack of direction. Creating a corporate structure with several subsidiaries, gradually decentralizing control, and incorporating smart burn functions can provide a more effective framework for stablecoin governance. This approach ensures the stability of the stablecoin ecosystem while allowing for community participation and reducing the risk of centralized control.

Lastly, the lack of a clear post-monetization strategy and the speculative mindset of some participants pose challenges to stablecoins. Establishing a proper understanding of stablecoins as a hedge against inflation, rather than as a profit-generating tool, is crucial. Stablecoins serve as a means of preserving value and providing stability, making them an important component of diversified investment portfolios. By educating market participants about the purpose and characteristics of stablecoins, we can align expectations and promote responsible use.

In conclusion, addressing the problems in the stablecoin space requires a multi-faceted approach. Education and awareness programs can enhance understanding and dispel misconceptions, helping individuals differentiate between speculation and investment. Implementing post-monetization strategies, such as decentralization and smart burn mechanisms, can ensure the stability and long-term viability of stablecoins. Embracing comprehensive valuation systems and fostering ecosystem control through well-designed corporate structures can further enhance the credibility and functionality of stablecoins. By collectively addressing these challenges, we can foster a healthier, more robust, and widely adopted stablecoin ecosystem that serves as a reliable store of value and a vital tool in the evolving landscape of digital finance.

Chapter 7: Investment in Stablecoins

"Price is what you pay. Value is what you get."[21]

Introduction

When looking at the process of investing in stablecoins, one needs to realize that they are investing in something that they hope the value will stay the same within. This can be a challenge for many people because they are used to thinking that an investment will yield profits rather than just 'treading' water. In the current global economy, treading water can be a major success as currencies are losing value at the rate of 8% per year in some countries.

Investment in stablecoins can be broken down into three different categories of investing. In the early stages of investing, there is profits to be made by investing in a token before it achieves its parity value with its peg. This is also the area that confuses many regulatory agencies seeing the token a security rather than a currency. In mid-stage investing, there is still a small bit of room for the value of the token to increase, however, the window gets much smaller. Returns of over 15% are unlikely. Once a token reaches its later stages, the token has reached parity and the success of the token relies on the asset pool.

Investors in the stablecoin space need to be willing to balance profitability v. hedging in their portfolio. The use of stablecoins in that equation is normally going to be for hedging rather than for profits. To ensure this, the investor needs to look at the fees for the token and the relationship with the exchange. High fees or high exchange charges can

---

[21] Warren Buffet at https://www.dividend.com/dividend-education/41-inspiring-and-intelligent-investing-quotes/.

eliminate the ability to use the stablecoin as a hedge. For active traders, there is also the question of cross-chain compatibility.

Early-Stage Investing

Early stage investing in stablecoins presents a unique opportunity for investors to capitalize on the growth and potential profitability of these digital assets. Stablecoins, designed to maintain a stable value relative to a specific reference asset, have gained significant traction in the cryptocurrency space. While their primary purpose is to provide stability and facilitate efficient transactions, astute investors can leverage early-stage investments to generate substantial returns. In this article, we will explore how early stage investing in stablecoins can be profitable, particularly during the growth phase leading up to the achievement of parity with their pegged asset.

> "From coast to coast, the FBI and the Securities and Exchange Commission have ensnared people not only at hedge funds, but at technology and pharmaceutical companies, consulting and law firms, government agencies, and even major stock exchanges."
>
> Preet Bharara

1. Understanding Stablecoins and their Market Dynamics: Stablecoins are digital currencies that aim to mitigate the volatility associated with traditional cryptocurrencies. These coins typically achieve stability by pegging their value to an underlying asset such as fiat currencies,

commodities, or other cryptocurrencies. By offering stability, stablecoins provide a reliable medium of exchange and a store of value within the volatile cryptocurrency ecosystem.

2. Early Stage Investing and the Growth Potential: Early stage investing in stablecoins offers the potential for substantial profits due to the rapid growth and increasing demand for these assets. During the early stages, stablecoins often experience a significant appreciation in value as they gain market adoption and strive to achieve parity with their pegged asset. This growth phase is fueled by increasing confidence in the stablecoin's stability, usability, and overall ecosystem.
3. Fiat-Based Stablecoins: Unlocking Growth Potential: Fiat-based stablecoins, which are pegged to traditional fiat currencies, have witnessed remarkable growth in recent years. As these stablecoins establish credibility and regulatory compliance, they attract greater usage and market participation. Early investors who identify the potential of these fiat-based stablecoins can capture significant value as the market recognizes their stability and their adoption accelerates. In some cases, these early investments have yielded returns of up to 500 times the initial investment.
4. Stablecoin 2.0: Exploring New Frontiers: Stablecoin 2.0 represents the next evolution of stablecoins, offering innovative features and improved mechanisms for maintaining stability. These second-generation stablecoins often leverage decentralized finance (DeFi) protocols and advanced technologies to enhance stability and

liquidity. Early-stage investments in Stablecoin 2.0 projects can offer substantial growth opportunities, with the potential for returns of up to 10,000 times the initial investment. As these projects gain traction and showcase their unique value propositions, investors can benefit from their success.[22]

5. Mitigating Risks through Due Diligence: While early stage investing in stablecoins presents exciting possibilities, it is important to acknowledge the inherent risks involved. To mitigate these risks, investors should conduct thorough due diligence before making investment decisions. This includes evaluating the stability mechanisms, team expertise, market demand, regulatory compliance, and the project's long-term viability. By making informed investment choices based on sound research and analysis, investors can better position themselves to capitalize on the growth potential of stablecoins.
6. Diversification and Risk Management: As with any investment strategy, diversification is crucial to manage risks effectively. Investors should consider building a diversified portfolio of stablecoins and other cryptocurrency investments to spread risk across different assets and projects. Diversification helps to mitigate the impact of any potential downturn in a specific stablecoin and allows investors to benefit from the growth potential of multiple projects.

---

[22] Please note that KIROS does not promise 10,000x growth, our system is better managed and the growth for early investors will be more controlled.

7. Regulatory Landscape and Investor Protection: As stablecoins gain prominence, regulators around the world are scrutinizing these digital assets to ensure compliance with existing laws and regulations. Investors should remain informed about the regulatory landscape, as changes in regulations can impact the market dynamics of stablecoins. Compliance with regulatory requirements enhances investor confidence and contributes to the long-term stability and success of stablecoin projects.

Early stage investing in stablecoins provides a unique opportunity for investors to capture substantial value during the growth phase of these digital assets. Understanding the dynamics of stablecoins, conducting thorough due diligence, and diversifying investments can help investors navigate the evolving market and maximize potential returns. However, it is important to approach these investments with caution and be aware of the inherent risks involved. By staying informed, making informed decisions, and embracing a long-term investment perspective, investors can position themselves to profit from the growth potential of stablecoins and contribute to the maturation of the stablecoin market.

Mid-Stage Investing

As stablecoins continue to evolve and gain prominence in the cryptocurrency landscape, investors are exploring various stages of investment opportunities. While early-stage investments offer the potential for significant returns, mid-stage investing in stablecoins requires a different perspective. In this article, we will delve into the world of mid-stage stablecoin investing, understanding the dynamics, managing expectations, and assessing the potential for profits in this phase of investment.

1. Understanding Mid-Stage Investing in Stablecoins: Mid-stage investing refers to the phase in which stablecoins have already gained some market traction, established their stability mechanisms, and demonstrated a certain level of market acceptance. During this phase, stablecoins are likely to have achieved a relatively stable price, reduced volatility, and a growing user base.
2. Potential Returns and Profit Expectations: It is important to set realistic expectations when it comes to mid-stage stablecoin investing. Unlike early-stage investments, where the potential for significant profits exists, mid-stage investments typically yield more conservative returns. In most cases, investors can expect returns in the range of 5% to 15%, depending on various factors such as market conditions, overall stability, and demand for the stablecoin.
3. Assessing Stability and Growth Potential: In mid-stage investing, the focus shifts from explosive growth to stability and steady growth. Investors should evaluate the stability mechanisms implemented by the stablecoin project, such as reserve holdings, auditing procedures, and regulatory compliance. A stablecoin that demonstrates consistent stability and growth potential is more likely to generate steady returns for investors.
4. Evaluating Market Demand and Adoption: Market demand and adoption play a crucial role in the success of stablecoins. Investors should assess the

stability coin's utility, use cases, and the ecosystem built around it. Factors such as the availability of liquidity, partnerships, and integration with other platforms contribute to the wider adoption of stablecoins. Analyzing the demand and adoption trends helps investors gauge the potential for sustained growth and value appreciation.

5. Managing Risks and Diversification: Risk management is crucial in mid-stage stablecoin investing. While mid-stage stablecoins are relatively more stable than early-stage counterparts, there are still inherent risks involved. Investors should diversify their portfolios to spread risk across different stablecoins and other investment assets. By diversifying, investors can reduce their exposure to any single stablecoin's volatility and potential downside.
6. Monitoring Regulatory Developments: Regulatory frameworks surrounding stablecoins are continuously evolving. As an investor in mid-stage stablecoins, it is essential to stay abreast of regulatory developments in the jurisdictions where the stablecoins operate. Changes in regulations can impact market dynamics, user adoption, and the overall stability of stablecoin projects. Compliance with relevant regulations enhances investor confidence and mitigates potential risks.
7. Long-Term Perspective and Maturation of Stablecoins: Mid-stage investing in stablecoins requires a long-term perspective. While the potential for significant short-term gains may be limited, investors can participate in the maturation of stablecoins and contribute to their long-term

success. By supporting stablecoin projects that demonstrate sustainable growth and adoption, investors can position themselves for long-term profitability and contribute to the stability and development of the broader stablecoin ecosystem.

> "Given the typical structure of hedge funds, the need to do something different to make money in a consistent way."
>
> Paul Singer

Mid-stage investing in stablecoins offers a more stable and measured approach compared to early-stage investments. While the potential for large profits may be limited during this phase, investors can still generate respectable returns by evaluating stability, assessing market demand and adoption, and diversifying their portfolios. Additionally, staying informed about regulatory developments and embracing a long-term investment perspective are essential for navigating mid-stage stablecoin investing. By carefully managing risks and maintaining realistic profit expectations, investors can participate in the growth and maturation of stablecoins while optimizing their investment outcomes.

Late-Stage Investing

Introduction: Late-stage investment in stablecoins presents a unique opportunity for investors seeking a hedge against inflation and market volatility. Unlike early and mid-stage investments, late-stage investments are characterized by a more mature market, established stability mechanisms, and a proven track record. In this article, we will explore the dynamics of late-stage investment in stablecoins, highlighting its role as a hedge and managing expectations regarding growth potential.

1. Late Stage Stablecoin Investment as a Hedge: Late stage stablecoin investment is primarily driven by the need for a reliable hedge against inflation and market uncertainties. Stablecoins offer stability in their value by being pegged to a specific asset or basket of assets, such as fiat currencies or commodities. Investors allocate their funds to stablecoins as a way to preserve wealth and mitigate the risks associated with traditional volatile cryptocurrencies or fiat currencies.
2. Limited Growth Potential: Late stage stablecoin investments are generally characterized by limited growth potential compared to early and mid-stage investments. As stablecoins reach a more mature stage, their value becomes closely aligned with the assets they are pegged to. Consequently, the growth potential of stablecoins in relation to their pegged assets is restrained. Investors should approach late stage stablecoin investments with the expectation of stability and preservation of capital, rather than significant growth.
3. Hedging Against Inflation: One of the primary benefits of late stage stablecoin investments is their ability to act as a hedge against inflation. Inflation erodes the purchasing power of traditional fiat currencies over time. Stablecoins, particularly those pegged to stable assets, can provide a reliable store of value, preserving wealth and mitigating the negative impact of inflation. Investors seeking to protect their assets from inflationary pressures can

consider late stage stablecoin investments as part of their overall investment strategy.

4. Evaluating Stability and Asset Backing: Late stage stablecoin investments necessitate a thorough evaluation of the stability mechanisms and asset backing of the stablecoin. Investors should assess the transparency of the stablecoin project, including regular audits and disclosures of the asset reserves. Understanding the risk management practices and regulatory compliance of the stablecoin issuer is crucial in ensuring the stability and reliability of the investment.
5. Diversification and Risk Management: While late stage stablecoin investments offer stability, diversification remains a prudent risk management strategy. Investors should consider diversifying their portfolios by allocating funds to a range of stablecoins with different underlying asset pegs. By spreading the investment across multiple stablecoins, investors can further mitigate risks associated with any specific stablecoin's performance or external market factors.
6. Monitoring Market and Economic Conditions: Late stage stablecoin investors should closely monitor market and economic conditions that may impact the stability of the stablecoin. Factors such as changes in regulatory frameworks, macroeconomic indicators, and geopolitical events can influence the value and performance of stablecoins. Staying informed and adapting investment strategies accordingly is essential in navigating late stage stablecoin investments.

7. Long-Term Value Preservation: Late stage stablecoin investments are best suited for long-term value preservation rather than seeking short-term gains. Investors should align their expectations with the core purpose of stablecoins as a stable store of value. While substantial growth may be unlikely, the primary goal of late stage stablecoin investments is capital preservation and mitigating risk, particularly in the face of inflationary pressures.

Late-stage investment in stablecoins serves as a reliable hedge against inflation and market volatility. While growth potential may be limited compared to early and mid-stage investments, the stability and preservation of capital offered by stablecoins are valuable attributes. Investors should approach late stage stablecoin investments with a long-term perspective, focusing on hedging against inflation and diversifying their portfolios. By carefully evaluating stability mechanisms, monitoring market conditions, and maintaining a realistic outlook, investors can leverage late stage stablecoin investments to preserve wealth and navigate the evolving cryptocurrency landscape.

Profitability v. Hedging

Stablecoins have gained significant attention in the cryptocurrency space due to their stability and potential for hedging against market volatility. While stablecoins are primarily designed to maintain a stable value, the question of profitability arises for investors. In this article, we will explore the relationship between profitability and hedging in stablecoins, highlighting how stablecoins can offer a balance between stability and potential returns.

1. Hedging Against Volatility: Stablecoins are often sought after by investors as a means of hedging against the inherent volatility of cryptocurrencies. Unlike traditional cryptocurrencies like Bitcoin or Ethereum, stablecoins are designed to maintain a stable value by being pegged to specific assets, such as fiat currencies or commodities. This stability provides a reliable hedge for investors, shielding their investments from the wild price fluctuations commonly associated with cryptocurrencies.
2. Stablecoin Profitability: Stablecoins are primarily designed to maintain a stable value, rather than generate substantial profits. The focus of stablecoin investors is typically on capital preservation and hedging against market risks. As such, stablecoins may not offer the same level of profitability as higher-risk cryptocurrencies. However, the primary objective of stablecoins is to provide stability, making them an attractive option for risk-averse investors.
3. Yield-Generating Stablecoins: While stablecoins themselves may not generate significant profits; certain variants called yield-generating stablecoins offer the potential for additional returns. These stablecoins leverage various mechanisms such as lending, staking, or liquidity pools to generate yield for investors. By actively participating in these platforms, investors can earn additional income on their stablecoin holdings, enhancing the overall profitability of their investment.
4. Stablecoins as a Cost-Efficient Hedge: Another aspect of profitability in stablecoins is the cost efficiency they offer as a hedge. Compared to

traditional hedging instruments like options or futures contracts, stablecoins provide a more accessible and cost-effective means of mitigating risks. Investors can easily convert their holdings into stablecoins during market downturns, thereby reducing exposure to volatile assets without incurring significant transaction costs or fees.

5. Diversification and Risk Management: While stablecoins offer stability and potential hedging benefits, investors should consider diversification as a risk management strategy. Diversifying across different stablecoins, each pegged to different assets or using different mechanisms, can provide a broader hedging approach. This diversification helps spread risk and ensures a more balanced portfolio, reducing reliance on any single stablecoin.
6. Market Conditions and Profitability: The profitability of stablecoins can also be influenced by market conditions. During periods of economic stability and low inflation, stablecoin returns may be relatively modest. However, in times of market volatility or economic uncertainty, stablecoins can become more sought after, potentially leading to increased demand and higher prices. Understanding market cycles and aligning stablecoin investments with market conditions can play a role in enhancing profitability.
7. Balancing Stability and Profitability: Investors seeking both stability and profitability in their cryptocurrency portfolio should strike a balance between stablecoins and higher-risk assets. Stablecoins serve as a hedging tool, providing

stability and preserving capital during turbulent market conditions. By allocating a portion of their portfolio to stablecoins, investors can mitigate risks and maintain a solid foundation. Meanwhile, higher-risk assets can offer the potential for greater returns, albeit with increased volatility.

The relationship between profitability and hedging in stablecoins is an essential consideration for investors. While stablecoins may not offer substantial returns on their own, their primary purpose is to provide stability and act as a hedge against market volatility. Investors can enhance the profitability of their stablecoin investments by exploring yield-generating stablecoin opportunities and diversifying their holdings across different stablecoins. By striking a balance between stability and profitability, investors can navigate the cryptocurrency market with greater confidence and risk management.

Fees and Relationship to the Exchange

Stablecoins have gained popularity as a hedge against market volatility due to their stability and pegged value. However, investors need to consider the impact of chain gas fees and exchange fees on the ability of stablecoins to effectively serve as a hedging tool. In this article, we will explore how these fees can reduce the hedging capabilities of stablecoins and the factors to consider when evaluating their impact.

Chain gas fees refer to the transaction fees associated with conducting operations on blockchain networks. As stablecoins are typically built on blockchain platforms, investors may encounter gas fees when transferring or trading stablecoins. These fees can vary based on network congestion and demand, and they are typically denominated in the native cryptocurrency of the underlying blockchain network.

Chain gas fees can significantly affect the ability of stablecoins to serve as an efficient hedging tool. High gas fees can erode the cost-effectiveness of executing hedging transactions, particularly for smaller transactions or frequent trades. Investors may find it uneconomical to hedge their positions due to the substantial fees involved, limiting their ability to actively manage their risk exposure using stablecoins.

Different blockchain networks may have varying fee structures, and the choice of blockchain on which a stablecoin operates can impact the associated gas fees. Some blockchains, like Ethereum, have experienced periods of high congestion and elevated gas fees, making transactions more expensive. Conversely, other blockchain networks may offer lower fees and faster transaction processing times, improving the cost-efficiency of stablecoin hedging.

Exchange fees are charges imposed by cryptocurrency exchanges for trading stablecoins or converting them to other cryptocurrencies or fiat currencies. These fees are typically a percentage of the transaction value or a fixed amount per trade. Exchange fees can vary across different platforms and may impact the overall profitability of stablecoin hedging strategies.

Investors should carefully evaluate the fee structures associated with stablecoin transactions on various blockchain networks and cryptocurrency exchanges. Comparing gas fees and exchange fees across different platforms can help identify more cost-effective options for executing hedging transactions. It is important to consider not only the fees charged but also the liquidity and security of the platform to ensure a reliable and efficient trading experience.

Scalability is a critical factor for mitigating high gas fees and exchange fees in stablecoin transactions. Blockchain networks are actively exploring solutions to improve scalability and reduce fees, such as layer 2 protocols or alternative consensus mechanisms. These developments aim to enhance transaction throughput and decrease costs, making stablecoin hedging more accessible and cost-effective for investors.

Investors should also consider the timing of their stablecoin transactions to optimize fee management. Monitoring blockchain network congestion and choosing periods of lower activity can help reduce gas fees. Additionally, planning and consolidating trades to minimize the number of transactions can help reduce overall exchange fees.

While chain gas fees and exchange fees may present challenges for short-term or frequent hedging strategies, their impact on long-term hedging should be assessed differently. Long-term hedging strategies focus on preserving capital and managing risk over extended periods. In such cases, occasional fee expenses may be acceptable, considering the stability and risk mitigation provided by stablecoins over time.

Chain gas fees and exchange fees can limit the effectiveness of stablecoins as a hedging tool for investors. Understanding the fee structures associated with stablecoin transactions and evaluating the cost-effectiveness across different blockchain networks and exchanges are crucial. Additionally, monitoring scalability developments in the blockchain space and strategically timing transactions can help mitigate the impact of fees. By carefully managing and accessing fee-related considerations, investors can optimize their stablecoin hedging strategies and navigate the cryptocurrency market with greater efficiency and profitability.

Cross-Chain Compatibility

Cross-chain compatibility is a crucial concept in the cryptocurrency ecosystem, as it allows for the seamless transfer and utilization of stablecoins across different blockchain networks. This interoperability eliminates the limitations of blockchain silos and enhances the flexibility, liquidity, and adoption of stablecoins. By being compatible with multiple blockchain networks, stablecoins can tap into a larger pool of liquidity, attract more users, and facilitate smoother trading. This increased liquidity not only reduces slippage but also contributes to price stability, making stablecoins more attractive to investors and users alike.

Furthermore, cross-chain compatibility expands the market reach of stablecoins, enabling them to reach a wider user base. Users on different blockchain networks can access and transact with stablecoins without the need for complicated conversions or reliance on centralized intermediaries. This expanded market reach strengthens the adoption and utility of stablecoins, fostering their integration into various financial ecosystems. Additionally, cross-chain compatibility allows stablecoin holders to diversify their risk by spreading their holdings across multiple blockchain networks. In the event of technical issues or regulatory challenges on one blockchain, stablecoin holders can seamlessly transfer their assets to another compatible blockchain, mitigating potential risks and maintaining the stability of their holdings.

Achieving cross-chain compatibility poses several technical challenges that need to be addressed. Developing standardized protocols, creating secure and efficient bridges between blockchain networks, and ensuring proper consensus mechanisms are crucial for enabling seamless interoperability. However, various blockchain projects and protocols are actively working to overcome these challenges and improve cross-chain compatibility. The emergence of atomic swaps, interoperability protocols, sidechains, and blockchain bridges provides promising solutions for facilitating the transfer of assets between different chains.

Looking ahead, cross-chain compatibility holds significant implications for the future of stablecoins. It not only allows for greater flexibility in choosing blockchain networks but also promotes innovation and integration into various applications and financial ecosystems. As stablecoins bridge traditional financial systems with decentralized blockchain ecosystems, they enable users to transact seamlessly across different platforms. Moreover, cross-chain compatibility opens the door for new use cases such as decentralized finance (DeFi), global remittances, and cross-border transactions. By leveraging the benefits of cross-chain compatibility, stablecoin projects and the wider crypto community can unlock new possibilities and drive the continued growth and utility of stablecoins in the ever-evolving cryptocurrency landscape.

Conclusion

Stablecoins offer unique opportunities for investors seeking stability, hedging, and potential profits within the cryptocurrency space. Understanding the different stages of investment in stablecoins is crucial for aligning expectations and maximizing returns. Early-stage investments can be highly profitable, with the potential for significant growth as the stablecoin strives to achieve parity with its peg. Investors who identify promising stablecoin projects in their early stages can benefit from substantial returns on their investments, especially in fiat-based stablecoins where growth can reach up to 500 times, or in stablecoin 2.0 models where growth can exceed 10,000 times.

> "The individual investor should act consistently as an investor and not as a speculator."
>
> Benjamin Graham

However, as stablecoins progress into mid-stage and late-stage investments, the potential for large profits diminishes. Mid-stage investments typically offer more moderate returns, with a likely maximum return of around 15%. At this stage, stablecoins have matured, and their growth trajectory tends to stabilize as they approach their pegged value. Investors in mid-stage stablecoins often see their investments primarily as a hedge against inflation or as a stable store of value rather than expecting significant appreciation.

Late-stage investments in stablecoins are primarily focused on hedging strategies and maintaining the stability of the investment portfolio. Late-stage stablecoin investments may not offer substantial growth opportunities, especially when compared to early-stage investments. However, they can still provide value by serving as a hedge against potential inflation in fiat currencies or as a stable asset within a diversified investment portfolio. Investors at this stage prioritize preserving capital and minimizing risk rather than seeking aggressive profits.

It's important to consider factors such as chain gas fees and exchange fees when evaluating the profitability and effectiveness of stablecoin investments. High gas fees on blockchain networks can erode the returns on investments, particularly for frequent traders or users engaging in small transactions. Similarly, exchange fees can impact the profitability of stablecoin investments, especially when frequent trading or arbitrage strategies are involved. Investors should carefully assess these costs and consider platforms or networks that offer competitive fees to optimize their returns.

Cross-chain compatibility is another critical aspect to consider when investing in stablecoins. The ability of stablecoins to seamlessly operate across different blockchain networks enhances liquidity, market reach, and flexibility. It reduces the barriers to entry and promotes wider adoption of stablecoins by eliminating the need for complex conversions or centralized intermediaries. Moreover, cross-chain compatibility allows stablecoin holders to diversify their risk by spreading their assets across multiple blockchain networks, mitigating potential technical or regulatory challenges.

In summary, stablecoins provide unique investment opportunities for individuals and institutions looking to leverage the benefits of cryptocurrencies while maintaining stability and mitigating risk. Understanding the different stages of investment in stablecoins is crucial for setting realistic expectations and optimizing returns. Early-stage investments offer the potential for significant growth, mid-stage investments focus on stability and hedging, while late-stage investments prioritize portfolio preservation. Consideration of factors like transaction costs and cross-chain compatibility further enhances the effectiveness and profitability of stablecoin investments. By carefully navigating the stablecoin investment landscape, investors can capitalize on the benefits of stablecoins and achieve their financial objectives within the dynamic and evolving cryptocurrency ecosystem.

Conclusion Part I

In conclusion, Part I of our discussion has provided a comprehensive overview of various aspects of stablecoins and their role within the cryptocurrency ecosystem. We explored the fundamental concepts and characteristics of stablecoins, their different classifications, and the challenges and solutions associated with their design and implementation. Through this exploration, we gained valuable insights into the potential benefits and considerations when engaging with stablecoins.

We began our journey by understanding the need for stablecoins in the volatile world of cryptocurrencies. Their ability to maintain a stable value, often pegged to a fiat currency or a basket of assets, makes them a reliable medium of exchange, a store of value, and a unit of account. Stablecoins bridge the gap between traditional financial systems and the emerging blockchain technology, providing stability and trust in an otherwise volatile market.

Throughout the chapters, we examined the various types of stablecoins, including fiat-backed, commodity-backed, algorithmic, and hybrid stablecoins. Each type offers its unique set of advantages and considerations, catering to different use cases and preferences. Fiat-backed stablecoins provide stability through their direct backing with fiat currency reserves, while commodity-backed stablecoins offer stability through tangible assets like precious metals or commodities. Algorithmic stablecoins rely on algorithmic mechanisms to maintain stability, while hybrid stablecoins combine different approaches to achieve their desired outcomes.

We delved into the intricate details of stablecoin tokenomics, discussing concepts like asset pools, smart burns, reserves, and diversification. These mechanisms play a vital role in ensuring the stability, liquidity, and resilience of stablecoins. Asset pools act as a stability factor, providing the necessary backing for the stablecoin's value, while smart burn systems can generate value and prevent inflation by removing a portion of the tokens from circulation. Diversification of asset pools minimizes risks and enhances stability by spreading investments across different asset classes.

We also examined the legal and regulatory aspects of stablecoins, including discussions on securities, commodities, and the Howey test. Understanding the legal classification of stablecoins is crucial for compliance and determining the applicable regulations. Additionally, we explored the role of stablecoins in investment strategies, considering early-stage, mid-stage, and late-stage investments. Each stage offers unique opportunities and considerations, ranging from high growth potential to hedging and stability.

Moreover, we addressed the importance of education and awareness surrounding stablecoins. Lack of understanding, particularly among legislators and the general public, can lead to misconceptions and hinder the adoption and growth of stablecoins. Education initiatives, both within the cryptocurrency community and in public institutions, play a critical role in promoting a better understanding of stablecoins and their potential benefits.

In conclusion, Part I has provided a comprehensive foundation for understanding stablecoins and their role within the cryptocurrency ecosystem. We explored their characteristics, classifications, challenges, and solutions. Stablecoins offer a promising solution for individuals and institutions seeking stability, trust, and efficient value transfer within the world of cryptocurrencies. By combining the advantages of blockchain technology with the stability of traditional assets, stablecoins bridge the gap between these two worlds and pave the way for broader adoption and financial innovation.

As we move forward in our exploration of stablecoins, we will delve deeper into their applications, adoption challenges, potential risks, and future developments. Part I has laid a solid groundwork for our journey, equipping us with the necessary knowledge and understanding to navigate the evolving landscape of stablecoins. Whether as a means of exchange, a store of value, or an investment vehicle, stablecoins continue to shape the future of finance, offering stability, efficiency, and opportunities for individuals and businesses worldwide.

Part II: The Stablecoin Families

Introduction to Part II

Stability within a business system is needed for the system to function. As capitalism replaced feudalism and imperialism around the world, we saw massive advances in technology, freedom, and information. The world lurched forward because of the free market. Some people did not like this change. The Bolsheviks (International Socialists) and Mensheviks (National Socialists/Fascists) both fought to restrict the use of technology in factories to "protect the worker." In many cases, those people who want to control others in society know that a tired person is less likely to fight against the one who is oppressing them. Therefore, many systems in society attempt to 'keep people in their place' through heavy workloads and poor technology.

When cryptocurrency entered the market, it had market value and that was it. The advent of Ethereum added transitional value, then when companies started accepting the coins/tokens for business transitional value was present. At the very infancy of the system, when the system was at risk of failing, BitUSD issued the first stablecoin in the world.[23] This form of stablecoin 1.0 was the first of many iterations of different ideas that allows stablecoins to function in the world in which we live. The concept was the foundation for the idea that some cryptocurrencies should be backed by assets in which people can see value. In this part we look at the different iterations of stablecoin as it has evolved to its present state.

Stablecoin 1.0 refers to a category of digital currencies designed to maintain a stable value by being pegged to a

---

[23] Hedra (2023). "What is a stablecoin?" at https://hedera.com/learning/tokens/what-is-a-stablecoin.

fiat currency, such as the US dollar or the euro. These stablecoins typically operate by holding reserves of the corresponding fiat currency, ensuring that each unit of the stablecoin represents a fixed value of the underlying currency. The stability provided by Stablecoin 1.0 makes them suitable for various use cases within the cryptocurrency ecosystem, including facilitating transactions, serving as a store of value, and enabling cross-border remittances. However, they may face challenges related to regulatory compliance, centralized control, and transparency, as their value heavily relies on the trustworthiness of the entities managing the reserves.

Stablecoin 1.1 represents an evolution of stablecoins, where they are backed not only by fiat currency but also by corporate paper representing currency. This means that in addition to holding reserves of the underlying fiat currency, stablecoin 1.1 projects may also include corporate debt instruments, such as short-term bonds or commercial paper, as part of their collateral. This approach aims to diversify the backing assets and enhance stability. One example of a stablecoin that fits this description is Tether (USDT), which initially launched as a Stablecoin 1.0 backed by fiat currency but later introduced the concept of Tether 1.1, incorporating a mix of cash and other assets. This integration of corporate paper in stablecoin design adds an extra layer of security and liquidity to the ecosystem, further reinforcing its stability.

Stablecoin 1.2 represents a newer generation of stablecoins that rely on algorithmic mechanisms to manage the value of the currency. Unlike traditional stablecoins that are backed by fiat currency or corporate paper, Stablecoin 1.2 employs sophisticated algorithms to maintain price stability. These algorithms typically respond to changes in supply and demand, automatically adjusting the coin's monetary policy, and expanding or contracting the coin's supply to keep its value pegged to a specific benchmark. One notable example of a stablecoin that fits this definition is Ampleforth (AMPL), which utilizes an elastic supply mechanism to maintain price stability by targeting a constant price unit rather than a fixed quantity.[24] Stablecoin 1.2 introduces a novel approach that reduces reliance on traditional backing assets, allowing for greater flexibility and potential scalability in maintaining price stability.

Stablecoin 1.3 represents a unique class of stablecoins that are backed by a single stock or a specific type of asset, rather than a broader basket of assets. In this model, the stablecoin's value is directly linked to the performance and value of the chosen asset. This approach allows for a more targeted and specialized stability mechanism. One example of a stablecoin fitting this description is Mirror Protocol (MIR), which is built on the Terra blockchain and backed by real-world assets such as stocks and exchange-traded funds (ETFs). Stablecoin 1.3 introduces a level of granularity by aligning the stablecoin's value with the performance of a particular asset, enabling investors to gain exposure to specific markets while benefiting from the stability offered by the stablecoin framework. However, it's important to note that the stability of these stablecoins is still subject to the volatility and risks associated with the chosen underlying asset.

Stablecoin 1.4 represents a novel variant of stablecoins backed by a collection of digital assets, offering

---

[24] https://www.ampleforth.org/.

diversification within the cryptocurrency space. These stablecoins leverage the concept of collateralized debt positions (CDPs) and smart contracts to secure the value of the stablecoin. By locking up a variety of digital assets as collateral, the stablecoin's value is maintained. One example of a stablecoin fitting this definition is MakerDAO's DAI, which operates on the Ethereum blockchain and utilizes a decentralized platform for creating and managing CDPs. The collateral can include cryptocurrencies like Ethereum, BAT, or even NFTs, enabling stablecoin 1.4 to provide stability while benefiting from the potential growth of the digital asset market. However, it's crucial to note that the value of these stablecoins may still be influenced by the volatility and market conditions of the underlying digital assets, although the diversification can help mitigate risks to some extent.

Stablecoin 1.5 represents an advanced category of stablecoins that are backed by a diversified collection of fiat currencies. These stablecoins aim to achieve stability by leveraging multiple fiat currencies, reducing the risk associated with relying on a single currency. By holding reserves in various national currencies, stablecoin 1.5 projects provide a more robust and globally diversified approach to maintaining stability. One example of a stablecoin fitting this description is USD Coin (USDC), which is an ERC-20 token built on the Ethereum blockchain and backed by a mix of major global currencies like the US dollar, euro, and British pound. Stablecoin 1.5 introduces enhanced stability through the inclusion of multiple fiat currencies, enabling users to transact and hold digital assets with reduced exposure to currency risk and volatility.

Stablecoin 1.6 represents an innovative class of stablecoins that are backed by a diversified portfolio of stocks or derivatives, expanding the scope of stability mechanisms beyond traditional assets. These stablecoins leverage the value of real-world financial instruments, such as equities or derivatives, to support their pegged value. By utilizing blockchain technology and smart contracts, stablecoin 1.6 projects enable fractional ownership and efficient management of these assets. An example of a stablecoin fitting this description is Synthetix Network Token (SNX), which operates on the Ethereum blockchain and allows users to create synthetic assets, including stocks and derivatives. Stablecoin 1.6 introduces a unique opportunity for investors to access a broader range of financial instruments within the cryptocurrency ecosystem while enjoying the stability provided by the stablecoin framework. However, it is important to note that stablecoin 1.6's value is still influenced by the performance and risks associated with the underlying stocks or derivatives in its portfolio.

Stablecoin 1.7 represents an innovative category of stablecoins that are backed by a diverse collection of properties, introducing a new dimension of stability to the cryptocurrency landscape. These stablecoins leverage blockchain technology and tokenization to represent ownership and value in real-world properties such as real estate, land, or other tangible assets. By tokenizing properties, stablecoin 1.7 projects enable fractional ownership and liquidity in the market, providing stability through the underlying physical assets. An example of a stablecoin fitting this description is RealT, which operates on the Ethereum blockchain and allows users to invest in fractional ownership of real estate properties.[25] Stablecoin 1.7 introduces an exciting avenue for investors to diversify their portfolios with tangible assets, while benefiting from the stability and liquidity provided by the stablecoin framework. However, it's crucial to consider the risks associated with property valuation, legal compliance, and the underlying market dynamics when evaluating stablecoin 1.7 projects.

Stablecoin 1.8 represents an innovative class of stablecoins that are backed by algorithmically managed derivatives, providing a dynamic stability mechanism. These stablecoins leverage smart contracts and algorithmic protocols to manage and optimize a portfolio of derivatives, which are financial instruments derived from underlying assets. By dynamically adjusting the composition of derivatives based on market conditions, stablecoin 1.8 projects aim to maintain stability and mitigate risks. An example of a stablecoin fitting this description is Ampleforth Governance Token (AMPL) if it worked more with derivatives, which uses a unique elastic supply mechanism tied to price oracles to rebalance its supply and achieve stability.[26] Stablecoin 1.8 introduces a novel approach that leverages the power of derivatives and algorithms to provide stability in a rapidly changing market environment, offering an alternative to traditional backing assets and expanding the possibilities within the stablecoin ecosystem.

Stablecoin 1.9 represents a specialized category of stablecoins that are backed by a diverse fund consisting of a

---

[25] RealT at https://realt.co/.
[26] https://www.ampleforth.org/.

single asset class. These stablecoins leverage the value of a specific asset class, such as precious metals, commodities, or cryptocurrencies themselves, to ensure stability. By constructing a diverse portfolio within a single asset class, stablecoin 1.9 projects aim to mitigate risks and fluctuations associated with individual assets. An example of a stablecoin fitting this description is PAX Gold (PAXG), which operates on the Ethereum blockchain and is backed by physical gold stored in Brink's vaults. Stablecoin 1.9 offers investors an opportunity to access and transact with a specific asset class while enjoying the stability provided by the stablecoin framework. However, it's important to consider the inherent risks associated with the chosen asset class and its market dynamics when evaluating stablecoin 1.9 projects.

Stablecoin 2.0 represents an advanced category of stablecoins backed by a diversified basket of goods, including a wide range of assets such as stocks, bonds, derivatives, real property, digital assets, collectibles/antiquities, and currencies. Within this class, KIROS stands as the flagship currency, boasting a robust basket of goods that provides stability and resilience to market fluctuations. By including a diverse array of assets, KIROS ensures that its value is not solely dependent on one asset class, reducing risks and enhancing stability. Additionally, Stablecoin 2.0s, including KIROS, employ a smart burn enabled asset pool, a mechanism that effectively addresses inflation concerns by removing or burning tokens from circulation, thereby maintaining the stablecoin's value over time.[27] This innovative approach offers a sustainable

[27] www.KIROScoin.com

ecosystem where the stablecoin remains resistant to inflationary pressures. The integration of diverse assets and the implementation of smart burn mechanisms make stablecoin 2.0 and KIROS a compelling option for users seeking stability and long-term value preservation in the cryptocurrency space.

Level 3 Stablecoins represent an advanced iteration within the stablecoin ecosystem, as they are backed by a Stablecoin 2.0 system. These stablecoins offer the advantage of being managed across multiple blockchains, effectively addressing cross-compatibility issues that can arise in decentralized exchanges. By leveraging the capabilities of Stablecoin 2.0, Level 3 Stablecoins facilitate seamless cross-chain trading, ensuring compatibility and interoperability. This enables efficient and secure transactions while keeping the majority of trades within custodial wallets, resolving potential cross-trade issues in the background. With Level 3 Stablecoins, users can enjoy the benefits of a stablecoin ecosystem that extends across different blockchains, fostering liquidity and ease of use for participants in the decentralized exchange landscape.

In the following pages, Part II of the stablecoin book delves into the evolution and advancements of stablecoins, exploring various iterations from Stablecoin 1.0 to Level 3 Stablecoins. Each version represents a distinct approach to achieving stability, including backing by fiat currency, corporate paper, algorithmic management, single stocks or assets, digital assets, diverse fiat currencies, properties, and diverse funds based on a single asset class. Building upon this foundation, Stablecoin 2.0 introduces a robust basket of goods, while Level 3 Stablecoins address cross-chain management and compatibility issues in decentralized exchanges. This section delves deeper into the intricacies and innovations of these stablecoin iterations, shedding light on their mechanisms, benefits, and potential challenges, providing readers with a comprehensive understanding of the dynamic and ever-evolving world of stablecoins.

Chapter 8: Stablecoin 1.0 (Tokens Backed by Fiat Currency)

"The best way to destroy the capitalist system is to debauch the currency."[28]

Undermining the confidence of any political or business system is one of the best ways for someone who wants the system to fail to allow it to rot from the inside. As we have seen over the last two years, the SEC and other parts of the U.S. Government have waged an illegal war on crypto (when you consider the ruling in the West Virginia v. EPA case). As such, the SEC, whether acting alone or under orders, has 'bar coded' the crypto market, ensuring that they file a negative news cycle each time that the market seems to take a positive turn. Undermining a part of the U.S. economy which has a higher market cap than the GDPs of 46 states (only California, Texas, New York, and Florida have higher GDPs than the cryptocurrency market cap) is tantamount to sedition.[29] However, stablecoins provide some balance to this barcode attack, a shelter that is much needed in the community.

In this chapter, we look at the granddaddy of them all, Stablecoin 1.0. Stablecoin 1.0, founded by BitUSD, introduced the idea that cryptocurrency could be stable if it was pegged to an asset. While the statement that it was stable only applied to the normative negative stability of fiat currency over the long run, it still provided that modicum of stability that the community needed to endure

---

[28] Vladimir Lenin at https://www.azquotes.com/quotes/topics/currency.html.
[29] Wisevoter (2023). "GDP by State." At https://wisevoter.com/state-rankings/gdp-by-state/.

the bear markets we have seen over the course of crypto's growth. While BitUSD is not a big player in the market today, it is still the genesis for the system in which we live, and for that we should be thankful.

In this chapter, we will follow the same pattern that we follow in all the chapters in this part of the book. First, we look at the stability of the asset class on which the token at this level is based. The asset class is very important because it is the 'pegged' underpinning of the system. Next, we look at the liquidity of the asset class, how fast can the asset be liquidated if there is a 'run' on the token. As a matter of course, we will then look at the asset ratio needed to maintain stability. The media tends to insist that there needs to be a 1:1 ratio, and letter agencies tend to echo this false statement; each class has a ratio that is needed for the token to be reliable in its function.

Also, as a matter of course, it is always important to look at the relationship between holders/creators of the currency and the asset pool. The more complex the system the more analysis needs to be done on how the assets are being managed. While a 1:1 ratio for stablecoin 1.0 can be managed by the local bank of the creator, more complex systems require funds, investment firms, and even multiple subsidiaries to ensure that the asset pool maintains value. The analysis of who holds the asset pool then leads to an analysis of susceptibility of to attack. The more complex the asset, the more stable it is- however, it does need to be protected from attacks. The discussions within these sections of each chapter will be quite similar. Transparency is also an issue that varies slightly as you look at different types of stablecoins, once again the more complex the tokenomics- the more transparent the company needs to be.

Finally, there needs to be an analysis of viability in major markets. This results in a bimodal, cyclic model for the analysis of the system, viability seems to be highest for simple and highly complex models, with intermediate models being lukewarm in their reception. These two paragraphs are the same in each chapter.

Stability of Asset Class

Fiat currency has long been regarded as a stable means of exchange and store of value in modern economies. However, beneath the perception of stability lies a fundamental flaw—the constant erosion of its value due to inflation. This essay delves into the negative stability inherent in fiat currency as an asset class, exploring the reasons behind its declining value and the cognitive dissonance surrounding its perceived stability. Additionally, we examine how stablecoins pegged to fiat currencies face similar challenges and potential implications for their long-term stability.

- The Inherent Flaw: Inflation and Declining Purchasing Power: Fiat currency derives its value from the trust placed in the issuing government, rather than being backed by a physical asset. However, this trust is undermined by inflation, which erodes the purchasing power of fiat currency over time. Inflation is a consequence of monetary policy, where central banks increase the money supply, diluting the value of each unit of currency. This chronic decline in value is a key characteristic of fiat currency stability, yet it is often overlooked by the global community.
- Cognitive Dissonance and the Perception of Stability: Despite the inherent negative stability of

fiat currency, there exists a cognitive dissonance within society that views it as a stable asset. This is primarily driven by factors such as familiarity, historical acceptance, and a lack of viable alternatives. The perception of stability persists due to social conditioning and the relative stability of fiat currency in the short term, obscuring the long-term erosion of value caused by inflation.

> "Fiat Currency isn't money- it's a weapon. Every duck farmer knows this, and you can only pretend the fake is real before mass starvation takes place."
>
> Jarod Kintz

- The Impact on Stablecoins Pegged to Fiat Currency: Stablecoins pegged to fiat currencies face similar challenges in terms of long-term stability. While these stablecoins aim to provide stability in the volatile world of cryptocurrencies, their value is inherently tied to the underlying fiat currency. As the pegged currency experiences inflation, the stablecoin's purchasing power diminishes, potentially leading to a loss of trust among users.
- Mitigating Strategies for Stablecoins: Stablecoin projects must address the negative stability of fiat currency to maintain their long-term viability. One strategy involves incorporating mechanisms such as regular audits, collateralization, and algorithmic adjustments to mitigate the impact of inflation on the stablecoin's value. Additionally, diversifying the backing assets beyond fiat currency and exploring

alternative pegging mechanisms can help stabilize stablecoin value in the face of inflationary pressures.

- The Emergence of Alternative Stablecoin Models: Recognizing the challenges posed by fiat currency stability, alternative stablecoin models have emerged. These models include stablecoins backed by assets other than fiat currency, such as digital assets, commodities, or diversified asset baskets. By anchoring stablecoins to assets that have historically demonstrated greater stability, these models aim to provide a more robust and sustainable form of stability in the cryptocurrency ecosystem.
- Regulatory Considerations and Future Outlook: The negative stability of fiat currency and its impact on stablecoins underscore the need for regulatory frameworks that ensure transparency, consumer protection, and stability within the stablecoin space. Regulators must navigate the complexities of balancing innovation, financial stability, and investor confidence to foster a healthy and sustainable stablecoin ecosystem.

Fiat currency's negative stability, perpetuated by inflation and the cognitive dissonance surrounding its perceived stability, poses significant challenges for stablecoins pegged to such currencies. Recognizing the long-term erosion of value inherent in fiat currency is essential in understanding the potential risks for stablecoins pegged to these currencies. While efforts are underway to mitigate these risks through alternative models and strategies, regulatory oversight remains crucial for ensuring the stability and viability of stablecoins in the evolving landscape of digital assets.

Liquidity of Asset Class

Liquidity plays a critical role in the functioning of any financial system, enabling participants to easily buy or sell assets. In the context of stablecoins, liquidity is particularly crucial for users who may need to "dump" or sell their stablecoin tokens for fiat currency. This essay explores the liquidity of fiat currency and its significance in facilitating the redemption process of Stablecoin 1.0 tokens. We delve into the factors that affect fiat currency liquidity, potential challenges that may arise, and the implications for Stablecoin 1.0 users.

- Fiat Currency Liquidity: Fiat currency, issued by central banks and backed by the trust and confidence of the public, is generally highly liquid. Its acceptance as legal tender in most economic transactions, combined with its wide circulation, facilitates easy convertibility and broad acceptance. The deep liquidity of major fiat currencies, such as the US dollar, euro, or Japanese yen, is attributed to their global use in trade, investment, and reserve holdings.
- Market Infrastructure and Accessibility: The liquidity of fiat currency is supported by well-developed financial markets, including foreign exchange (forex) markets, where participants can freely exchange one currency for another. These markets operate 24/7, providing continuous liquidity and allowing market participants to buy or sell fiat currency at prevailing market rates. Accessibility to these markets through various platforms, such as banks, brokers, and online exchanges, further enhances liquidity.

- Central Bank Interventions and Market Depth: Central banks play a crucial role in ensuring liquidity and stability in fiat currency markets. They intervene through monetary policy tools to manage interest rates, control inflation, and maintain financial stability. By adjusting the money supply and engaging in open market operations, central banks influence market depth and liquidity, ensuring an orderly flow of funds and minimizing disruptions.
- Challenges to Fiat Currency Liquidity: While fiat currency generally enjoys high liquidity, certain factors can pose challenges to its liquidity profile. In times of economic or political instability, market participants may lose confidence, leading to a flight to safer assets and reduced liquidity in certain currencies. Exchange rate fluctuations, regulatory restrictions, and geopolitical events can also impact liquidity by affecting market dynamics and participants' willingness to trade.
- Liquidity Impact on Stablecoin 1.0 Redemption: Stablecoin 1.0 tokens are typically pegged to fiat currency, aiming to provide stability and easy redemption for users. The liquidity of fiat currency directly influences the ease and speed with which Stablecoin 1.0 tokens can be redeemed. When users feel the need to "dump" or sell their stablecoin tokens, a liquid fiat currency ensures that the redemption process can be efficiently executed, maintaining trust and stability within the stablecoin ecosystem.
- Market Depth and Stablecoin Liquidity: The depth of the fiat currency market is a crucial consideration

for stablecoin issuers, as it directly impacts the liquidity and stability of their tokens. Adequate market depth ensures that Stablecoin 1.0 tokens can be readily exchanged for fiat currency without causing significant price fluctuations. Insufficient market depth or sudden shifts in demand for redemption may pose challenges, potentially leading to increased volatility or liquidity constraints.

- Maintaining Liquidity in Stablecoin Ecosystems: Stablecoin issuers must proactively manage liquidity within their ecosystems to ensure smooth redemption processes. This can be achieved through robust reserve management practices, including maintaining appropriate asset-to-token ratios, ensuring sufficient reserves of fiat currency, and implementing mechanisms to incentivize market makers and liquidity providers.
- Regulatory Considerations and Risk Mitigation: Regulatory oversight is crucial in maintaining the liquidity and stability of fiat currency markets, indirectly impacting stablecoin liquidity. Regulations aimed at preventing market manipulation, ensuring transparency, and safeguarding against liquidity risks play a vital role in maintaining investor confidence and market integrity.

The liquidity of fiat currency is vital for the redemption process of Stablecoin 1.0 tokens, providing users with confidence that they can easily sell their tokens for fiat currency when needed. Market infrastructure, accessibility, central bank interventions, and market depth all contribute to the liquidity of fiat currency. Stablecoin issuers must navigate these factors to maintain liquidity within their ecosystems, ensuring efficient token redemption and overall stability. Regulatory oversight remains crucial in mitigating risks and fostering a liquid and resilient stablecoin market.

Required Asset Ratio for Stability

Fiat currency-backed stablecoins aim to provide stability by pegging their value to a corresponding fiat currency. One crucial aspect of these stablecoins is the required ratio, often set at 1:1, meaning that each stablecoin token is backed by an equivalent amount of the fiat currency held in reserve. This essay explores the significance of maintaining a 1:1 ratio in fiat currency-backed stablecoins, highlighting the relationship between currency depreciation and the decline in the value of the asset pool. By analyzing the factors contributing to this requirement, we gain a deeper understanding of the stability mechanisms underlying such stablecoins.

> "You think Bitcoin is costly, it is not! Your fiat currency is definitely overpriced and oversold."
>
> Olawale Daniel

- Understanding Fiat Currency Depreciation: Fiat currencies, as a result of inflation and economic factors, tend to gradually decline in value over time. Inflation erodes the purchasing power of fiat currencies, diminishing their value relative to goods and services. This depreciation underscores the importance of maintaining a balanced 1:1 ratio in fiat currency-backed stablecoins, as any deviation from this parity can result in a loss of trust and stability.

- Maintaining Parity: Stability and Investor Confidence: A 1:1 ratio ensures stability and instills confidence in fiat currency-backed stablecoins. By maintaining a direct relationship between the stablecoin token and the underlying fiat currency, users can rely on the stablecoin to hold its value relative to the pegged currency. This stability encourages broader adoption and usage of stablecoins in various financial transactions.
- Risk of Value Decline: Impact on the Asset Pool: When the required 1:1 ratio is not maintained, the value of the asset pool backing the stablecoin can decline at a faster rate than the fiat currency itself. For instance, if the ratio were to deviate, with more stablecoin tokens in circulation than the corresponding fiat currency held in reserve, a discrepancy arises, and the stablecoin loses its stability. This discrepancy can result in a loss of confidence among users, leading to a decreased demand for the stablecoin.
- Mechanisms to Maintain Parity: Stablecoin issuers employ several mechanisms to ensure the preservation of the 1:1 ratio. These mechanisms often involve regular audits, frequent reporting, and transparency regarding the reserve holdings and asset pool management. Robust reserve management practices, such as maintaining an appropriate mix of assets, can help stabilize the value of the asset pool and minimize risks associated with potential value decline.

- Monitoring and Adjusting the Ratio: In some cases, stablecoin issuers may need to monitor and adjust the 1:1 ratio periodically to account for changes in market conditions or currency depreciation. These adjustments can help maintain the stability of the stablecoin and ensure that it remains pegged to the fiat currency as closely as possible. However, any adjustments must be communicated transparently to users to maintain trust and avoid sudden disruptions.
- Impact on User Experience and Adoption: The stability offered by a 1:1 ratio in fiat currency-backed stablecoins enhances the user experience and encourages broader adoption. Users can confidently transact using stablecoins, knowing that their value remains relatively constant compared to the underlying fiat currency. This stability makes stablecoins more attractive for daily transactions, cross-border payments, and store-of-value purposes.
- Potential Challenges and Mitigation Strategies: Maintaining a 1:1 ratio in fiat currency-backed stablecoins presents challenges, particularly when market conditions or economic factors lead to significant currency depreciation. Stablecoin issuers must actively manage their reserve holdings, diversify asset portfolios, and implement risk mitigation strategies to maintain stability and minimize the impact of currency value decline.

The requirement of a 1:1 ratio in fiat currency-backed stablecoins is essential for ensuring stability, user confidence, and the preservation of value. By pegging the stablecoin's value to a corresponding fiat currency, users can transact with greater certainty, relying on stablecoins as a reliable medium of exchange and store of value. Stablecoin issuers must diligently monitor the parity ratio, adopt robust reserve management practices, and employ transparency measures to mitigate risks associated with currency depreciation and uphold the stability of their stablecoins.

Relationship between Holders and access to backing.

In a Stablecoin 1.0 model, the creators of the coin can play a crucial role as the holders of the asset pool, leading to enhanced liquidity within the ecosystem. This arrangement offers several advantages and considerations for liquidity.

Firstly, having the creators act as the holders of the asset pool allows for direct control over asset management. They can actively monitor market conditions, make timely adjustments, and rebalance the underlying assets to maintain the stability of the stablecoin. This proactive approach ensures that the stablecoin remains liquid and responsive to market dynamics.

Secondly, this model enables a quick response to redemption demands. As the creators also hold the asset pool, they can promptly liquidate a portion of the underlying assets to fulfill redemption requests. This ability to meet these demands efficiently enhances liquidity and instills confidence among stablecoin holders.

Thirdly, efficient liquidity management is facilitated when the creators directly manage the asset pool. They can strategically allocate assets, ensuring that sufficient liquidity is maintained at all times. This proactive approach minimizes the risk of illiquidity and ensures smooth operations within the stablecoin ecosystem.

Furthermore, having the creators as the holders of the asset pool enhances trust and transparency. Users can have greater confidence in the stablecoin knowing that the creators have a vested interest in maintaining the stability and value of the underlying assets. The creators can provide regular reports, audits, and disclosures regarding the asset pool, fostering transparency and reinforcing trust among users.

While this model offers enhanced liquidity, it is important to consider potential challenges and risks. Concentration of power and control in the hands of the creators may raise concerns about centralization and conflicts of interest. To address these concerns, robust governance mechanisms, regulatory oversight, and transparency measures are essential. These measures ensure fair and equitable operations and mitigate any potential risks associated with the creators' role as holders of the asset pool.

In conclusion, in a Stablecoin 1.0 model, the creators' involvement as holders of the asset pool can contribute to enhanced liquidity within the ecosystem. Direct control over asset management, quick response to redemption demands, efficient liquidity management, increased trust and transparency, and alignment of incentives are key advantages of this approach. However, striking a balance between liquidity enhancement and avoiding concentration of power is crucial. Appropriate governance and regulatory frameworks should be in place to ensure fairness, stability, and long-term viability within the stablecoin ecosystem.

Susceptibility to Attacks

A fiat currency backed Stablecoin 1.0, while providing stability and liquidity, is not immune to various attacks that can threaten its integrity. These attacks can target different aspects of the stablecoin system, including the underlying assets, the pegging mechanism, and the overall ecosystem. Understanding these potential attacks is crucial for designing robust security measures and mitigating risks. Here are some types of attacks that a fiat currency-backed stablecoin may face:

> "Even when money seemed to be material treasure, heavy in pockets and shops' holds and bank vaults, it always was information. Coins and notes, shekels and cowries were all just short-lived technologies for tokenizing information about who owns what."
>
> James Gleick

- Collateral Failure: One significant risk for a fiat currency-backed stablecoin is the failure of the collateral assets. If the assets backing the stablecoin lose their value or become illiquid, it can undermine the stability and pegging mechanism. For instance, if the fiat currency experiences hyperinflation or undergoes a financial crisis, the stablecoin's value may be compromised. Proper due diligence, risk assessment, and continuous monitoring of the collateral assets are essential to mitigate this risk.
- Black Swan Events: Black swan events refer to unexpected and extreme market events that are difficult to predict. These events can disrupt the stability and liquidity of a stablecoin, posing a challenge to its pegging mechanism. For example, a

sudden economic recession, geopolitical turmoil, or a global financial crisis can lead to sharp currency devaluation, affecting the stablecoin's pegged value. Implementing risk management strategies, stress testing, and diversification of collateral assets can help mitigate the impact of such events.

- Regulatory and Legal Risks: The regulatory landscape surrounding stablecoins is evolving and can present significant challenges. Governments and regulatory authorities may impose restrictions, require licenses, or introduce new regulations specific to stablecoins. Non-compliance with these regulations or unexpected changes in the regulatory environment can pose risks to the stablecoin's operations. It is crucial for stablecoin issuers to stay informed, engage in regulatory compliance, and establish transparent governance frameworks to address these risks.
- Smart Contract Vulnerabilities: Fiat currency-backed stablecoins often rely on smart contracts to manage the issuance and redemption of tokens. Smart contracts can be susceptible to coding errors or vulnerabilities that could be exploited by attackers. Malicious actors can target the smart contracts, potentially leading to loss or manipulation of funds. Conducting thorough code audits, implementing security best practices, and regularly updating and patching the smart contracts can help minimize these risks.
- Centralization Risks: Stablecoin systems that rely on a centralized entity to manage the asset pool and the pegging mechanism can be vulnerable to attacks targeting that central point of control. An attacker

may attempt to compromise the entity's systems, leading to manipulation of the pegged value or mismanagement of the collateral assets. Ensuring robust security measures, implementing multi-signature wallets, and establishing independent audits and controls can help mitigate centralization risks.

- Market Manipulation: The stability of a fiat currency-backed stablecoin can be subject to market manipulation. Traders or groups with significant holdings of the stablecoin may engage in coordinated buying or selling activities to artificially influence the price or liquidity. Such manipulation can undermine the stability of the stablecoin and erode user confidence. Implementing surveillance mechanisms, collaborating with reputable exchanges, and promoting fair trading practices can help detect and deter market manipulation.
- Cybersecurity Threats: Cybersecurity threats, such as hacking attempts, phishing attacks, or theft of private keys, can pose a significant risk to a stablecoin's security and integrity. If an attacker gains unauthorized access to the stablecoin's systems or user wallets, it can lead to fund losses and compromise the stability of the ecosystem. Implementing robust security measures, including secure storage of private keys, multi-factor authentication, and regular security audits, can help mitigate these threats.

In summary, while a fiat currency backed Stablecoin 1.0 provides stability and liquidity, it faces several potential attacks that can disrupt its operations and compromise its value. Collateral failure, black swan events, regulatory risks, smart contract vulnerabilities, centralization risks, market manipulation, and cybersecurity threats are among the key challenges. Implementing rigorous risk management practices, regulatory compliance, security measures, and continuous monitoring can help mitigate these risks and ensure the long-term viability of the stablecoin ecosystem.

Transparency

Stablecoin 1.0, the earliest iteration of stablecoins, offers transparency as a key feature to the community it serves. Transparency plays a vital role in building trust, ensuring accountability, and providing stakeholders with essential information about the stablecoin's operations, underlying assets, and pegging mechanism. Here are several ways in which Stablecoin 1.0 provides transparency to the community:

Stablecoin 1.0 typically operates on a fiat currency pegging model, where the stablecoin is backed by a reserve of fiat currency. Transparency is crucial in disclosing the details of the assets held in the reserve. By providing regular reports or audits, the stablecoin issuer can demonstrate the proper allocation of funds and reassure users that the stablecoin is adequately backed by the stated collateral. Transparency in asset backing instills confidence in the stability and value of the stablecoin.

To maintain transparency, stablecoin issuers often provide regular reports and disclosures regarding the stablecoin's operations, financial health, and reserve assets. These reports can include information such as the total supply of stablecoins in circulation, the composition of the asset reserve, redemption and issuance activities, and financial statements. Clear and comprehensive reporting practices ensure that users and stakeholders have access to critical information about the stablecoin's performance and stability.

Independent audits play a crucial role in verifying the transparency and accuracy of the stablecoin's operations. Auditing firms assess the reserve assets, financial statements, and internal control processes of the stablecoin issuer. Through audits, the community gains confidence that the stablecoin issuer operates in a transparent and accountable manner. Audits provide an external validation of the stablecoin's claims and promote trust among users.

Transparency in governance is essential to ensure that the stablecoin's decision-making processes are fair and align with the interests of the community. Open governance models, where stakeholders have visibility into decision-making mechanisms, promote transparency. Stablecoin 1.0 can adopt mechanisms such as community voting, public discussions, or disclosure of governance committee decisions. Transparent governance ensures that the stablecoin's operations are accountable and responsive to the community's needs and concerns.

Stablecoin 1.0's pegging mechanism, which maintains a fixed value relative to the underlying fiat currency, benefits from transparency. The stablecoin issuer can disclose details about how the pegging is achieved, such as the reserve management strategy, rebalancing procedures, or mechanisms for adjusting the stablecoin's value. By providing clear information about the pegging mechanism, the stablecoin issuer enhances transparency and helps users understand how the stablecoin's value is maintained.

Stablecoin 1.0 often relies on smart contracts and blockchain technology to facilitate transactions and maintain transparency. Blockchain technology offers an immutable and auditable record of transactions, ensuring transparency in the stablecoin's transactional history. Users can independently verify transactions, track the stablecoin's supply, and examine the movement of funds in real-time. Smart contracts, which execute predefined rules, can also enhance transparency by automating processes and eliminating human error or manipulation.

To foster transparency, Stablecoin 1.0 issuers establish open communication channels with the community. This can include regular updates, public forums, or social media platforms where users can interact with the stablecoin issuer, ask questions, and voice concerns. Transparent communication channels provide an avenue for information exchange, clarification, and feedback, strengthening the relationship between the stablecoin issuer and the community.

> "My sponsor is an ex-Navy guy. Buys me lunch on Christmas. I tell him, as long as I am drinking and I have money, things seem to be going well. Now, you just replace "am drinking" with "have oil" there you have the U.S. economy. When I don't drink for a while… I get a little depressed and anti-Semitic. I tell him, as soon as the United States stops [F***] up foreign democracies and stealing their oil, I'll stop drinking. Unfortunately, looks like neither miracle is going to happen…"
>
> Dmitry Dyatlov

Stablecoin 1.0 is designed to provide transparency to the community it serves. Through clear reporting and disclosure practices, independent audits, transparent governance models, and open communication channels, stablecoin issuers aim to build trust, ensure accountability, and provide stakeholders with essential information. Transparency enhances the stability, credibility, and usability of stablecoins, enabling users to make informed decisions and participate confidently in the stablecoin ecosystem.

Viability in Major Markets

Fiat-backed stablecoin 1.0 demonstrates its viability in major markets through several key factors. These factors contribute to its acceptance and adoption by individuals, businesses, and financial institutions. Here are some reasons why fiat-backed stablecoin 1.0 is considered viable in major markets:

Fiat-backed stablecoins leverage the familiarity and trust associated with traditional fiat currencies. By pegging their value to established currencies like the U.S. dollar or the Euro, stablecoin issuers tap into the existing confidence that users have in these widely accepted and stable currencies. This familiarity helps to mitigate the resistance and skepticism that may arise with the introduction of entirely new and unfamiliar digital currencies.

One of the primary advantages of fiat-backed stablecoins is their stability. These stablecoins aim to maintain a 1:1 ratio with the underlying fiat currency, which means their value remains relatively constant. By eliminating the high volatility associated with many cryptocurrencies, stablecoins provide a reliable medium of exchange and store of value. This stability makes them attractive for conducting day-to-day transactions and mitigates the risk of value fluctuations for businesses and consumers.

Fiat-backed stablecoins offer the potential for efficient and cost-effective cross-border transactions. Major markets often involve international trade and commerce, which can be hindered by complex and costly cross-border payment systems. Stablecoins provide a digital means of transferring value across borders, bypassing traditional banking systems and reducing transaction fees and processing times. This increased efficiency makes stablecoins a viable option for global business transactions, especially in regions with limited banking infrastructure.

> "If you are thinking the price of Bitcoin is going up, you must also realize that the purchasing power of Bitcoin is increasing, hence you are guaranteed of freedom. You must decide not to allow the fiat currency to be your mental intermediary when thinking about the world."
>
> Olawale Daniel

Fiat-backed stablecoins often operate within regulatory frameworks, ensuring compliance with legal requirements in major markets. By adhering to regulatory standards, stablecoin issuers enhance trust and credibility among users and businesses. Compliance with anti-money laundering (AML) and know-your-customer (KYC) regulations provides an added layer of security and reduces the risk of illicit activities associated with cryptocurrencies. Regulatory compliance also opens up opportunities for collaboration with financial institutions, as stablecoin issuers demonstrate a commitment to operating within established legal frameworks.

Major markets often grapple with issues of financial inclusion, where certain segments of the population have limited access to traditional banking services. Fiat-backed stablecoins can help bridge this gap by providing an inclusive financial system accessible to anyone with a smartphone and internet connection. By eliminating barriers such as geographical restrictions and high transaction costs, stablecoins enable individuals in underserved areas to participate in the digital economy and gain access to financial services.

Fiat-backed stablecoins can integrate with existing financial infrastructure, enabling seamless interoperability with traditional banking systems and payment networks. This integration allows for the easy conversion between stablecoins and fiat currencies, facilitating liquidity and usability. Stablecoins can be integrated into payment gateways, online platforms, and mobile wallets, making them compatible with existing financial tools and infrastructure used in major markets.

The viability of fiat-backed stablecoins in major markets is driven by market demand and support from industry players. As businesses and individuals increasingly recognize the benefits of stablecoins, the demand for these digital assets grows. This demand attracts industry support, including partnerships with financial institutions, payment processors, and technology providers. The involvement of established market players enhances the credibility and acceptance of stablecoins within major markets.

In conclusion, fiat-backed stablecoin 1.0 demonstrates its viability in major markets through familiarity, stability, cross-border capabilities, regulatory compliance, financial inclusion, integration with existing infrastructure, and market demand. These factors contribute to the growing acceptance and adoption of stablecoins as a reliable and efficient means of conducting transactions and accessing financial services. As stablecoin ecosystems continue to evolve and mature, their viability in major markets is likely to strengthen, opening up new opportunities for individuals, businesses, and economies worldwide.

Conclusion

In conclusion, Stablecoin 1.0, with its fiat currency-backed model, has emerged as a significant milestone in the evolution of digital currencies. It introduced the concept of stability to the volatile cryptocurrency market by pegging the value of stablecoins to established fiat currencies. While Stablecoin 1.0 may not be without its limitations and challenges, it has played a crucial role in providing a modicum of stability and trust in an otherwise unpredictable ecosystem.

One of the key advantages of Stablecoin 1.0 is its ability to address the inherent volatility and speculative nature of cryptocurrencies. By pegging the value to fiat currencies, stablecoins offer users a familiar reference point and a reliable store of value. This stability makes stablecoins more suitable for day-to-day transactions and provides a hedge against the price fluctuations that plague many other digital assets.

Furthermore, Stablecoin 1.0 has demonstrated its viability in major markets through various factors. Its use of fiat currencies as the underlying collateral leverages familiarity and trust, gaining acceptance from individuals, businesses, and financial institutions. The stability and reduced volatility of stablecoins make them an attractive option for conducting cross-border transactions efficiently and cost-effectively.

The regulatory compliance embraced by Stablecoin 1.0 issuers enhances trust and credibility in major markets. By adhering to legal frameworks and implementing stringent measures such as AML and KYC, stablecoin issuers ensure compliance with regulations, reducing the risk of illicit activities. This compliance also fosters collaboration with traditional financial institutions, opening up avenues for broader adoption and integration within existing financial systems.

Stablecoin 1.0's potential for enhanced liquidity stems from its creators often being the holders of the underlying asset pool in escrow. This arrangement not only ensures transparency and accountability but also facilitates liquidity management. The ability of stablecoin creators to hold the asset pool enhances the availability of funds for potential buybacks, contributing to the stability and resilience of the ecosystem.

Transparency plays a pivotal role in the stablecoin community. Stablecoin 1.0 models strive to provide comprehensive and accessible information to stakeholders, ensuring informed decision-making. By disclosing the asset pool composition, auditing processes, and issuance and redemption mechanisms, stablecoin issuers foster trust and confidence among users. This transparency helps establish Stablecoin 1.0 as a credible and reliable alternative to traditional financial instruments.

While Stablecoin 1.0 has demonstrated its viability, it is essential to acknowledge its limitations and potential risks. The 1:1 ratio between stablecoins and fiat currencies is crucial for maintaining stability. Any deviation from this ratio can undermine the trust and value of stablecoins, potentially leading to loss of confidence in the entire system. As fiat currencies are subject to inflationary pressures and decline in value over time, stablecoin issuers must diligently manage the asset pool to ensure the necessary parity and protect against value erosion.

Moreover, stablecoin ecosystems, including Stablecoin 1.0, are not immune to potential attacks and vulnerabilities. Security risks, including hacking attempts, regulatory challenges, or market manipulation, can threaten the stability and integrity of stablecoins. Robust security measures, adherence to best practices, and ongoing monitoring and audits are crucial for mitigating these risks and safeguarding the ecosystem.

Stablecoin 1.0 represents a significant step in the development of stablecoin technology. It has provided stability, reliability, and transparency to the cryptocurrency market, paving the way for broader adoption and integration within major markets. While challenges remain, the ongoing advancements in stablecoin technology continue to refine and enhance the stability and usability of digital currencies. Stablecoin 1.0 serves as a foundation, setting the stage for the emergence of subsequent iterations and innovations in the stablecoin ecosystem. As the cryptocurrency landscape evolves, stablecoins will likely continue to play a pivotal role in reshaping the future of finance and providing a reliable bridge between the traditional and digital worlds.

Chapter 9: Stablecoin 1.1 (Tokens Backed by Fiat Currency and Corporate Paper)

"Money is the worst currency that ever grew among mankind. This sacks cities, this drives men from their homes, this teaches and corrupts the worthiest mins to turn to base deeds."[30]

Early on, those who use digital money realized that using fiat currency on its own as a method for providing value to the digital system was a folly. Intrinsically, we were moving away from physical currency when Satoshi invented Bitcoin. The movement to 'link' digital currency wholly to a single fiat currency was antithetical to the concept of cryptocurrency (which is why many government agencies are currently attacking cryptocurrency). This led to a company called Tether to create a cryptocurrency which was backed by the US dollar and also corporate paper.[31] This allowed the ecosystem to be profitable but still have a level of stability within the system.

In this chapter, we will follow the same pattern that we follow in all the chapters in this part of the book. First, we look at the stability of the asset class on which the token at this level is based. The asset class is very important because it is the 'pegged' underpinning of the system. Next, we look at the liquidity of the asset class, how fast can the asset be liquidated if there is a 'run' on the token. As a matter of course, we will then look at the asset ratio needed to maintain stability. The media tends to insist that there needs to be a 1:1 ratio, and letter agencies tend to echo this

---

[30] Sophocies at https://www.azquotes.com/quotes/topics/currency.html.
[31] Messari.io (2023). "Tether." At https://messari.io/asset/tether/profile.

false statement; each class has a ratio that is needed for the token to be reliable in its function.

> "Paper money is liable to be abused, has been, is, and forever will be abused, in very country in which it is permitted."
>
> Thomas Jefferson

Also, as a matter of course, it is always important to look at the relationship between holders/creators of the currency and the asset pool. The more complex the system the more analysis needs to be done on how the assets are being managed. While a 1:1 ratio for stablecoin 1.0 can be managed by the local bank of the creator, more complex systems require funds, investment firms, and even multiple subsidiaries to ensure that the asset pool maintains value. The analysis of who holds the asset pool then leads to an analysis of susceptibility of to attack. The more complex the asset, the more stable it is- however, it does need to be protected from attacks. The discussions within these sections of each chapter will be quite similar. Transparency is also an issue that varies slightly as you look at different types of stablecoins, once again the more complex the tokenomics- the more transparent the company needs to be. Finally, there needs to be an analysis of viability in major markets. This results in a bimodal, cyclic model for the analysis of the system, viability seems to be highest for simple and highly complex models, with intermediate models being lukewarm in their reception.

Stablecoin 1.1 has a higher level of complexity than Stablecoin 1.0, but it is still one of the least complex stablecoins on the market. This allows for an ease of understanding that facilitates an ease of use. This is one of the reasons that regulators are having a difficult time attacking Stablecoin 1.1, because people can understand that if stablecoin 1.1 is not stable, then neither is the economy on which it is based. This leads to a 'catch 22' for the government- either they accept this advancement for human currency, or they admit that the government managed fiat currency experiment of the last 110 years is a failure, and it is time to move on to mathematical based currencies- either way, cryptocurrency wins.

Stability of Asset Class

Stablecoin 1.1 represents an evolution in the world of stablecoins by introducing a hybrid approach to backing the token with both fiat currency and corporate paper. This new model aims to enhance stability compared to Stablecoin 1.0, which relies solely on fiat currency as its underlying asset. In this discussion, we will explore the reasons why Stablecoin 1.1 offers improved stability and resilience by incorporating corporate paper into its backing.

One of the primary reasons why Stablecoin 1.1 is considered more stable than Stablecoin 1.0 is diversification. By including corporate paper as part of its collateral, Stablecoin 1.1 spreads the risk across multiple assets. While fiat currency is subject to inflationary pressures and value erosion over time, corporate paper represents a different asset class with its own dynamics. The inclusion of corporate paper reduces the reliance on a single asset type, mitigating the impact of potential fluctuations in the value of fiat currency alone.

Corporate paper refers to short-term debt obligations issued by corporations, such as commercial paper or corporate bonds. These instruments typically have a fixed maturity and provide a steady income stream to investors. By including high-quality corporate paper in the backing of Stablecoin 1.1, the stability of the overall system is enhanced. The value of corporate paper is influenced by factors such as the creditworthiness of the issuing company, prevailing interest rates, and market demand. The inclusion of these factors introduces a layer of stability and potential appreciation to the stablecoin's underlying assets.

Another advantage of incorporating corporate paper into the stablecoin's backing is the potential for increased liquidity. While fiat currency is generally highly liquid, corporate paper can provide an additional avenue for liquidity management. Corporate bonds and commercial paper can be traded in secondary markets, offering an opportunity for stablecoin issuers to access liquidity quickly if needed. This liquidity buffer can help address potential redemption demands or market fluctuations, enhancing the stability of the stablecoin ecosystem.

Furthermore, the inclusion of corporate paper brings potential income generation to the stablecoin ecosystem. Corporate bonds, for instance, offer interest payments to bondholders. This income stream can be utilized to bolster the stability of the stablecoin by contributing to the maintenance of the necessary parity with the underlying assets. The ability to generate income from corporate paper holdings can act as a counterbalance to the gradual erosion of value in fiat currency, further enhancing the stability and sustainability of Stablecoin 1.1.

Risk mitigation is another key aspect that Stablecoin 1.1 aims to address through its diversified backing. While fiat currency is subject to various risks such as inflation, economic fluctuations, and political instability, corporate paper introduces a different risk profile. The inclusion of high-quality corporate paper helps mitigate the risk associated with a single asset class, providing stability to the stablecoin ecosystem. The creditworthiness of the issuing companies, coupled with thorough risk assessment and due diligence, contributes to a more robust risk management framework for Stablecoin 1.1.

Moreover, the combination of fiat currency and corporate paper as backing assets in Stablecoin 1.1 offers a more resilient ecosystem in the face of external shocks and market volatility. As different asset classes respond differently to market conditions, the inclusion of corporate paper provides a potential cushion during times of economic uncertainty. For instance, during periods of economic downturn or currency devaluation, the stablecoin ecosystem can benefit from the stability and potential appreciation of corporate paper holdings, helping to maintain the value and stability of the stablecoin.

> "The commercial paper market, when that dries up, you know, that's just like sucking the blood out of the economic body of the United States."
>
> Howard Warren Buffet

In summary, Stablecoin 1.1, with its backing of both fiat currency and corporate paper, represents an advancement in stablecoin design by incorporating diversification, increased liquidity, potential income generation, risk mitigation, and resilience to external shocks. By combining the stability of fiat currency with the dynamics of corporate paper, Stablecoin 1.1 aims to provide a more robust and reliable stablecoin ecosystem. The inclusion of corporate paper diversifies the underlying assets, mitigating risks associated with fiat currency alone and enhancing the stability, liquidity, and sustainability of the stablecoin. Stablecoin 1.1 paves the way for a new generation of stablecoins that offer improved stability, trust, and value preservation for users in the evolving digital economy.

Liquidity of Asset Class

Stablecoin 1.1, with its innovative incorporation of corporate paper as a backing system, introduces a new dimension to the stability and liquidity of the stablecoin ecosystem. While corporate paper adds diversification and stability to the backing assets, it does have implications for the overall liquidity of Stablecoin 1.1. In this discussion, we will explore the impact of corporate paper on the liquidity of Stablecoin 1.1, highlighting its slightly lower level of liquidity compared to fiat-backed stablecoins. However, we will also emphasize that Stablecoin 1.1 maintains necessary liquidity by ensuring that it never requires more than 50% liquidity, except in extreme circumstances such as the company's bankruptcy.

Liquidity is a critical aspect of any stablecoin, as it determines the ease and speed with which the stablecoin can be bought or sold in the market. It ensures that users can efficiently convert their stablecoins into other assets or currencies when needed. In the case of Stablecoin 1.1, the inclusion of corporate paper in the backing system introduces certain considerations that impact the liquidity dynamics.

Corporate paper, such as commercial paper and corporate bonds, represents debt obligations issued by corporations. While these instruments can provide a steady income stream and diversify the backing assets, they may have lower liquidity compared to fiat currency. Trading corporate paper generally involves participating in secondary markets where buyers and sellers interact to establish market prices. These secondary markets may have specific trading hours, limited market participants, and potentially higher transaction costs, which can impact the overall liquidity of corporate paper-backed stablecoins.

As a result, Stablecoin 1.1 may have a slightly lower level of liquidity compared to stablecoins solely backed by fiat currency. The limited liquidity of corporate paper-backed stablecoins can be attributed to the characteristics of the corporate paper market itself. However, it is important to note that this does not imply that Stablecoin 1.1 lacks the necessary liquidity to function effectively.

To ensure sufficient liquidity, Stablecoin 1.1 is designed to maintain a certain level of reserves that can be readily converted into fiat currency or other highly liquid assets. The reserves held by the stablecoin issuer act as a liquidity buffer, ensuring that there are sufficient funds available to meet redemption demands and market dynamics. While the liquidity may be slightly lower than fiat-backed stablecoins, Stablecoin 1.1 manages its liquidity by maintaining a prudent reserve ratio.

The reserve ratio for Stablecoin 1.1 is typically set at a level that ensures stability and meets regulatory requirements. The exact ratio may vary depending on the specific stablecoin issuer's risk management framework and the nature of the corporate paper holdings. However, it is generally recommended that the reserve ratio should be above a certain threshold, such as 50%, to ensure sufficient liquidity for the stablecoin.

By requiring a reserve ratio of at least 50%, Stablecoin 1.1 ensures that it can maintain necessary liquidity to meet normal redemption demands and market dynamics. This means that at any given time, the stablecoin issuer should hold at least 50% of the stablecoin's value in highly liquid assets or fiat currency. This provides confidence to the stablecoin holders that they can readily convert their stablecoins into the backing assets if desired.

It is important to note that the requirement for a reserve ratio above 50% provides a safety net even in the event of extreme circumstances such as the stablecoin issuer's bankruptcy. In such cases, the reserve assets can be used to fulfill redemption requests, ensuring that stablecoin holders can recover a significant portion of their investment. This aspect highlights the importance of robust risk management practices and transparency in the stablecoin ecosystem.

> "To be truly successful in business companies need to have a corporate mission that is bigger than business."
>
> Marc Benioff

While the slightly lower liquidity of Stablecoin 1.1 may pose certain considerations, it is crucial to understand that the stablecoin issuer's ability to manage and maintain the necessary liquidity is fundamental to its stability and utility. Through prudent reserve management and risk mitigation strategies, Stablecoin 1.1 strikes a balance between diversification, stability, and liquidity, providing users with a reliable and trustworthy stablecoin option.

In conclusion, Stablecoin 1.1's utilization of corporate paper as a backing system introduces a slightly lower level of liquidity compared to fiat-backed stablecoins. However, the stablecoin maintains necessary liquidity by ensuring that it never requires more than 50% liquidity, except in extreme scenarios like the issuer's bankruptcy. The inclusion of corporate paper diversifies the underlying assets, enhancing stability and reducing reliance on a single asset class. The prudent management of reserves and risk mitigation strategies are crucial for maintaining the necessary liquidity and ensuring the stability and utility of Stablecoin 1.1 in the broader cryptocurrency ecosystem.

Required Asset Ratio for Stability

Stablecoin 1.1, with its unique backing system that incorporates corporate paper, introduces an interesting dynamic to the stability of the stablecoin ecosystem. Unlike traditional stablecoins that require a 100% backing to maintain stability, Stablecoin 1.1 operates within a zone of movement that allows for a slightly lower backing ratio of around 80%. This is made possible by the stabilizing diversity provided by corporate paper, which contributes to long-term stability despite short-term fluctuations.

The inclusion of corporate paper as part of the backing assets in Stablecoin 1.1 brings inherent advantages in terms of diversification. Corporate paper represents debt obligations issued by corporations and provides an additional layer of stability to the stablecoin ecosystem. By incorporating this asset class, Stablecoin 1.1 reduces its reliance on a single asset, such as fiat currency, and gains exposure to the creditworthiness and performance of corporate issuers.

The diversification offered by corporate paper allows Stablecoin 1.1 to operate within a zone of movement in the short term. While the token may experience some fluctuations in value on a day-to-day basis, these movements are generally within an acceptable range and do not significantly impact its overall stability. This zone of movement provides flexibility and allows the stablecoin to adapt to market dynamics without jeopardizing its long-term stability.

The underlying principle behind Stablecoin 1.1's zone of movement is the understanding that short-term fluctuations and market volatility are inherent to any financial system, including stablecoins. By accepting this reality, Stablecoin 1.1 acknowledges that maintaining absolute stability in the short term may not be feasible or necessary. Instead, it focuses on maintaining a balance between stability and flexibility, ensuring that the stablecoin remains neutral or, at the very least, less negatively impacted by market fluctuations over the long term.

The zone of movement in Stablecoin 1.1 is influenced by the diverse nature of the backing assets. While the value of individual corporate paper holdings may fluctuate due to factors such as changes in credit ratings or market conditions, the overall impact on the stablecoin's stability is mitigated by the collective performance of the asset class. The stability of corporate paper, coupled with the backing assets' diversification, helps stabilize the stablecoin's value over time.

The lower backing ratio of around 80% in Stablecoin 1.1 also contributes to the long-term stability of the stablecoin. Unlike stablecoins that require a full 100% backing, Stablecoin 1.1 benefits from the flexibility provided by the slightly lower ratio. This flexibility allows the stablecoin issuer to allocate a portion of the funds towards strategic investments, liquidity management, or operational expenses while still maintaining a robust level of backing to support stability.

It is important to note that the backing ratio of 80% does not imply any compromise on the stability or reliability of Stablecoin 1.1. The stablecoin issuer carefully manages the backing assets to ensure that they remain liquid, creditworthy, and readily convertible. The reserve management strategies employed by the issuer are designed to maintain the necessary balance between liquidity and stability, taking into account the unique characteristics of corporate paper as an asset class.

While Stablecoin 1.1 may exhibit some movement on day-to-day charts, its long-term stability and neutrality with the market make it an attractive option for users seeking a stable and reliable digital asset. The diversity offered by corporate paper, along with the slightly lower backing ratio, allows Stablecoin 1.1 to navigate short-term fluctuations while maintaining stability in the broader context. This approach recognizes the inherent nature of market dynamics and provides a robust framework for sustainable stability.

Stablecoin 1.1's unique composition, which incorporates corporate paper as part of its backing system, allows for a zone of movement in the short term while maintaining long-term stability. The diversification provided by corporate paper contributes to the stablecoin's ability to weather market fluctuations, reducing reliance on a single asset class. The slightly lower backing ratio of around 80%, coupled with prudent reserve management strategies, ensures the necessary balance between liquidity and stability. Stablecoin 1.1's approach provides a compelling model for stability in the stablecoin ecosystem, offering users a reliable and resilient digital asset for various financial transactions.

Relationship between Holders and access to backing.

Ensuring transparency and accountability is crucial in the stablecoin ecosystem to build trust among users and maintain the stability of the token. In the case of Stablecoin 1.1, which is backed by a combination of fiat currency and corporate paper, it is advisable for the holders of the stablecoin to involve a third party to hold at least 50% of the asset pool. This arrangement provides several benefits, including independent verification, increased transparency, and reliable reporting on the holdings and their utilization.

One of the primary reasons for involving a third party to hold a significant portion of the asset pool is to establish independent verification of the holdings. By entrusting a reputable third-party custodian with a substantial stake in the asset pool, the stability of the stablecoin can be validated by an unbiased entity. The third-party custodian acts as a neutral party that verifies the existence and quantity of the assets held as backing for the stablecoin. This independent verification adds a layer of trust and confidence in the stability of the stablecoin, assuring users that the asset pool is accurately represented.

Moreover, having a third party hold a significant portion of the asset pool enhances transparency in the ecosystem. The custodian can provide regular reports on the composition and status of the assets held, including detailed information on the holdings and how they are being utilized. This transparency helps stakeholders, including users and regulators, understand the stability and risk profile of the stablecoin. It allows for better assessment of the backing assets, ensuring their quality, liquidity, and compliance with regulatory standards. Transparency also mitigates the possibility of mismanagement or misuse of the asset pool, promoting accountability within the stablecoin ecosystem.

The involvement of a third-party custodian holding at least 50% of the asset pool also ensures reliable reporting on the holdings and their utilization. The custodian can provide periodic reports that detail the assets held, any changes in the composition of the pool, and how the assets are being managed. These reports can include information on investment strategies, risk management practices, and compliance measures undertaken by the custodian. This level of reporting enhances the overall understanding of the stablecoin's stability and risk management framework, promoting informed decision-making among users and regulators.

Furthermore, the presence of a third-party custodian helps mitigate potential conflicts of interest that may arise when the stablecoin creators or holders directly manage the entire asset pool. By involving a neutral third party, the risk of self-dealing or manipulation of the asset pool is minimized. The custodian acts as a guardian of the stablecoin's stability, ensuring that the asset pool remains secure and well-managed. This arrangement fosters trust and confidence in the stablecoin, as users can be assured that the stablecoin's creators have taken measures to mitigate conflicts of interest and prioritize the stability and integrity of the ecosystem.

Involving a third party to hold at least 50% of the asset pool in Stablecoin 1.1 also aligns with regulatory expectations and requirements. Regulatory bodies often emphasize the need for independent verification and transparency in financial systems, including stablecoins. By adhering to these principles and involving a third-party custodian, stablecoin issuers demonstrate their commitment to regulatory compliance and foster a cooperative relationship with regulatory authorities. This cooperation enhances the overall credibility and legitimacy of the stablecoin in the eyes of regulators, which can lead to a more favorable regulatory environment for its adoption and use.

In conclusion, involving a third party to hold at least 50% of the asset pool in Stablecoin 1.1 brings several advantages to the ecosystem. It ensures independent verification of the holdings, increases transparency through regular reporting, and promotes accountability and trust among users. By involving a reputable third-party custodian, stablecoin issuers can demonstrate their commitment to transparency, regulatory compliance, and responsible management of the stablecoin's stability. This approach contributes to the long-term viability and acceptance of Stablecoin 1.1 in the market, establishing it as a reliable and trustworthy digital asset for various financial transactions.

Susceptibility to Attacks

Stablecoin 1.1, backed by a combination of fiat currency and corporate paper, faces various potential attacks that could impact its stability and functionality within the cryptocurrency ecosystem. Understanding these types of attacks is essential for implementing robust security measures and mitigating risks. In this discussion, we will explore some common attack vectors and their potential implications for Stablecoin 1.1.

1. Collusion Attacks: One possible attack on Stablecoin 1.1 involves collusion among the stakeholders. Collusion can occur when participants

with significant influence over the stablecoin's asset pool conspire to manipulate or exploit the system. For example, if a large percentage of the asset pool is controlled by a small group of individuals or entities, they may collude to create artificial scarcity or manipulate the value of the backing assets. This type of attack undermines the stability and trustworthiness of the stablecoin, eroding its value and reliability.

2. Market Manipulation: Stablecoin 1.1 is not immune to market manipulation attacks, which involve intentional actions to manipulate the price or demand for the stablecoin. Malicious actors can engage in practices such as pump-and-dump schemes, where they artificially inflate the price of the stablecoin by creating false buying pressure, only to sell off their holdings at a profit, causing a sharp price decline. Market manipulation attacks can undermine the stability and liquidity of Stablecoin 1.1, leading to significant losses for holders and a loss of confidence in the stablecoin.
3. Regulatory Crackdown: Stablecoin 1.1 operates within a regulatory landscape that is subject to evolving rules and regulations. A regulatory crackdown, triggered by concerns over compliance, financial stability, or investor protection, can pose a significant risk to the stablecoin's operations. If regulatory authorities impose stringent regulations or outright bans on stablecoins, it can restrict the ability to maintain a stable value or access certain markets. Such regulatory actions can create uncertainty, erode trust, and limit the viability of Stablecoin 1.1 in major markets.

4. Security Breaches: Stablecoin 1.1, like any digital asset, is susceptible to security breaches that can result in the loss or theft of funds. Hackers may target the stablecoin's infrastructure, wallets, or exchanges to gain unauthorized access and compromise the system's integrity. A security breach could lead to a loss of funds, undermine user confidence, and negatively impact the stablecoin's stability. Implementing robust security measures, including strong encryption, multi-factor authentication, and regular audits, is crucial to protect Stablecoin 1.1 from these types of attacks.
5. Asset Mismatch: Stablecoin 1.1's stability relies on the underlying assets maintaining their value and liquidity. However, an asset mismatch can occur when the value of the backing assets deviates significantly from the stablecoin's value. For instance, if the value of the corporate paper or fiat currency backing the stablecoin experiences a sharp decline, it may lead to an asset shortfall. This mismatch can threaten the stablecoin's stability and erode confidence in its value.
6. Regulatory Non-Compliance: Non-compliance with regulatory requirements can expose Stablecoin 1.1 to legal risks and regulatory actions. Failure to adhere to anti-money laundering (AML) and know-your-customer (KYC) regulations or other financial regulations may result in penalties, fines, or even the suspension of the stablecoin's operations. Regulatory non-compliance can also tarnish the stablecoin's reputation, making it less attractive to users and hindering its adoption.

To mitigate these potential attacks, stablecoin issuers and participants should implement robust security measures, including strong governance frameworks, regular audits, and comprehensive risk management practices. Engaging in third-party audits and verifications can enhance transparency and provide additional assurance to users. Collaborating with regulatory authorities and adhering to regulatory requirements can help ensure compliance and minimize regulatory risks. Furthermore, promoting decentralization, diversification of the asset pool, and open-source development can enhance the stability and resilience of Stablecoin 1.1. By understanding and addressing these potential attack vectors, the stablecoin ecosystem can work towards building a secure and stable financial infrastructure that fosters trust, promotes widespread adoption, and facilitates seamless transactions.

Transparency

Transparency is a crucial aspect of stablecoin 1.1 as it helps to build trust, foster confidence, and ensure the stability of the ecosystem. In this discussion, we will explore the need for transparency in stablecoin 1.1 and address some of the complaints raised regarding Tether's transparency policy.

Transparency plays a pivotal role in stablecoin 1.1 for several reasons. First and foremost, stablecoins aim to provide a reliable and stable digital representation of real-world assets. Users rely on the stability of these tokens to preserve the value of their holdings and facilitate seamless transactions. Transparency helps to ensure that the stablecoin's value is adequately backed and that the assets supporting it are of high quality and sufficient in quantity.

By providing transparent information about the composition and management of the asset pool, stablecoin 1.1 instills confidence in users, investors, and regulators. Users want to know that the assets backing the stablecoin are held in a secure manner, that the stablecoin issuer has mechanisms in place to verify the assets' existence and value, and that there are regular audits to validate these claims. Transparent reporting allows users to assess the stability and reliability of the stablecoin and make informed decisions about its usage.

> "In todays era of volatility, there is no other way but to re-invent. The only sustainable advantage you can have over others is agility. That is because nothing else is sustainable everything ese you create, somebody else will replicate."
>
> Jeff Bezos

Transparency also acts as a safeguard against potentially fraudulent activities and mismanagement. By disclosing information about the asset pool and the parties responsible for its custody, stablecoin 1.1 can detect and prevent unauthorized activities. Users can verify the authenticity and accuracy of the stablecoin's claims, ensuring that the issuer is acting in the best interest of the community. Transparent reporting acts as a deterrent, as issuers are less likely to engage in questionable practices when their actions are subject to scrutiny.

Furthermore, transparency promotes market integrity and fair competition. In a rapidly evolving cryptocurrency market, various stablecoin projects compete for users' attention and trust. Transparent reporting helps distinguish credible projects from those with questionable practices. Users can evaluate the level of transparency and make informed choices based on their preferences and risk appetite. Market participants can also assess the stability and risk profile of different stablecoins, promoting healthy competition and innovation.

While transparency is a vital aspect of stablecoin 1.1, it is essential to address concerns raised regarding Tether's transparency policy. Tether, one of the prominent stablecoins in the market, has faced criticism for its lack of transparency and alleged opacity in its operations. Critics argue that Tether's claims of having a 1:1 backing with fiat currencies lack sufficient evidence and transparency.

The complaints against Tether primarily revolve around the limited disclosure of its reserves, the absence of regular audits, and questions regarding the nature of the assets backing the stablecoin. Transparency requires clear and verifiable information, and the lack thereof can erode trust and raise doubts about the stability of the stablecoin.

To address these concerns, stablecoin 1.1 should prioritize transparency in its operations. Regular and independent audits of the asset pool should be conducted by reputable third-party firms to validate the claims of asset backing and ensure compliance with the stated ratio. Disclosing the identity of the custodians, providing detailed reports on the asset allocation, and implementing robust governance mechanisms are essential steps in enhancing transparency.

Moreover, stablecoin 1.1 should consider adopting open-source development practices. Open-source code allows independent developers and auditors to scrutinize the stablecoin's underlying technology, ensuring that it operates as intended and that there are no hidden vulnerabilities or risks. This level of transparency promotes accountability and provides the community with a deeper understanding of the stablecoin's operations.

Transparency is a fundamental requirement for stablecoin 1.1 to establish trust, ensure stability, and foster confidence among users, investors, and regulators. It enables users to assess the stability and reliability of the stablecoin, protects against fraudulent activities, promotes fair competition, and enhances market integrity. While concerns have been raised regarding Tether's transparency policy, it is important for stablecoin 1.1 to address these concerns by implementing robust reporting mechanisms, conducting regular audits, and embracing open-source development practices. By prioritizing transparency, stablecoin 1.1 can strengthen its position in the market and contribute to the development of a more trustworthy and resilient stablecoin ecosystem.

Viability in Major Markets

Stablecoin 1.1 has emerged as a successful solution in various markets, including credit card companies, due to its unique characteristics and benefits. In this discussion, we will explore the reasons behind the high levels of success that stablecoin 1.1 has achieved in these markets.

One of the key factors contributing to the success of stablecoin 1.1 in credit card companies is its stability. Stablecoins are designed to maintain a stable value, often pegged to a fiat currency or a basket of assets. This stability makes stablecoin 1.1 an attractive option for credit card companies as it minimizes the risk associated with price volatility. Credit card companies can seamlessly integrate stablecoin 1.1 into their payment systems, providing customers with a reliable and consistent means of transacting without worrying about sudden value fluctuations.

Moreover, stablecoin 1.1 offers faster settlement times compared to traditional payment methods. Credit card transactions typically involve intermediaries, such as banks and payment processors, which can introduce delays and increase transaction costs. Stablecoin 1.1, being built on blockchain technology, enables peer-to-peer transactions with near-instant settlement times, bypassing the need for intermediaries. This efficiency not only enhances the user experience but also reduces costs for credit card companies, making stablecoin 1.1 an attractive alternative payment method.

The global nature of stablecoin 1.1 also contributes to its success in credit card companies and other markets. Traditional cross-border transactions often involve lengthy processes, high fees, and currency exchange complexities. Stablecoin 1.1 allows for frictionless cross-border payments, eliminating the need for multiple intermediaries and reducing transaction costs. Credit card companies can leverage stablecoin 1.1's borderless nature to expand their customer base and offer more accessible payment options to individuals and businesses worldwide.

Another advantage of stablecoin 1.1 for credit card companies is increased privacy and security. Blockchain technology, on which stablecoins are built, provides a transparent and immutable ledger of transactions. This transparency helps prevent fraud, ensures transaction integrity, and builds trust among users. Additionally, stablecoin 1.1 transactions can be executed with pseudonymity, protecting users' personal information. This privacy-enhancing feature aligns with the evolving demands for data protection and security in the digital age.

Furthermore, stablecoin 1.1's compatibility with smart contract functionality opens up new possibilities for credit card companies. Smart contracts are self-executing contracts with predefined rules and conditions encoded within the blockchain. They can automate various aspects of payment processing, such as escrow services, multi-signature requirements, and programmable triggers. Credit card companies can leverage stablecoin 1.1's smart contract capabilities to streamline their operations, enhance payment security, and offer innovative financial products to their customers.

In addition to these market-specific advantages, stablecoin 1.1 aligns with the broader trends of decentralization and financial inclusion. The traditional financial system often excludes individuals without access to banking services or those in regions with unstable currencies. Stablecoin 1.1 provides an inclusive digital payment solution, enabling individuals to participate in the global economy without the need for a bank account. Credit card companies can tap into this growing demand for inclusive financial services by integrating stablecoin 1.1 into their offerings, expanding their customer base, and fostering financial inclusion.

It is worth noting that the success of stablecoin 1.1 in credit card companies and other markets also relies on regulatory compliance and industry partnerships. Stablecoin projects need to navigate regulatory frameworks to ensure legal compliance, address concerns regarding money laundering and terrorist financing, and establish trust with regulators. Collaborations with established financial institutions and payment processors help bridge the gap between traditional finance and the emerging stablecoin ecosystem, fostering adoption and driving success.

In conclusion, stablecoin 1.1 has witnessed high levels of success in credit card companies and various markets due to its stability, faster settlement times, global accessibility, privacy and security features, compatibility with smart contracts, and alignment with decentralization and financial inclusion trends. By leveraging these advantages, stablecoin 1.1 offers credit card companies an innovative payment solution that enhances user experience, reduces costs, expands market reach, and promotes financial inclusion. However, continued regulatory compliance and strategic partnerships remain essential for sustained growth and broader adoption of stablecoin 1.1 in the credit card industry and beyond.

Conclusion

Stablecoin 1.1 has emerged as a promising solution in the realm of digital currencies and financial transactions. Its success in various markets, including credit card companies, can be attributed to its stability, efficiency, global accessibility, privacy, and security features. Stablecoin 1.1's ability to maintain a stable value, often backed by a combination of fiat currency and corporate paper, provides users with a reliable means of transacting while minimizing the risks associated with price volatility. The faster settlement times and reduced costs offered by stablecoin 1.1 make it an attractive option for credit card companies seeking to enhance their payment systems and improve the user experience for their customers.

The borderless nature of stablecoin 1.1 enables frictionless cross-border transactions, bypassing the complexities and inefficiencies of traditional cross-border payments. This feature presents a significant opportunity for credit card companies to expand their customer base globally and offer accessible payment options to individuals and businesses worldwide. Moreover, the transparency and security provided by blockchain technology contribute to stablecoin 1.1's appeal, ensuring transaction integrity, preventing fraud, and protecting user data. These factors align with the evolving demands for privacy and security in the digital age.

Furthermore, stablecoin 1.1's compatibility with smart contract functionality unlocks new possibilities for credit card companies, allowing for the automation of payment processes and the development of innovative financial products. By harnessing the power of smart contracts, credit card companies can streamline their operations, enhance payment security, and deliver enhanced services to their customers. However, it is important to note that the success of stablecoin 1.1 in credit card companies and other sectors hinges on regulatory compliance and strategic partnerships. Adherence to regulatory frameworks and collaboration with established financial institutions and payment processors are crucial for ensuring legal compliance, building trust, and fostering broader adoption.

Overall, stablecoin 1.1 presents a compelling proposition for credit card companies, offering stability, efficiency, global accessibility, privacy, and security in an increasingly digitized financial landscape. Its ability to address key pain points in traditional payment systems positions stablecoin 1.1 as a transformative force in the industry. As the stablecoin ecosystem continues to evolve, it is imperative to maintain a balance between innovation, regulatory compliance, and industry collaboration to unlock the full potential of stablecoin 1.1 and drive its continued success in credit card companies and the broader market.

Chapter 10: Stablecoin 1.2 (Algorithm Based Tokens)

"Governments create money and spend it into the economy by running budget deficits. The paper currency in your pocket is technically a government debt."[32]

The concept of government debt is a dangerous concept. Because if people realized that any money that we have is just a government obligation to pay us, and that we would be paid in the self-same currency that is a promise for the government to pay us, it becomes known that it is all a charade. Money is worth something because we believe that it is worth something, if the people stop believing that the money is worth anything, then the money becomes worth nothing. Stablecoin 1.2 moves beyond the attachment that we have to fiat currency and the promise of a better life that the government offers us on a generation-by-generation basis. The promises of the government are not much better than the promises of a campaigning politician, mainly because the government is always campaigning for our support. The introduction of algorithms into the cryptocurrency stablecoin space allows for the system to asset the holder of the currency in determining the value, which in turn cuts the human out of the equation to allow for a better system.

In this chapter, we will follow the same pattern that we follow in all the chapters in this part of the book. First, we look at the stability of the asset class on which the token at this level is based. The asset class is very important

---

[32] Michael Hudson at https://www.azquotes.com/quotes/topics/currency.html.

because it is the 'pegged' underpinning of the system. Next, we look at the liquidity of the asset class, how fast can the asset be liquidated if there is a 'run' on the token. As a matter of course, we will then look at the asset ratio needed to maintain stability. The media tends to insist that there needs to be a 1:1 ratio, and letter agencies tend to echo this false statement; each class has a ratio that is needed for the token to be reliable in its function.

> "Money is only a tool that will take you wherever you wish, but it will not replace you as the driver."
>
> Ayn Rand

Also, as a matter of course, it is always important to look at the relationship between holders/creators of the currency and the asset pool. The more complex the system the more analysis needs to be done on how the assets are being managed. While a 1:1 ratio for stablecoin 1.0 can be managed by the local bank of the creator, more complex systems require funds, investment firms, and even multiple subsidiaries to ensure that the asset pool maintains value. The analysis of who holds the asset pool then leads to an analysis of susceptibility of to attack. The more complex the asset, the more stable it is- however, it does need to be protected from attacks. The discussions within these sections of each chapter will be quite similar. Transparency is also an issue that varies slightly as you look at different types of stablecoins, once again the more complex the tokenomics- the more transparent the company needs to be. Finally, there needs to be an analysis of viability in major markets. This results in a bimodal, cyclic model for the analysis of the system, viability seems to be highest for simple and highly complex models, with intermediate models being lukewarm in their reception.

Overall, the advancement of using computer technology to balance out the inflation paradox was a major leap forward in the idea of stablecoins. This can often be seen as the turning point between level 1 and level 2 stablecoins. While not truly a level 2 stablecoin, stablecoin 1.2 is still one of the 'founding fathers' of the system that will allow everyone access to their wealth at a better rate, without the worry of some human making a mistake and devaluing the currency before it has a chance.

Stability of Asset Class

Stablecoins have emerged as a viable solution in the cryptocurrency ecosystem, offering stability and mitigating the price volatility associated with cryptocurrencies. While stablecoin 1.0, backed solely by fiat currency, has provided a stable value proposition, stablecoin 1.2 introduces a novel approach by incorporating algorithms to balance the value of the token. This essay aims to delve into the stability of stablecoin 1.2 and compare it with stablecoin 1.0 (fiat stablecoins) and stablecoin 1.1 (fiat and corporate paper stablecoins). By analyzing their respective mechanisms, underlying assets, and risk factors, we can gain a comprehensive understanding of stablecoin 1.2's stability in the cryptocurrency landscape.

Stability Mechanisms of Stablecoin 1.2:

Stablecoin 1.2 utilizes algorithmic management to maintain price stability. Algorithms dynamically adjust the supply and demand of the stablecoin based on market conditions and predetermined rules. This dynamic approach enables stablecoin 1.2 to adapt to changing market forces and stabilize the token's value. By employing algorithmic balancing, stablecoin 1.2 aims to achieve stability without relying solely on traditional backing assets like fiat currency or corporate paper.

Comparison with Stablecoin 1.0 (Fiat Stablecoins):

Stablecoin 1.0, backed by fiat currency, offers stability by pegging the token's value to a specific fiat currency, such as the US dollar. While fiat stablecoins have been successful in providing stability in the cryptocurrency market, they still face inherent challenges due to the declining value of fiat currencies over time. In contrast, stablecoin 1.2's algorithmic approach allows for more dynamic adjustments and potential for greater resilience to inflationary pressures.

Moreover, stablecoin 1.2's algorithmic balancing mechanism enables it to respond to market fluctuations and maintain price stability even during times of high volatility. This flexibility distinguishes stablecoin 1.2 from stablecoin 1.0, as the former can adapt its supply and demand dynamics in real-time, ensuring a stable value for users. However, the algorithmic approach also introduces its own set of risks and challenges, including the need for robust algorithms, accurate data feeds, and potential vulnerabilities to algorithmic manipulation.

Comparison with Stablecoin 1.1 (Fiat and Corporate Paper Stablecoins):

Stablecoin 1.1, backed by a combination of fiat currency and corporate paper, provides an enhanced level of stability compared to stablecoin 1.0. The inclusion of corporate paper assets diversifies the underlying asset pool, reducing the reliance on fiat currency alone. This diversification can mitigate the impact of inflation on the stablecoin's value and provide additional stability to users.

> "Money is based on two universal principles:
>
> a. Universal Convertibility: With money as an alchemist, you can turn land into loyalty, justice into health, and violence into knowledge.
> b. Universal Trust: with money as a go-between any two people can cooperate on any project."
>
> Noah Harari

While stablecoin 1.1's combination of fiat currency and corporate paper offers increased stability, stablecoin 1.2's algorithmic balancing mechanism takes stability to another level. Stablecoin 1.2's algorithmic approach allows for more agile adjustments to supply and demand, offering potential benefits in responding to market fluctuations and maintaining price stability. However, stablecoin 1.2's reliance on algorithms introduces a different set of challenges, including the need for transparent and auditable algorithms, robust risk management, and potential algorithmic risks.

Risk Factors and Challenges:

Despite the advantages of stablecoin 1.2's algorithmic stability, there are potential risks and challenges to consider. Algorithmic stability relies on accurate and timely data feeds, making data quality and reliability crucial for maintaining stability. Additionally, the transparency and auditability of the algorithms employed in stablecoin 1.2 are essential for user trust and confidence. Without proper transparency and oversight, algorithmic stability may be vulnerable to manipulation or inaccuracies.

Furthermore, stablecoin 1.2's algorithmic stability may also face challenges in extreme market conditions or unforeseen events. Rapid market fluctuations or external shocks could strain the algorithm's ability to maintain stability, potentially leading to temporary deviations from the pegged value. Robust risk management protocols and stress-testing mechanisms are crucial to address these challenges and ensure the stability of stablecoin 1.2 in various market scenarios.

Stablecoin 1.2 introduces a novel approach to stability in the cryptocurrency market, utilizing algorithmic balancing to maintain the value of the token. While stablecoin 1.0 (fiat stablecoins) and stablecoin 1.1 (fiat and corporate paper stablecoins) have provided stability, stablecoin 1.2's algorithmic approach offers greater flexibility, adaptability, and potential resilience to inflationary pressures. However, algorithmic stability also comes with its own set of challenges, including data reliability, algorithm transparency, and vulnerability to market shocks. Robust risk management, transparent algorithms, and accurate data feeds are essential for the successful implementation of stablecoin 1.2's algorithmic stability. As the cryptocurrency ecosystem evolves, stablecoin 1.2 represents an innovative solution that can contribute to a more stable and efficient digital financial system.

Liquidity of Asset Class

Stablecoin 1.2, an algorithm-based stablecoin, offers a unique approach to achieving price stability in the cryptocurrency market. While algorithmic balancing introduces advantages in terms of flexibility and responsiveness to market dynamics, it also raises concerns regarding liquidity. This essay aims to explore the liquidity issues associated with stablecoin 1.2 and analyze the factors that contribute to its liquidity profile. By understanding the challenges and potential solutions, stakeholders can evaluate the viability and effectiveness of stablecoin 1.2 as a liquid digital asset.

Understanding Liquidity in Algorithm-Based Stablecoin 1.2: Liquidity refers to the ability to convert an asset into cash or other assets quickly and without significant impact on its price. In the context of stablecoins, liquidity plays a crucial role in maintaining the stability of the token's value and enabling smooth transactions. However, algorithm-based stablecoin 1.2 may face unique liquidity challenges due to its dynamic nature.

Algorithmic balancing relies on adjusting the supply and demand of stablecoin 1.2 based on market conditions. While this dynamic approach can help stabilize the token's value, it also introduces potential liquidity constraints. Unlike stablecoin 1.0 or 1.1, which typically maintain a fixed backing asset or combination of assets, stablecoin 1.2's algorithmic balancing may lead to fluctuations in token supply and demand, impacting liquidity.

Factors Influencing Liquidity in Stablecoin 1.2:

- Market Depth: The depth of the market, represented by the volume of buy and sell orders, affects the liquidity of stablecoin 1.2. A liquid market with a significant number of participants and substantial trading activity allows for efficient transactions and reduces the risk of slippage when converting stablecoins.
- Algorithm Efficiency: The effectiveness and accuracy of the algorithm employed in stablecoin 1.2's balancing mechanism are critical. A well-designed and properly functioning algorithm can help maintain stability and prevent sudden price movements that may hinder liquidity. Conversely, algorithmic errors or inefficiencies could impact stability and liquidity.
- Demand-Supply Imbalance: Sudden shifts in market conditions or fluctuations in user demand for stablecoin 1.2 may lead to imbalances in the token's supply and demand. If the algorithm fails to respond adequately to these imbalances, it could result in liquidity challenges, such as insufficient liquidity for users looking to buy or sell stablecoin 1.2.
- External Market Factors: Stablecoin 1.2's liquidity is also influenced by external market factors. Changes in overall market sentiment, regulatory developments, or economic events can impact liquidity conditions. Increased volatility or uncertainty in the broader cryptocurrency market may have spillover effects on stablecoin 1.2's liquidity.

Addressing Liquidity Challenges:

- Market Making and Incentives: Stablecoin 1.2 issuers can employ market-making strategies, incentivizing liquidity providers to maintain a liquid market for the token. Market makers play a crucial role in improving liquidity by continuously quoting buy and sell prices, narrowing spreads, and absorbing imbalances in supply and demand.
- Liquidity Pools: Implementing liquidity pools, similar to those used in decentralized finance (DeFi), can enhance stablecoin 1.2's liquidity. Liquidity providers contribute their assets to the pool, enabling users to trade stablecoin 1.2 with minimal slippage. Incentives, such as yield farming or staking rewards, can attract liquidity providers to participate in these pools.
- Partnerships and Integrations: Collaborations with established cryptocurrency exchanges and platforms can expand stablecoin 1.2's liquidity ecosystem. Integration into popular trading venues and decentralized exchanges (DEXs) provides users with greater access to liquidity, improving overall market depth.
- Transparent Governance: Transparent governance models, where token holders participate in decision-making processes, can enhance confidence in stablecoin 1.2 and attract more users. Open discussions about liquidity management strategies, risk mitigation measures, and audits can help establish trust and ensure the stability of the algorithm-based stablecoin.

While stablecoin 1.2's algorithmic balancing mechanism introduces advantages in terms of flexibility and responsiveness, it also presents unique challenges to liquidity. Market depth, algorithm efficiency, demand-supply imbalances, and external market factors are crucial factors influencing stablecoin 1.2's liquidity profile. By implementing market-making strategies, liquidity pools, forging partnerships, and adopting transparent governance, stablecoin 1.2 can address liquidity challenges and enhance its overall market liquidity. Evaluating and improving liquidity is essential to ensure the stability and widespread adoption of algorithm-based stablecoins in the evolving cryptocurrency ecosystem.

Required Asset Ratio for Stability

Stablecoin 1.2 introduces a novel approach to achieving stability by leveraging algorithms to dynamically balance the token's value. Unlike stablecoin 1.0, which requires a 100% backing ratio to maintain stability, stablecoin 1.2 benefits from algorithmic adjustments that allow for a lower asset ratio of approximately 70%. This essay aims to delve into the efficiencies offered by stablecoin 1.2's algorithms, exploring how they enable a more optimized and responsive system while maintaining stability.

> "Programming is the art of algorithm design and the craft of debugging errant code."
>
> Ellen Ullman

Algorithmic Balancing for Enhanced Efficiency: Stablecoin 1.2's algorithms play a crucial role in continuously adjusting the asset ratio to maintain stability. These algorithms employ various mechanisms, such as supply and demand monitoring, rebalancing, and collateral adjustments, to ensure that the token's value remains pegged to the desired target, typically a fiat currency or a combination of assets.

By dynamically managing the asset ratio, stablecoin 1.2 achieves a delicate balance between maintaining stability and optimizing efficiency. The algorithms monitor market conditions, price movements, and liquidity to determine when and how to adjust the asset composition. This adaptability allows stablecoin 1.2 to respond swiftly to market fluctuations and maintain a more efficient system compared to stablecoin 1.0.

Efficiency Advantages of Lower Asset Ratio:

- Reduced Capital Requirement: Stablecoin 1.2's lower asset ratio provides an efficiency advantage by requiring less capital to maintain stability. In stablecoin 1.0, the 100% backing ratio necessitates holding an equivalent amount of assets to support the token's value. Stablecoin 1.2, with its lower asset ratio, allows for a more capital-efficient system, enabling creators to allocate resources effectively.
- Improved Liquidity: The lower asset ratio in stablecoin 1.2 can contribute to enhanced liquidity. By requiring a smaller pool of assets to back the token's value, stablecoin 1.2 may free up additional resources that can be deployed to enhance liquidity mechanisms. This increased liquidity facilitates smoother transactions, reduced slippage, and improved market depth, making stablecoin 1.2 a more attractive choice for users and market participants.
- Greater Flexibility: Stablecoin 1.2's algorithms provide greater flexibility in adjusting the asset

composition. Unlike stablecoin 1.0, which relies on a fixed asset backing, stablecoin 1.2 can dynamically adapt to changing market conditions. The algorithms continuously analyze the token's supply and demand, market sentiment, and price movements to determine the optimal asset ratio. This flexibility allows stablecoin 1.2 to better respond to market dynamics, optimizing stability while maximizing efficiency.

- Enhanced Risk Management: The algorithmic balancing of stablecoin 1.2 introduces risk management benefits. The dynamic asset ratio adjustments enable the system to respond to potential risks or vulnerabilities promptly. The algorithms can mitigate potential imbalances or overexposure to certain assets by rebalancing the composition, thereby reducing the risk of instability. This proactive risk management approach enhances the resilience and robustness of stablecoin 1.2.

Considerations and Challenges:

While the lower asset ratio in stablecoin 1.2 offers efficiency advantages, it is essential to consider potential challenges and risks associated with algorithm-based stablecoins. The reliance on algorithms introduces the need for continuous monitoring and auditing to ensure algorithmic integrity, safeguard against potential vulnerabilities or manipulation, and maintain user trust. Transparency in algorithm design and implementation is critical to building confidence in the stability and efficiency of stablecoin 1.2.

Furthermore, algorithmic balancing requires robust technical infrastructure and expertise to execute effectively.

The accuracy and reliability of the algorithms heavily influence stablecoin 1.2's performance. Therefore, thorough testing, auditing, and ongoing development are necessary to ensure the algorithms operate as intended and align with the stability goals of the stablecoin.

Stablecoin 1.2's algorithm-based approach redefines the asset ratio required for stability, allowing for a more efficient system compared to stablecoin 1.0. The algorithms' ability to dynamically adjust the asset composition ensures stability while optimizing capital allocation, enhancing liquidity, providing flexibility, and improving risk management. While challenges exist in terms of algorithmic integrity and technical infrastructure, diligent monitoring, auditing, and transparency measures can mitigate these concerns. Stablecoin 1.2 represents a significant evolution in stablecoin design, offering an efficient alternative to traditional fiat-backed stablecoins while maintaining the desired stability.

Relationship between Holders and access to backing.

Stablecoin 1.2 introduces a revolutionary approach to asset management by employing algorithms to balance the value of the token dynamically. This algorithm-based stability mechanism opens up a range of possibilities for determining who holds the asset pool. As long as the algorithm, or the individual responsible for executing its decisions, has access to the necessary funds, various entities can hold the asset pool. This essay explores the potential holders for the asset pool of stablecoin 1.2, highlighting the benefits and considerations associated with each option.

1. Creators of Stablecoin 1.2: The creators of stablecoin 1.2 can choose to hold the asset pool

themselves. This option provides direct control and oversight over the assets, ensuring immediate execution of the algorithm's decisions. By holding the asset pool, the creators take responsibility for maintaining the stability and integrity of the stablecoin. However, this approach requires sufficient financial resources and expertise to manage the assets effectively.

2. Financial Institutions: Established financial institutions, such as banks or investment firms, can serve as holders of the asset pool for stablecoin 1.2. These institutions have the necessary infrastructure, expertise, and regulatory compliance to manage and safeguard the assets. Their involvement can enhance the stability and credibility of the stablecoin, attracting institutional investors and fostering broader adoption. Collaborating with financial institutions also provides an additional layer of oversight and transparency.
3. Trust Companies: Trust companies specialize in managing assets on behalf of others. They can act as custodians of the asset pool for stablecoin 1.2, ensuring secure storage and efficient execution of asset transfers. Trust companies are subject to regulatory frameworks and often offer advanced security measures to safeguard the assets. By involving a trusted third party, stablecoin 1.2 creators can enhance transparency, provide independent oversight, and instill confidence in users.
4. Decentralized Autonomous Organizations (DAOs): DAOs are community-driven organizations governed by smart contracts and decentralized

decision-making. These entities can hold the asset pool for stablecoin 1.2, distributing control and decision-making power among token holders. DAOs provide a transparent and inclusive framework where stakeholders collectively participate in managing the assets. This approach aligns with the decentralized nature of cryptocurrencies and fosters community involvement in the stablecoin ecosystem.

5. Asset Management Firms: Asset management firms specialize in optimizing investment portfolios and can bring their expertise to manage the asset pool for stablecoin 1.2. These firms employ professional fund managers who can make informed decisions based on market analysis and risk assessment. Involving asset management firms adds a layer of expertise and diversification to the asset pool, potentially increasing stability and attracting institutional investors.

When selecting a holder for the asset pool, several considerations come into play. Transparency and accountability are crucial, ensuring that the selected entity operates with integrity and adheres to regulatory requirements. The chosen holder should have robust security measures in place to protect the assets from potential threats and attacks.

The benefits of involving various holders for the asset pool of stablecoin 1.2 are multifaceted. Firstly, it allows for a diversified approach to asset management, potentially reducing risk and enhancing stability. Different entities bring unique perspectives, expertise, and risk management strategies to the table. Secondly, involving external entities such as financial institutions or trust companies enhances the perceived trustworthiness and reliability of the stablecoin, encouraging wider adoption and attracting more users.

Additionally, the choice of holder can influence the transparency and governance of the stablecoin ecosystem. Collaborating with financial institutions or DAOs can promote community participation and democratic decision-making. Engaging asset management firms can leverage their expertise in optimizing the asset pool's performance, potentially generating higher returns.

Stablecoin 1.2's algorithmic stability mechanism provides flexibility in determining who holds the asset pool. Creators, financial institutions, trust companies, DAOs, and asset management firms are among the potential holders. Each option offers distinct benefits and considerations, ranging from direct control and oversight to enhanced security, transparency, and expertise. The selection of a suitable holder for the asset pool should prioritize transparency, accountability, and robust security measures to ensure the stablecoin's long-term stability and trustworthiness. By carefully considering the available options, stablecoin 1.2 can leverage the strengths of different entities to create a resilient and efficient asset management framework.

Susceptibility to Attacks

As stablecoin 1.2 relies on algorithmic mechanisms to maintain its stability, it becomes crucial to assess and address potential attacks that may target this type of stablecoin. While algorithm-based stablecoins offer unique advantages, they also face specific vulnerabilities that malicious actors could exploit. This essay explores various types of attacks that stablecoin 1.2, and similar algorithmic stablecoins, may encounter, and discusses potential mitigation strategies to enhance their security and resilience.

- Exploiting Algorithmic Vulnerabilities: Algorithmic stablecoins are susceptible to attacks that aim to manipulate or exploit vulnerabilities within the underlying algorithms. These attacks could involve exploiting price or liquidity oracle discrepancies, algorithmic biases, or vulnerabilities in the code. For example, attackers may attempt to flood the market with trades or create false signals to trigger cascading effects and disrupt the stablecoin's balance. Ensuring rigorous code auditing, continuous testing, and employing robust security measures can help mitigate these risks. Regular monitoring and prompt response to any algorithmic anomalies or suspicious activities can also minimize the impact of potential attacks.
- Liquidity Attacks: Liquidity attacks are a common concern for stablecoins, including algorithmic-based ones. These attacks aim to exploit liquidity imbalances or lack of depth in the trading markets. Attackers may attempt to initiate large trades or create artificial market pressure to manipulate prices and disrupt the stablecoin's peg. To mitigate liquidity attacks, stablecoin 1.2 should strive to maintain sufficient liquidity across different trading platforms and establish strategic partnerships with reputable liquidity providers. Implementing circuit breakers or trading halts during periods of extreme market volatility can also help prevent rapid price fluctuations due to liquidity attacks.

- Collateral Attacks: Algorithmic stablecoins that rely on collateral reserves face the risk of collateral attacks. These attacks involve targeting the assets held as collateral and aiming to undermine their value or security. Attackers may engage in short-selling or spreading negative rumors to manipulate the value of the collateral assets, leading to potential instability of the stablecoin. Implementing rigorous due diligence processes when selecting and managing collateral assets is vital. Diversification of the collateral pool and regular assessments of the assets' performance, creditworthiness, and market conditions can help mitigate the risks associated with collateral attacks.
- Oracle Exploitation: Oracles play a crucial role in algorithmic stablecoins by providing external data, such as market prices, for the algorithms to make informed decisions. However, oracle exploitation attacks involve manipulating or compromising the oracles to provide inaccurate or misleading data. Attackers may attempt to feed false price information, leading to mispricing of the stablecoin and potentially triggering instability. Implementing multi-oracle solutions, using reputable and decentralized oracle networks, and employing data verification mechanisms can help mitigate the risks of oracle exploitation attacks. Regular audits and monitoring of oracle performance can also ensure the integrity of the data used by the algorithm.
- Governance Attacks: Algorithmic stablecoins often have governance mechanisms that allow token holders to participate in decision-making processes. However, these governance systems can be targeted

in governance attacks, where attackers aim to gain control over the decision-making power or disrupt the governance process. Attackers may attempt to accumulate a significant portion of the tokens, launch a coordinated voting attack, or exploit vulnerabilities in the governance smart contracts. Implementing robust governance protocols, ensuring broad token distribution, and conducting regular security audits of the governance infrastructure can help safeguard against governance attacks.

As algorithmic stablecoins like stablecoin 1.2 gain prominence, understanding and addressing potential attack vectors is essential to ensure their stability and security. By anticipating and mitigating attacks that target algorithmic vulnerabilities, liquidity, collateral, oracles, and governance systems, stablecoin 1.2 can enhance its resilience and maintain its pegged value over the long term. Implementing comprehensive security measures, conducting regular audits, fostering transparency, and engaging with the broader community can further bolster trust in algorithmic stablecoins and support their widespread adoption in the financial ecosystem.

Transparency

Transparency is a critical aspect of stablecoins, providing users with confidence in the stability and integrity of the system. In the case of stablecoin 1.2, which relies on algorithmic mechanisms, transparency becomes paramount to maintain trust among users. This essay explores the important factors for transparency in stablecoin 1.2, with a particular focus on the algorithm and its moves. By ensuring transparency in algorithmic operations, stablecoin

1.2 can enhance accountability, foster trust, and promote a healthier ecosystem.

- Algorithmic Operations and Disclosed Parameters: Transparency in stablecoin 1.2 begins with a clear understanding of the underlying algorithm and its operations. The algorithm's design, parameters, and key variables should be openly disclosed to the community. This includes the formulae used for determining the stablecoin's value, rebalancing mechanisms, and any specific rules governing the algorithm's behavior. By sharing this information, stablecoin 1.2 promotes transparency and allows users to assess the algorithm's reliability and stability.
- Real-Time Reporting and Public Audits: To ensure transparency, stablecoin 1.2 should provide real-time reporting on its algorithmic moves and actions. This includes regular updates on the rebalancing process, collateral utilization, and changes in asset allocations. Real-time reporting allows users to track the stability of the stablecoin, understand the algorithm's decision-making process, and evaluate its effectiveness. Additionally, conducting periodic public audits by reputable third-party firms can further enhance transparency and instill confidence in the system's integrity.
- Governance and Community Involvement: Stablecoin 1.2 can foster transparency through a robust governance framework that encourages community involvement. This involves allowing token holders to participate in decision-making processes, such as voting on proposed changes to the algorithm or collateral assets. Transparent

governance mechanisms, with clear rules and processes, ensure that decisions are made collectively and in the best interest of the stablecoin's ecosystem. Regular updates on governance activities, including voting outcomes and proposals, should be publicly accessible to maintain transparency.

- Algorithmic Risk Disclosures: Transparency in stablecoin 1.2 extends to addressing potential risks associated with the algorithmic operations. Clear and comprehensive risk disclosures should be provided, outlining the potential vulnerabilities, limitations, and scenarios in which the stablecoin's stability may be challenged. This includes disclosure of potential attacks, algorithmic biases, liquidity risks, and dependencies on external data sources. By openly discussing these risks, stablecoin 1.2 enables users to make informed decisions and assess the potential implications for their investments.
- External Audit and Regulatory Compliance: To reinforce transparency, stablecoin 1.2 should undergo external audits by reputable auditing firms to validate the accuracy of its reporting, compliance with regulations, and adherence to disclosed parameters. Independent audits help ensure that the stablecoin's operations align with stated objectives, providing an additional layer of verification and accountability. Furthermore, transparent engagement with regulatory authorities, adhering to relevant regulations, and maintaining compliance with financial standards can strengthen trust in

stablecoin 1.2's operations and foster wider acceptance.

Transparency is a fundamental pillar for stablecoin 1.2, particularly in its algorithmic operations. By providing clarity on algorithmic moves, real-time reporting, public audits, inclusive governance, risk disclosures, and regulatory compliance, stablecoin 1.2 can foster transparency, accountability, and user trust. This commitment to transparency enables users to understand the stablecoin's operations, assess its stability, and make informed decisions. As stablecoin 1.2 continues to evolve, maintaining and enhancing transparency will be crucial for its long-term success in the broader financial landscape.

Viability in Major Markets

Stablecoin 1.2, with its algorithmic approach to maintaining stability, has gained significant attention in the cryptocurrency space. This essay explores how Stablecoin 1.2 has successfully met the demands of various markets and gained acceptance among users and businesses. By offering enhanced stability, efficiency, and versatility, Stablecoin 1.2 has emerged as a reliable and popular choice for individuals, merchants, and financial institutions.

One of the key reasons for the market acceptance of Stablecoin 1.2 is its ability to provide stability in volatile market conditions. Unlike traditional cryptocurrencies, Stablecoin 1.2's algorithmic design allows it to automatically adjust its value based on predefined parameters, ensuring that it remains closely pegged to a specific asset or basket of assets. This stability makes Stablecoin 1.2 an attractive option for users seeking to minimize the impact of market volatility on their transactions and holdings.

Stablecoin 1.2's algorithmic stability also makes it a preferred choice for cross-border transactions. By avoiding the need for traditional banking intermediaries and their associated fees and delays, Stablecoin 1.2 enables fast and cost-effective international transfers. The stability of the coin ensures that the value sent remains consistent throughout the transaction, reducing currency exchange risks and providing a seamless experience for individuals and businesses engaged in global commerce.

Stablecoin 1.2 has witnessed substantial adoption by merchants across various industries. Its stability, coupled with low transaction fees and near-instant settlement, make it an attractive payment solution for businesses. Stablecoin 1.2's compatibility with existing payment systems and easy integration through application programming interfaces (APIs) further accelerates its acceptance among merchants. This widespread adoption in the merchant community contributes to the ecosystem's liquidity and facilitates the broader acceptance of Stablecoin 1.2 in everyday transactions.

Financial institutions have also recognized the value and potential of Stablecoin 1.2. Its stability, transparency, and compliance with regulatory frameworks have paved the way for partnerships between Stablecoin 1.2 providers and traditional banks or financial service providers. These collaborations enable users to access Stablecoin 1.2 through their existing banking infrastructure, providing them with a seamless experience and offering the benefits of cryptocurrency while maintaining the oversight and stability of traditional financial systems.

Stablecoin 1.2's versatility has contributed to its success in the market. Its stable value and compatibility with smart contract platforms have opened up a wide range of use cases, such as decentralized finance (DeFi), tokenized assets, and lending/borrowing protocols. This flexibility allows users to explore investment opportunities, earn interest on their holdings, and participate in decentralized applications (dApps) built on blockchain networks. The ability to generate passive income or access alternative investment options has attracted a growing number of users to Stablecoin 1.2.

Stablecoin 1.2 has emerged as a stable, efficient, and versatile option in the cryptocurrency market. Its algorithmic stability, which ensures value preservation even in volatile markets, has gained acceptance among individuals, merchants, and financial institutions. The ability to facilitate cross-border transactions, the adoption by merchants, collaboration with traditional financial institutions, and the wide range of use cases have further contributed to the market success of Stablecoin 1.2. As the demand for stable and efficient digital currencies continues to grow, Stablecoin 1.2's ability to meet these requirements positions it as a promising player in the evolving landscape of digital finance.

Conclusion

In conclusion, Stablecoin 1.2 represents a significant advancement in the world of digital finance, offering a paradigm shift in terms of stability, efficiency, and versatility. Through its algorithmic approach, Stablecoin 1.2 has successfully addressed the challenges faced by traditional fiat-backed stablecoins and introduced new possibilities for individuals, businesses, and financial institutions. This conclusion highlights the key aspects and implications of Stablecoin 1.2, cementing its position as a transformative force in the cryptocurrency ecosystem.

First and foremost, Stablecoin 1.2 has revolutionized stability within the cryptocurrency space. By leveraging algorithmic mechanisms, it provides a more robust and resilient pegging system, mitigating the risks associated with price fluctuations and market volatility. The ability of Stablecoin 1.2 to maintain a relatively stable value, often hovering around the 70% asset backing ratio, enhances confidence and trust among users. This stability makes Stablecoin 1.2 an attractive medium of exchange and store of value, offering an alternative to traditional fiat currencies that are subject to inflation and geopolitical uncertainties.

Furthermore, Stablecoin 1.2 has demonstrated remarkable efficiency, particularly in cross-border transactions. With its algorithmic stability and compatibility with blockchain technology, it enables seamless and cost-effective peer-to-peer transfers, eliminating the need for intermediaries and associated fees. The ability to transact with Stablecoin 1.2 in a borderless manner not only enhances financial inclusion but also unlocks new opportunities for global commerce, empowering businesses, and individuals to engage in transactions without the traditional barriers of time, cost, and complexity.

The acceptance and adoption of Stablecoin 1.2 by merchants have played a pivotal role in its success. By embracing Stablecoin 1.2 as a payment solution, merchants benefit from lower transaction fees, faster settlement times, and reduced exposure to currency risks. The compatibility of Stablecoin 1.2 with existing payment systems and the ease of integration through APIs have facilitated its integration into mainstream commerce, opening doors for broader acceptance and usage. The widespread adoption by merchants further enhances the liquidity and utility of Stablecoin 1.2, solidifying its position as a viable and convenient payment option.

Collaboration between Stablecoin 1.2 providers and traditional financial institutions has been instrumental in driving its market success. Financial institutions recognize the value of Stablecoin 1.2's stability, compliance with regulations, and transparency, leading to partnerships that bridge the gap between the digital currency ecosystem and the traditional financial system. These collaborations allow users to access Stablecoin 1.2 through their existing banking infrastructure, expanding its reach and enabling a seamless transition between fiat currencies and digital assets. The involvement of financial institutions also contributes to regulatory compliance, ensuring the long-term sustainability and trustworthiness of Stablecoin 1.2.

In conclusion, Stablecoin 1.2 has ushered in a new era of stability, efficiency, and versatility within the cryptocurrency landscape. Its algorithmic mechanisms have provided a robust and reliable solution to the challenges faced by traditional fiat-backed stablecoins. The ability of Stablecoin 1.2 to offer stability, efficient cross-border transactions, merchant acceptance, and collaboration with financial institutions highlights its potential to reshape the future of digital finance. As the demand for stable and efficient digital currencies continues to grow, Stablecoin 1.2 stands as a testament to innovation, offering users a more reliable and secure financial ecosystem.

Chapter 11: Stablecoin 1.3 (Tokens based on a Stock or a Single Asset)

"It's not like I'm Anti-China. I just think that it's ridiculous that we allow them to do what they're doing to this country, with the manipulations of the currency…"[33]

Market manipulation is a very concerning problem in the global economy. For years we had to worry about China attacking the US economy, now we have the problem of actors inside the economy undermining the stability of the US Dollar. The more stablecoin is used, and is transparent, the more that we can fight back against these negative actors. In this chapter we look at the final level one stablecoin, stablecoin 1.3. Basing stablecoins on stocks (or other singular assets) was ca logical progression of the stablecoin evolution, however, it was an evolutionary dead end. While this approach had promise, the promise was realized in the diversified accounts of the level two stablecoins.

In this chapter, we will follow the same pattern that we follow in all the chapters in this part of the book. First, we look at the stability of the asset class on which the token at this level is based. The asset class is very important because it is the 'pegged' underpinning of the system. Next, we look at the liquidity of the asset class, how fast can the asset be liquidated if there is a 'run' on the token. As a matter of course, we will then look at the asset ratio needed to maintain stability. The media tends to insist that there needs to be a 1:1 ratio, and letter agencies tend to echo this

---

[33] Donald Trump at https://www.azquotes.com/quotes/topics/currency.html.

false statement; each class has a ratio that is needed for the token to be reliable in its function.

Also, as a matter of course, it is always important to look at the relationship between holders/creators of the currency and the asset pool. The more complex the system the more analysis needs to be done on how the assets are being managed. While a 1:1 ratio for stablecoin 1.0 can be managed by the local bank of the creator, more complex systems require funds, investment firms, and even multiple subsidiaries to ensure that the asset pool maintains value. The analysis of who holds the asset pool then leads to an analysis of susceptibility of to attack. The more complex the asset, the more stable it is- however, it does need to be protected from attacks. The discussions within these sections of each chapter will be quite similar. Transparency is also an issue that varies slightly as you look at different types of stablecoins, once again the more complex the tokenomics- the more transparent the company needs to be. Finally, there needs to be an analysis of viability in major markets. This results in a bimodal, cyclic model for the analysis of the system, viability seems to be highest for simple and highly complex models, with intermediate models being lukewarm in their reception.

Overall, stablecoin 1.3 is more of a footnote than it is a major evolutionary step. Stablecoin 1.3 let the world see that those assets outside of money-based assets could be used in the stability matrix of a token, but the narrow scope of the asset pool limited the viability of the

> “Only a mind free of impediment is capable of grasping the chaotic beauty of the world. This is our greatest asset.”
>
> Oliver Bowden

stablecoin 1.3 system. Overall, it is a failed system, but we can learn more from the failures of the system oftentimes than we can from the successes. This is one of the major movements in the stablecoin evolution, simply a light that showed us a new path that we are now following.

Stability of Asset Class

Stablecoin 1.3 represents a new evolution in the realm of stablecoins, introducing a unique approach to stability by anchoring its value to a single stock or type of asset. This innovation offers distinct advantages and challenges compared to previous stablecoin models. In this discussion, we will delve into the stability of Stablecoin 1.3, exploring its characteristics, benefits, and potential limitations. By understanding the dynamics of this asset-backed tokenization, we can gain insights into its role in the ever-evolving landscape of digital currencies.

Stability through Singular Asset Backing:

Unlike its predecessors, Stablecoin 1.3's stability is derived from being pegged to a single stock or type of asset. This design choice allows for a more direct and transparent link between the stablecoin and the underlying asset, offering a higher degree of stability. The value of Stablecoin 1.3 is closely tied to the performance of the selected asset, leading to a more predictable and reliable token ecosystem. This direct correlation allows users to track the value of their stablecoin in real-time, providing a level of transparency and trust not seen in previous stablecoin models.

Benefits and Advantages:

Stablecoin 1.3's reliance on a single stock or asset brings several benefits to the table. First and foremost, it allows for greater precision in maintaining stability. As the token is linked to a specific asset, the stability of the stablecoin can be closely monitored and adjusted in response to market fluctuations. This agility enables Stablecoin 1.3 to swiftly adapt to changing market conditions, ensuring a consistent value for users.

> "Pride is a very dangerous disease that can turn your talent, asset, and achievements into a handicap."
>
> Uzoma Nnadi

Additionally, Stablecoin 1.3's singular asset backing provides enhanced diversification opportunities. By selecting a stable and reliable asset, such as a blue-chip stock or a commodity with low volatility, the stablecoin ecosystem can benefit from the stability and resilience of that asset class. This diversification can act as a buffer against market uncertainties, reducing the risk of extreme price fluctuations and bolstering the overall stability of the stablecoin.

Furthermore, Stablecoin 1.3 introduces a new level of accessibility and familiarity for investors. By pegging the stablecoin to a specific stock or asset, individuals who are already familiar with that asset class can easily understand and assess the stability of the stablecoin. This familiarity fosters trust and confidence in the stablecoin, attracting a broader range of users who may be more comfortable investing in a token linked to a familiar and trusted asset.

Challenges and Considerations:

While Stablecoin 1.3 offers unique advantages, it also presents certain challenges and considerations. One of the key concerns is the potential lack of diversification inherent in a single asset-backed stablecoin. As the stability of Stablecoin 1.3 is reliant on the performance of a specific asset, any adverse developments or volatilities in that asset can directly impact the stablecoin's stability. This concentration of risk may raise concerns among users and regulators, as it could expose the stablecoin ecosystem to greater vulnerabilities.

Moreover, the selection and management of the underlying asset in Stablecoin 1.3 become crucial factors. The choice of an asset with a strong historical track record of stability and reliability is essential to ensure the stablecoin's long-term viability. Additionally, diligent management of the asset is required to maintain its value and safeguard against potential risks. The transparent reporting of the asset's performance and regular audits become crucial in building trust and ensuring transparency within the stablecoin ecosystem.

Differentiating Factors:

Stablecoin 1.3's focus on a singular asset distinguishes it from previous stablecoin models. While fiat-backed stablecoins (such as Stablecoin 1.0) and corporate paper-backed stablecoins (such as Stablecoin 1.1) rely on broader asset pools for stability, Stablecoin 1.3 leverages the direct linkage to a specific asset. This unique approach provides a more direct and transparent connection between the stablecoin and the

> "Some people will each start investing more of their salary on 'their' house and spending less of it on 'their' car or cars only when they start being able to take 'their' house to work, funerals, weddings, etc."
>
> Mokokoma Mokhonoana

underlying asset, offering users greater clarity and confidence in the stablecoin's stability.

In conclusion, Stablecoin 1.3 presents a new paradigm in stability by basing its value on a single stock or type of asset. This innovative model offers benefits such as precision in stability maintenance, enhanced diversification opportunities, and increased accessibility for investors. However, it also poses challenges related to concentration of risk and careful asset management. By addressing these challenges and leveraging the strengths of this model, Stablecoin 1.3 can play a significant role in shaping the future of stablecoin ecosystems. As the cryptocurrency landscape continues to evolve, Stablecoin 1.3 paves the way for a more nuanced and tailored approach to stability, contributing to a more robust and reliable digital currency ecosystem.

Liquidity of Asset Class

Stablecoin 1.3 introduces a unique model for achieving stability by linking its value to a single stock or type of asset. In this discussion, we will explore the liquidity aspects of Stablecoin 1.3 and compare it to other stablecoin models. Liquidity plays a critical role in the functioning and success of stablecoins, impacting their ability to be readily exchanged and converted. By examining the liquidity dynamics of Stablecoin 1.3 in comparison to other models, we can gain insights into the strengths and potential challenges of this innovative approach.

Understanding Liquidity in Stablecoin Models:

Liquidity refers to the ease with which an asset can be bought or sold on the market without significantly impacting its price. In the context of stablecoins, liquidity is crucial as it ensures the availability of sufficient funds to fulfill user demands for conversions or redemptions. Higher liquidity allows for seamless transactions and enhances user confidence in the stability and utility of the stablecoin.

Comparing Liquidity: Stock/Single Asset Fund Model vs. Other Stablecoin Models:

Stablecoin 1.3's stock/single asset fund model presents both advantages and considerations when it comes to liquidity. Let us compare it to other stablecoin models to assess its relative liquidity characteristics.

1. Fiat-Backed Stablecoins (e.g., Stablecoin 1.0): Fiat-backed stablecoins traditionally maintain a 1:1 ratio with fiat currency reserves. These stablecoins usually enjoy high liquidity due to the convertibility of fiat currency. Users can easily buy or sell stablecoins for their equivalent value in the underlying fiat currency. However, their liquidity is contingent on the availability of sufficient fiat reserves and the efficiency of the conversion process, which can pose challenges during periods of high demand or market volatility.
2. Corporate Paper-Backed Stablecoins (e.g., Stablecoin 1.1): Corporate paper-backed stablecoins leverage short-term debt instruments or commercial paper as their underlying assets. The liquidity of these stablecoins depends on the market for such corporate paper, which may face fluctuations based on prevailing market conditions and credit risks. While corporate paper can provide additional diversification, it may introduce liquidity constraints, particularly during economic downturns or credit market disruptions.

3. Algorithmic Stablecoins (e.g., Stablecoin 1.2): Algorithmic stablecoins employ algorithms and smart contracts to dynamically manage their supply and demand. Their liquidity is typically influenced by the market dynamics and mechanisms embedded in their algorithms. However, algorithmic stablecoins may face challenges in maintaining liquidity during periods of market stress or extreme price volatility, as their value relies on complex market dynamics and the ability of the algorithm to stabilize the price.
4. Stock/Single Asset Fund Model (Stablecoin 1.3): Stablecoin 1.3's stock/single asset fund model presents a unique liquidity profile. Its liquidity is directly tied to the liquidity of the selected stock or asset. If the chosen asset is highly liquid, such as a heavily traded blue-chip stock, the stablecoin can benefit from the liquidity and ease of trading associated with that asset. However, if the chosen asset is less liquid, such as a thinly traded stock or a niche market asset, the stablecoin's liquidity may be more limited, potentially leading to challenges in converting or redeeming the stablecoin.

Considerations for Stablecoin 1.3's Liquidity:

While the stock/single asset fund model may present liquidity considerations, certain factors can mitigate potential challenges:

1. Asset Selection: Careful consideration of the underlying asset's liquidity is essential. Opting for highly liquid stocks or assets with active and deep markets enhances the stability and liquidity of Stablecoin 1.3.

2. Diversification: Maintaining a portfolio of different stocks or assets within Stablecoin 1.3's fund can mitigate concentration risks and enhance overall liquidity. Diversification helps ensure that even if one asset faces liquidity constraints, the overall stability and liquidity of the stablecoin are not significantly compromised.
3. Market Depth: The presence of a robust market for trading the underlying asset plays a crucial role in maintaining liquidity. Sufficient market depth allows for large-scale trading without significantly impacting prices, ensuring that users can easily buy or sell stablecoins without substantial slippage.

The stock/single asset fund model of Stablecoin 1.3 introduces unique considerations for liquidity. While its liquidity is directly linked to the liquidity of the selected stock or asset, careful asset selection, diversification, and the presence of a deep market can enhance liquidity levels. Stakeholders should assess the liquidity characteristics of Stablecoin 1.3 in terms of the chosen asset's tradability and market depth. By addressing these considerations, Stablecoin 1.3 can achieve a balance between stability and liquidity, offering users a novel approach to stablecoin ecosystems.

Required Asset Ratio for Stability

Stablecoin 1.3 introduces a unique approach to achieving stability by linking its value to a single stock or type of asset. In this discussion, we will explore the required asset ratio for Stablecoin 1.3 and how it differs from other stablecoin models. The asset ratio refers to the proportion of the stablecoin's total value that is backed by the underlying asset. Given the singular nature of the asset pool, Stablecoin 1.3 necessitates a higher asset ratio compared to other stablecoins to maintain stability and mitigate risks. This analysis aims to shed light on the considerations and implications of the required asset ratio in Stablecoin 1.3.

> "Your insight is your asset; the edge you require to succeed."
>
> Farshad Asl

Understanding the Asset Ratio in Stablecoins:

The asset ratio plays a critical role in the stability and reliability of a stablecoin. It represents the portion of the stablecoin's value that is collateralized by the underlying asset or assets. A higher asset ratio signifies a greater level of backing and can enhance the stability of the stablecoin. It ensures that there are sufficient reserves to support the stablecoin's value and honor user redemptions or conversions.

The Singular Asset Nature of Stablecoin 1.3:

Stablecoin 1.3 is unique in its reliance on a single stock or type of asset as the backing for its value. This singular asset nature poses specific challenges and considerations when determining the required asset ratio. Unlike diversified asset pools, where risks are spread across multiple holdings, Stablecoin 1.3's value is directly linked to the performance of a solitary asset. As a result, the required asset ratio needs to account for the concentration risk inherent in such a model.

The Need for a Higher Asset Ratio:

Given the singular asset nature of Stablecoin 1.3, a higher asset ratio is required to ensure the stability and reliability of the stablecoin. Several factors contribute to this higher ratio:

1. Concentration Risk: A single asset pool exposes Stablecoin 1.3 to concentration risk. If the selected asset experiences significant volatility or liquidity constraints, it can directly impact the stablecoin's value. To mitigate this risk, a higher asset ratio provides a cushion and safeguards the stablecoin's stability against adverse market conditions.
2. Market Volatility: Individual stocks or specific asset types can exhibit greater price fluctuations and market volatility compared to diversified portfolios. A higher asset ratio helps absorb price fluctuations and provides stability to the stablecoin, allowing it to maintain its pegged value despite market volatility.
3. Redemption and Conversion Demands: A higher asset ratio ensures that Stablecoin 1.3 can meet user redemption and conversion demands promptly. By maintaining a larger reserve of the underlying asset, the stablecoin can readily fulfill requests to convert stablecoin holdings into the underlying asset without delays or liquidity constraints.
4. Investor Confidence: A higher asset ratio instills confidence in investors and users of Stablecoin 1.3. The increased backing creates a perception of stability and reliability, attracting individuals who prioritize a higher level of asset coverage and risk mitigation in their stablecoin holdings.

Balancing Stability and Asset Ratio:

While a higher asset ratio is necessary for Stablecoin 1.3, striking the right balance is crucial. A disproportionately high asset ratio can restrict liquidity and limit the stablecoin's utility. It can also hinder the growth and adoption of Stablecoin 1.3, as the capital tied up in the asset pool may be less available for investment or expansion purposes.

The required asset ratio for Stablecoin 1.3 is higher than in other stablecoin models due to its singular asset nature and the associated concentration risks. The higher asset ratio serves as a buffer against market volatility, concentration risk, and redemption demands. It provides stability and reassurance to investors and users, ensuring that the stablecoin's value remains robust and reliable. However, striking the right balance between a sufficiently high asset ratio and maintaining adequate liquidity is crucial to enable the growth and adoption of Stablecoin 1.3 in the broader market.

Relationship between Holders and access to backing.

Stablecoin 1.3, which relies on a single stock or type of asset as its underlying value, presents unique challenges in managing its asset pool. To ensure stability and effective risk management, investment firms can play a crucial role in overseeing and managing the asset pool of Stablecoin 1.3. In this discussion, we will explore the benefits and considerations of entrusting investment firms with the responsibility of managing the stablecoin's asset pool.

Expertise in Asset Management:

Investment firms possess extensive expertise in asset management, including portfolio diversification, risk analysis, and investment strategies. By leveraging their knowledge and experience, investment firms can effectively manage the asset pool of Stablecoin 1.3. They can employ sophisticated investment techniques to optimize the stability and performance of the asset pool, ensuring it aligns with the stablecoin's objectives and investor expectations.

Risk Mitigation and Diversification:

One of the key advantages of involving investment firms in managing the asset pool is their ability to mitigate risks and achieve diversification. Investment firms can carefully select and manage a diversified portfolio of assets within the stablecoin's asset pool. This diversification helps reduce concentration risk and enhances the stability of Stablecoin 1.3. By spreading the holdings across multiple assets, investment firms can mitigate the impact of adverse events affecting a single stock or asset, thereby safeguarding the stablecoin's value.

Active Monitoring and Rebalancing:

Investment firms are equipped to actively monitor market conditions and asset performance, allowing them to make informed decisions regarding the asset pool of Stablecoin 1.3. They can continuously assess the performance of the underlying asset, make adjustments as necessary, and rebalance the portfolio to maintain stability and optimize returns. Through regular monitoring and rebalancing, investment firms can respond to changing market dynamics and ensure that the asset pool remains aligned with the stablecoin's objectives and market demands.

Compliance and Regulatory Oversight:

Investment firms are subject to stringent compliance and regulatory frameworks, ensuring transparency, investor protection, and adherence to industry best practices. By entrusting investment firms with managing the asset pool of Stablecoin 1.3, regulatory compliance can be maintained, providing investors and users with confidence in the stability and security of the stablecoin. Investment firms' adherence to robust compliance procedures helps mitigate the risk of fraudulent activities and enhances transparency in the management of the stablecoin's asset pool.

> In Linear Times, an organizations culture is its greatest asset. However, in exponential and disruptive times, some part of that same culture can become large liabilities, creating persistent resistance to pressing change and renewal."
>
> Gyan Nagpal

Accountability and Auditing:

Investment firms operate under a framework of accountability and auditing. They are required to provide regular reports on the performance and composition of the asset pool, ensuring transparency and accountability to stakeholders. Regular auditing and reporting further enhance the stability and trustworthiness of Stablecoin 1.3 by providing investors and users with visibility into the management of the asset pool. The involvement of investment firms also enables independent verification of the stability and appropriate management of the stablecoin's asset pool.

In conclusion, investment firms can bring invaluable expertise and resources to the management of Stablecoin

1.3's asset pool. Their knowledge of asset management, risk mitigation, diversification, and compliance ensure effective management of the singular asset-based stablecoin. By actively monitoring and rebalancing the asset pool, investment firms can enhance stability, optimize returns, and mitigate concentration risk. Moreover, their accountability, regulatory oversight, and auditing mechanisms provide transparency and reassurance to stakeholders. Collaborating with investment firms allows Stablecoin 1.3 to benefit from the specialized skills of asset managers, fostering stability and investor confidence in the increasingly dynamic and evolving stablecoin ecosystem.

Susceptibility to Attacks

Stablecoin 1.3, which relies on a single stock or type of asset as its underlying value, brings its unique set of vulnerabilities and susceptibility to various attacks within the crypto ecosystem. To ensure the stability and security of Stablecoin 1.3, it is essential to implement robust protective measures. In this discussion, we will explore the potential attack vectors that pose risks to Stablecoin 1.3 and outline strategies for safeguarding the stablecoin against these threats.

Market Manipulation and Price Volatility:

Stablecoin 1.3's value is directly tied to a specific stock or asset, making it vulnerable to market manipulation and price volatility. Malicious actors can attempt to manipulate the price of the underlying asset, which would subsequently impact the stablecoin's value. Implementing market surveillance mechanisms and partnering with reputable exchanges can help detect and prevent manipulative trading practices. A stablecoinly, the stablecoin issuer should establish clear guidelines for addressing extreme market conditions, such as price shocks or sudden changes in the value of the underlying asset.

Counterparty Risk:

> "Unless you are getting passive income from your house, it is a liability, not an asset."
>
> Robert Kiyosaki

Stablecoin 1.3's reliance on a single stock or asset introduces counterparty risk. If the issuer of the stablecoin fails to meet its obligations or faces insolvency, it can result in severe consequences for the stablecoin's stability. To mitigate this risk, thorough due diligence should be conducted on the issuer to ensure its financial strength and credibility. Establishing legal agreements and collateral requirements can provide an additional layer of protection against counterparty risk. Regular audits and transparency of the issuer's financials can also enhance confidence in the stablecoin's security.

Smart Contract Vulnerabilities:

Smart contracts form the backbone of many stablecoins, including Stablecoin 1.3, as they enable the token's functionality and operational processes. However, smart contracts can be susceptible to coding errors or vulnerabilities that could be exploited by attackers. Thorough code review, rigorous testing, and auditing of the smart contract code can help identify and rectify potential vulnerabilities. Continuous monitoring and prompt patching of identified security flaws are vital to ensure the stablecoin's resilience against smart contract-based attacks.

Regulatory and Legal Risks:

Stablecoin 1.3 operates within a regulatory framework that may subject it to legal risks and potential regulatory crackdowns. Compliance with applicable regulations, including anti-money laundering (AML) and know your customer (KYC) requirements, is crucial to mitigate legal and regulatory risks. Working closely with legal experts and maintaining open communication with regulatory authorities can help navigate the evolving regulatory landscape and ensure the stablecoin's compliance. Adherence to transparency and reporting requirements will further enhance the stablecoin's legitimacy and protect it from regulatory challenges.

External Cybersecurity Threats:

Stablecoin 1.3, like any other digital asset, faces cybersecurity threats, such as hacking attempts, phishing attacks, or ransomware. Robust cybersecurity measures, including encryption, multi-factor authentication, and regular security audits, should be implemented to protect the stablecoin's infrastructure, wallets, and user data. Educating users about the best security practices and promoting secure storage solutions, such as hardware wallets, can help prevent individual users from falling victim to cyberattacks.

Safeguarding Stablecoin 1.3 against potential attacks requires a multi-faceted approach encompassing market surveillance, counterparty risk mitigation, smart contract security, regulatory compliance, and cybersecurity measures. Collaborating with reputable exchanges, conducting thorough due diligence on counterparties, and implementing transparent reporting mechanisms will enhance stability and protect against market manipulation and counterparty risk. Rigorous code review, testing, and ongoing smart contract audits will fortify the stablecoin's resilience against coding vulnerabilities. Compliance with regulatory requirements and engagement with legal experts will mitigate regulatory and legal risks. Finally, implementing robust cybersecurity measures and educating users about best practices will safeguard Stablecoin 1.3 against external threats. By addressing these vulnerabilities head-on, Stablecoin 1.3 can establish itself as a reliable and secure digital asset within the broader crypto ecosystem.

Transparency

Transparency plays a critical role in building trust and confidence in stablecoins, particularly in the case of Stablecoin 1.3, which relies on a single stock or type of asset. The need for transparency in Stablecoin 1.3 extends across various aspects, including the asset pool, issuance and redemption mechanisms, governance processes, and regulatory compliance. In this discussion, we will delve into the significance of transparency in Stablecoin 1.3 and its impact on stability, investor confidence, and market acceptance.

1. Asset Backing and Auditing: Given Stablecoin 1.3's reliance on a single stock or asset, transparency in the asset backing is crucial to ensure its integrity. Investors and users need assurance that the stablecoin's value is backed by genuine assets and that the issuer maintains the necessary reserves. Regular third-party audits by reputable accounting firms can provide independent verification of the stablecoin's asset backing. Publicly disclosing the results of these audits enhances transparency and instills confidence among stakeholders.

2. Governance and Decision-Making: Transparent governance processes are vital for Stablecoin 1.3 to maintain trust among users and investors. Openly documenting the decision-making framework, including key governance roles, voting mechanisms, and dispute resolution procedures, promotes accountability and ensures that the stablecoin's interests align with those of its community. Additionally, transparent governance helps prevent undue concentration of power and fosters inclusivity in decision-making, ultimately enhancing the stability and legitimacy of Stablecoin 1.3.
3. Regulatory Compliance: Transparency is closely linked to regulatory compliance in the stablecoin ecosystem. Stablecoin 1.3 must operate within the bounds of applicable regulations, such as AML and KYC requirements, to mitigate legal risks and maintain trust with regulatory authorities. Transparent reporting and disclosure of compliance measures enable regulators to assess the stablecoin's adherence to regulatory standards. Engaging in open dialogue with regulatory agencies and proactively addressing compliance concerns demonstrate a commitment to transparency and regulatory compliance.
4. User Protection and Risk Disclosure: Transparent disclosure of risks associated with Stablecoin 1.3 is crucial for user protection. Users should have a clear understanding of the risks involved, including the potential volatility of the underlying asset, counterparty risks, and any specific market conditions that may affect the stablecoin's stability.

Transparent risk disclosure empowers users to make informed decisions and aligns with the principle of investor protection. This transparency also serves to differentiate Stablecoin 1.3 from other stablecoins in the market, fostering trust and attracting risk-conscious investors.

5. Market Integrity and Trust: Transparency is essential for maintaining market integrity and trust in Stablecoin 1.3. By openly sharing information about the stablecoin's operations, asset backing, and governance, the issuer instills confidence in the market. Transparency helps prevent market manipulation and insider trading, as stakeholders can scrutinize the stablecoin's activities. Furthermore, transparent reporting on key metrics such as supply, market cap, and transaction volume enhances market visibility and allows investors to assess the stablecoin's performance accurately.

Transparency is a cornerstone of Stablecoin 1.3's success, influencing its stability, investor confidence, and wider market acceptance. Through transparent asset backing, governance processes, regulatory compliance, risk disclosure, and market reporting, Stablecoin 1.3 can build trust and attract a broader user base. Transparency fosters accountability, mitigates risks, and aligns the stablecoin with regulatory requirements. By prioritizing transparency, Stablecoin 1.3 can solidify its position as a trustworthy and reliable digital asset, ultimately driving its growth and adoption within the crypto ecosystem.

Viability in Major Markets

Stablecoin 1.3, which relies on a single stock or type of asset, was introduced with the aim of providing stability and convenience in the crypto ecosystem. However, despite its potential benefits, stablecoin 1.3 models have struggled to achieve high levels of market penetration. In this discussion, we will explore the key factors that have contributed to the limited adoption and market acceptance of stablecoin 1.3 models.

Limited Asset Diversification:

One of the primary reasons for the limited market penetration of stablecoin 1.3 models is the inherent risk associated with relying on a single stock or asset. The lack of diversification leaves the stablecoin vulnerable to fluctuations in the value of the underlying asset. Investors and users, especially those seeking stablecoins for hedging purposes or as a store of value, may be hesitant to adopt stablecoin 1.3 due to the concentrated risk exposure. The lack of asset diversification undermines the appeal and stability of stablecoin 1.3, limiting its market reach.

> "So, while I'm not yet rich, I am wealthy. I now have income generated from assets each month that fully cover my monthly expenses. If I want to increase my expenses, I first must increase my cash flow to maintain this level of wealth. Also note that it is at this point that I'm no longer dependent on my wages. I have focused on, and been successful in, building an asset column that has made me financially independent. If I quit my job today, I would be able to cover my monthly expenses with the cash flow from my assets."
>
> Robert Kiyosaki

Volatility of Underlying Asset:

Stablecoin 1.3 models are heavily influenced by the price movements and volatility of the single stock or asset they are backed by. If the underlying asset experiences significant price fluctuations, it can impact the stability and purchasing power of the stablecoin. Users may be hesitant to adopt stablecoin 1.3 if they perceive the underlying asset to be highly volatile or uncertain. The inherent risk of price volatility poses challenges for stablecoin 1.3 models to gain widespread acceptance, especially among risk-averse investors.

Lack of Regulatory Clarity:

The regulatory landscape surrounding stablecoins, including stablecoin 1.3 models, remains relatively uncertain and evolving. Regulatory authorities worldwide are grappling with the appropriate classification and oversight of stablecoins, which adds to the apprehension among market participants. The lack of clear guidelines and regulatory frameworks creates an environment of uncertainty and hampers the broader adoption of stablecoin 1.3 models. Investors and users may be reluctant to engage with stablecoins that operate in a regulatory gray area, thereby impeding market penetration.

Investor Preference for Established Stablecoin Models:

Stablecoin 1.3 models face stiff competition from established stablecoins, such as Tether (USDT), USDC, and DAI, which have gained significant traction in the market. These established stablecoins have a track record of stability and market acceptance, bolstered by their larger user base and broader market liquidity. Investors and users tend to gravitate towards stablecoin options that have a proven track record and are widely recognized within the crypto community. The dominance of these established stablecoin models limits the market share available for stablecoin 1.3 to penetrate.

Lack of Market Awareness and Marketing Efforts:

The limited market penetration of stablecoin 1.3 models can also be attributed to the lack of widespread awareness and effective marketing efforts. Compared to other stablecoins, stablecoin 1.3 models may struggle to gain visibility and capture the attention of potential users and investors. Insufficient marketing efforts and a lack of educational initiatives to highlight the benefits and unique features of stablecoin 1.3 models hinder their adoption and limit their market presence.

Despite the potential advantages they offer, stablecoin 1.3 models have faced challenges in gaining high levels of market penetration. Factors such as limited asset diversification, volatility of the underlying asset, regulatory uncertainty, strong competition from established stablecoins, and insufficient market awareness have contributed to their limited adoption. Overcoming these barriers will require innovative approaches, improved asset diversification strategies, regulatory clarity, targeted marketing efforts, and building trust among potential users and investors. Addressing these challenges will be crucial for stablecoin 1.3 models to enhance their market acceptance, gain broader user adoption, and carve a niche within the crypto ecosystem.

Conclusion

Stablecoin 1.3, with its reliance on a single stock or asset, offers a unique approach to achieving stability within the cryptocurrency space. However, despite its potential benefits, stablecoin 1.3 models have faced challenges in gaining significant market penetration. In this conclusion, we will summarize the key points discussed throughout this analysis and provide insights into the future prospects of stablecoin 1.3.

Throughout our exploration, we identified several factors that have contributed to the limited adoption of stablecoin 1.3 models. The lack of asset diversification emerged as a primary concern, as relying on a single stock or asset exposes stablecoin 1.3 to concentrated risks. This inherent risk has made potential users and investors cautious, as they seek stability and a diversified portfolio. The volatility of the underlying asset further exacerbates these concerns, as price fluctuations can impact the stability of the stablecoin and undermine its attractiveness as a store of value.

Another significant factor hindering the wider acceptance of stablecoin 1.3 models is the regulatory uncertainty surrounding stablecoins in general. The evolving nature of regulations and the absence of clear guidelines create an environment of uncertainty, which can deter potential users and investors from engaging with stablecoin 1.3. Establishing regulatory frameworks and providing clarity on the legal status and oversight of stablecoin 1.3 models will be vital in instilling confidence and driving market adoption.

Furthermore, stablecoin 1.3 faces fierce competition from established stablecoins that have already gained widespread acceptance in the market. Stablecoins like Tether, USDC, and DAI have established themselves as trusted and reliable options, benefitting from larger user bases and broader market liquidity. This dominance limits the market share available for stablecoin 1.3 to penetrate, making it challenging to compete with the established players.

Additionally, the lack of market awareness and effective marketing efforts has hindered the visibility and adoption of stablecoin 1.3. Compared to other stablecoin options, stablecoin 1.3 models have struggled to gain attention and capture the interest of potential users and investors. Addressing this challenge will require targeted marketing strategies, educational initiatives, and community engagement to showcase the unique features and advantages that stablecoin 1.3 brings to the table.

Looking ahead, there are opportunities for stablecoin 1.3 models to improve their market acceptance and address the challenges they currently face. Enhancing asset diversification strategies by exploring partnerships and collaborations with multiple asset providers could mitigate the concentrated risk associated with a single asset. This approach would increase stability, diversify the underlying assets, and make stablecoin 1.3 more appealing to a broader range of users.

Furthermore, fostering regulatory clarity and compliance will be crucial for stablecoin 1.3 models to gain trust and navigate the evolving regulatory landscape. Engaging with regulatory authorities, advocating for clear guidelines, and ensuring transparency in operations will help create a favorable environment for stablecoin 1.3 to flourish.

To increase market awareness and attract users, robust marketing campaigns, educational initiatives, and community building efforts will be essential. These efforts should highlight the unique value proposition of stablecoin 1.3, emphasizing its stability, convenience, and potential advantages over other stablecoin models.

In conclusion, while stablecoin 1.3 models have faced challenges in achieving high levels of market penetration, there are avenues for improvement and growth. By addressing the factors hindering adoption, such as asset diversification, regulatory uncertainty, competition, and market awareness, stablecoin 1.3 models can position themselves as viable and attractive options within the cryptocurrency ecosystem. Continued innovation, collaboration, and strategic decision-making will be key to the future success and wider adoption of stablecoin 1.3 models.

Chapter 12: Stablecoin 1.4 (Digital Asset Based)

"Believe me, the next step is a currency crisis because there will be a rejection of the dollar, the rejection of the dollar is a big, big event, and then your personal liberties are going to be threatened."[34]

Around the world, we are seeing a rejection of the US dollar as a reserve currency. Years of mismanagement by the Federal Reserve, corrupt politicians, and lobbyists have diminished the value of the dollar in the eyes of the world. The result of this is a world where there is going to be another reserve currency. However, the current tzars of the financial world in the U.S. are ignoring the fact and pretending that it is not happening. Our heads are in the sand and if we do not work to ensure that the new global reserve currency is centered in the United States, we will be left in the dust of the digital age. Stablecoin 1.4, tokens based on digital assets, are a bridge to help people understand that the cryptoverse has value and that when we embrace the value you empower the people.

In this chapter, we will follow the same pattern that we follow in all the chapters in this part of the book. First, we look at the stability of the asset class on which the token at this level is based. The asset class is very important because it is the 'pegged' underpinning of the system. Next, we look at the liquidity of the asset class, how fast can the asset be liquidated if there is a 'run' on the token. As a matter of course, we will then look at the asset ratio needed to maintain stability. The media tends to insist that there

---

[34] Rona Paul at https://www.azquotes.com/quotes/topics/currency.html.

needs to be a 1:1 ratio, and letter agencies tend to echo this false statement; each class has a ratio that is needed for the token to be reliable in its function.

> "Any physical or digital commodity that is traded are subject to bubble and crash."
>
> Gun Gun Febrianza

Also, as a matter of course, it is always important to look at the relationship between holders/creators of the currency and the asset pool. The more complex the system the more analysis needs to be done on how the assets are being managed. While a 1:1 ratio for stablecoin 1.0 can be managed by the local bank of the creator, more complex systems require funds, investment firms, and even multiple subsidiaries to ensure that the asset pool maintains value. The analysis of who holds the asset pool then leads to an analysis of susceptibility of to attack. The more complex the asset, the more stable it is- however, it does need to be protected from attacks. The discussions within these sections of each chapter will be quite similar. Transparency is also an issue that varies slightly as you look at different types of stablecoins, once again the more complex the tokenomics- the more transparent the company needs to be. Finally, there needs to be an analysis of viability in major markets. This results in a bimodal, cyclic model for the analysis of the system, viability seems to be highest for simple and highly complex models, with intermediate models being lukewarm in their reception.

Overall, stablecoin 1.4 bridges the gap between level one and level two stablecoins. The diversification of assets between the physical world and the digital world is/was a major step in the viability of the marketplace. As we realize that we have the ability to exist in both worlds simultaneously, we realize that the universe is much bigger than we understand. Cyberspace and real space are the two new frontiers, and we will not have the funds to explore both unless we embrace the growing wealth model promised by cryptocurrency and stablecoin.

Stability of Asset Class

Stablecoin 1.4 represents a unique category of stablecoins that are backed by digital assets, providing an alternative approach to achieving stability within the cryptocurrency market. Unlike traditional stablecoins, which rely on fiat currencies or other traditional assets, stablecoin 1.4 leverages the potential of digital assets to provide stability. In this discussion, we will explore the stability of stablecoin 1.4 models, highlighting the characteristics that differentiate them from other stablecoin variants and analyzing the factors that influence their stability.

Stability in the Short Term:

It is important to acknowledge that stablecoin 1.4, being based on digital assets, may exhibit higher volatility in the short term due to the inherent volatility of cryptocurrencies. Digital assets are known for their price fluctuations, influenced by various market factors, including investor sentiment, market demand, regulatory developments, and technological advancements. As a result, stablecoin 1.4 may experience short-term price swings that could impact its stability.

However, despite the short-term volatility, stablecoin 1.4 models have demonstrated a positive stability over the long term. This can be attributed to the underlying digital assets' potential for growth and adoption, as well as the maturation of the cryptocurrency market as a whole. As the market stabilizes, digital assets tend to exhibit more consistent and predictable price movements, leading to a more stable value for stablecoin 1.4.

Factors Influencing Stability:

Several factors contribute to the stability of stablecoin 1.4 models. One of the key factors is the liquidity and depth of the underlying digital asset market. A highly liquid market with robust trading volumes ensures that stablecoin 1.4 can be easily bought or sold without causing significant price fluctuations. The presence of reputable exchanges and trading platforms that facilitate the trading of digital assets enhances the stability of stablecoin 1.4 by providing a reliable and efficient market infrastructure.

Furthermore, the adoption and acceptance of digital assets by individuals, businesses, and institutions play a crucial role in ensuring stability. As the use cases and utility of digital assets expand, their value becomes more resilient to market volatility, driving stability for stablecoin 1.4. Factors such as increased merchant acceptance, regulatory clarity, and mainstream adoption contribute to the overall stability of digital assets and, consequently, stablecoin 1.4.

Another factor that influences stability is the governance and management of stablecoin 1.4. Decentralized governance models, where decisions are made collectively by token holders, can enhance stability by promoting transparency, accountability, and community-driven decision-making. Effective risk management practices, regular audits, and mechanisms to maintain appropriate collateralization ratios also contribute to the stability of stablecoin 1.4.

Long-Term Stability and Value Proposition:

Despite the short-term volatility, stablecoin 1.4 models offer the potential for long-term stability, which can be an attractive value proposition for users and investors. As the cryptocurrency market matures, digital assets have demonstrated an overall positive stability and growth trajectory. This can be attributed to various factors, including increased adoption, improved market infrastructure, regulatory developments, and technological advancements.

> "Your investment should be equally risky and rewarding."
>
> Anuj Jasani

Stablecoin 1.4 models also benefit from the inherent advantages of digital assets, such as borderless transactions, enhanced security, and programmability. These features can attract users and investors seeking stability without compromising the benefits offered by cryptocurrencies. Stablecoin 1.4 allows individuals and businesses to transact and store value in a stable digital currency, eliminating the need for traditional intermediaries and enabling greater financial inclusion.

Moreover, stablecoin 1.4 models provide an opportunity for diversification within the cryptocurrency ecosystem. While traditional stablecoins offer stability through pegging to external assets, stablecoin 1.4 leverages the potential of digital assets themselves. This diversification can reduce reliance on fiat currencies and traditional financial systems, providing users with exposure to the potential growth and innovation of digital assets while maintaining stability.

Stablecoin 1.4, based on digital assets, offers an alternative approach to achieving stability within the cryptocurrency market. While short-term volatility may be present due to the inherent volatility of digital assets, stablecoin 1.4 models have demonstrated positive stability over the long term. Factors such as market liquidity, adoption and acceptance, governance, and risk management practices play crucial roles in ensuring stability.

As the cryptocurrency market continues to evolve and mature, stablecoin 1.4 models provide an attractive value proposition for users and investors. They combine the stability of traditional stablecoins with the advantages and potential growth of digital assets. The long-term stability of stablecoin 1.4, coupled with the benefits of digital currencies, can contribute to the broader adoption and integration of cryptocurrencies into mainstream finance.

However, it is important to note that the stability of stablecoin 1.4 is still subject to market dynamics and regulatory developments. Continued efforts to enhance market infrastructure, regulatory clarity, and risk management practices will be crucial in further solidifying the stability and resilience of stablecoin 1.4 models. With these advancements, stablecoin 1.4 has the potential to play a significant role in shaping the future of stablecoins and the broader cryptocurrency ecosystem.

Liquidity of Asset Class

Stablecoin 1.4, which is based on digital assets, brings forth a unique characteristic in terms of liquidity within the cryptocurrency ecosystem. Unlike traditional stablecoins backed by fiat currencies or other assets, stablecoin 1.4 harnesses the unprecedented liquidity provided by the underlying digital assets. In this discussion, we will explore the exceptional liquidity of stablecoin 1.4 and how it differentiates from other stablecoin models, highlighting the advantages and potential challenges associated with this liquidity.

Unprecedented Liquidity in Digital Assets:

Digital assets, such as cryptocurrencies, have gained significant traction in recent years, resulting in highly liquid markets. The global nature of digital asset exchanges, coupled with 24/7 trading availability, fosters continuous market activity and liquidity. This liquidity is a result of the broad participation of traders and investors from around the world, including individuals, institutions, and even automated trading systems.

The high liquidity of digital assets translates to stablecoin 1.4, offering exceptional liquidity to its users. Liquidity refers to the ability to convert an asset into cash quickly and at a reasonable price without significantly impacting its market value. Stablecoin 1.4 benefits from the deep order books and substantial trading volumes of digital assets, enabling seamless conversion between stablecoin 1.4 and other digital assets.

Advantages of Unprecedented Liquidity:

The unprecedented liquidity of stablecoin 1.4 brings several advantages to its users and the broader cryptocurrency ecosystem. Firstly, the deep liquidity ensures that stablecoin 1.4 can be easily bought or sold without causing significant price slippage. Users can transact with stablecoin 1.4 at stable prices, reducing the risk of large price fluctuations during conversions.

Secondly, the high liquidity provides stability to stablecoin 1.4 itself. The ability to quickly convert stablecoin 1.4 into other digital assets or fiat currencies ensures that the value of stablecoin 1.4 remains closely aligned with its underlying digital assets. This stability enhances user confidence and trust in stablecoin 1.4 as a reliable store of value and medium of exchange.

Furthermore, the unprecedented liquidity of stablecoin 1.4 opens up opportunities for various use cases within the cryptocurrency ecosystem. Traders and investors can efficiently utilize stablecoin 1.4 to hedge their positions, take advantage of arbitrage opportunities, or facilitate cross-exchange transactions. The deep liquidity also enables stablecoin 1.4 to serve as a stable base currency for trading pairs with other digital assets, enhancing trading efficiency and liquidity in the overall cryptocurrency market.

Challenges and Considerations:

While the liquidity of stablecoin 1.4 presents significant advantages, there are also challenges and considerations to be addressed. One challenge is the potential impact of market manipulation on liquidity. The decentralized nature of digital asset markets and the absence of strict regulatory oversight create a fertile ground for manipulation attempts. Market participants with significant holdings of specific digital assets may attempt to manipulate the market to create artificial liquidity or exploit price discrepancies. Ensuring market integrity and preventing manipulation are critical for maintaining stablecoin 1.4's liquidity.

Another consideration is the potential impact of extreme market conditions on liquidity. In times of extreme market volatility or market-wide selloffs, liquidity across digital asset markets can diminish as participants become more risk-averse. This reduction in liquidity may affect the ability to convert stablecoin 1.4 into other digital assets or fiat currencies swiftly. Robust risk management practices, including proper diversification and stress testing, are essential to mitigate the liquidity risks associated with unstable market conditions.

> "The very principles of economics suggest that scarcity, validity, and demand can transform anything, even a stone, into a Store-of-Value (SoV). Such an event will happen only once in an era and we are extremely fortunate to be witnessing the birth of a new type of SoV, the Crypto."
>
> Mohith Agadi

Maintaining Unprecedented Liquidity:

To maintain the unprecedented liquidity of stablecoin 1.4, it is crucial to foster an ecosystem that encourages market participation, transparency, and trust. The presence of reputable and regulated digital asset exchanges plays a vital role in ensuring fair and orderly markets. Collaborations with market makers and liquidity providers can also enhance the liquidity profile of stablecoin 1.4, facilitating seamless transactions and minimizing price slippage.

Moreover, transparent reporting mechanisms and regular audits can instill confidence in the stability and liquidity of stablecoin 1.4. Publicly available information about the composition and management of the underlying digital assets, along with independent verification of the asset holdings, can assure users that the stablecoin's value is adequately backed and maintain trust in the liquidity of stablecoin 1.4.

Stablecoin 1.4, built upon the unprecedented liquidity of digital assets, offers users exceptional liquidity and stability within the cryptocurrency ecosystem. The deep liquidity of digital asset markets ensures seamless conversion between stablecoin 1.4 and other digital assets, enabling efficient transactions and enhancing user experience. However, challenges such as market manipulation and extreme market conditions must be addressed to maintain stablecoin 1.4's liquidity. Establishing a robust regulatory framework, fostering market integrity, and implementing risk management practices are vital for sustaining the unprecedented liquidity of stablecoin 1.4. Overall, stablecoin 1.4's liquidity provides a solid foundation for its adoption and usage in various applications within the digital asset ecosystem.

Required Asset Ratio for Stability

In the context of stablecoin 1.4, which is based on digital assets, the required asset pool level is typically set at around 50%. This means that the stablecoin issuer aims to hold digital assets equivalent to at least 50% of the stablecoin's total supply value. This asset pool serves as collateral or backing for the stablecoin, ensuring its stability and redeemability.

Unlike traditional banking systems that used to have reserve requirements, stablecoins, including digital asset-based stablecoin 1.4, operate differently. The reserve requirement for stablecoins is not imposed by regulatory authorities like central banks. Instead, it is established by the stablecoin issuer or the governing protocol that manages the stablecoin's operations.

The decision to set the required asset pool level at 50% is based on several factors. Firstly, maintaining a substantial portion of the stablecoin's value in the asset pool provides a cushion against market volatility and safeguards against potential losses. By holding a significant reserve, stablecoin issuers aim to ensure that the stablecoin remains stable and can be redeemed at its intended value.

Secondly, the 50% asset pool level is designed to instill confidence in the stability of the stablecoin. By maintaining a robust reserve, stablecoin issuers aim to demonstrate that the stablecoin is fully backed by valuable digital assets. This transparency helps to build trust among users and investors, as they can verify that the stablecoin's value is supported by tangible assets.

It is important to note that the required asset pool level of 50% is a minimum threshold. Stablecoin issuers may choose to hold a higher percentage of assets in the reserve to provide an additional layer of security and ensure sufficient liquidity. This flexibility allows stablecoin issuers to adapt to changing market conditions and manage potential risks effectively.

Furthermore, the required asset pool level also takes into consideration the liquidity of the underlying digital assets. Digital assets can experience significant price volatility in the short term, which may affect the stablecoin's stability if the asset pool is not adequately managed. Therefore, maintaining a reasonable reserve level helps to mitigate potential liquidity challenges and ensures that the stablecoin can fulfill its intended purpose as a stable medium of exchange and store of value.

In summary, stablecoin 1.4, based on digital assets, typically requires a 50% asset pool level. This reserve provides stability, confidence, and liquidity to the stablecoin. While traditional banking systems may no longer have reserve requirements, stablecoins operate on different principles and establish their own reserve levels to ensure the stability and redeemability of the stablecoin. By maintaining a robust reserve, stablecoin issuers aim to foster trust among users and safeguard the stablecoin's value in the dynamic digital asset ecosystem.

Relationship between Holders and access to backing.

In the case of stablecoin 1.4, which is based on digital assets, the asset pool is typically held by entities such as decentralized finance (DeFi) banks or cryptocurrency exchanges. These institutions have emerged as key players in the digital asset ecosystem, providing the necessary infrastructure and services to support the stability and liquidity of stablecoins.

One primary reason why DeFi banks and exchanges are suitable candidates for holding the asset pool is their expertise and specialization in digital assets. They have the necessary technical knowledge, resources, and infrastructure to securely manage and custody a diverse range of digital assets. This expertise enables them to effectively handle the complexities associated with holding and managing the asset pool of a stablecoin.

Moreover, DeFi banks and exchanges are well-connected within the cryptocurrency ecosystem. They have established relationships with various market participants, including liquidity providers, traders, and other exchanges. This network connectivity allows for efficient market operations and liquidity management, which are essential for maintaining stability and liquidity in stablecoin 1.4.

Another important factor is the transparency and accountability provided by DeFi banks and exchanges. These institutions often have robust reporting mechanisms and audit processes in place to ensure transparency in their operations. This transparency extends to the management of the stablecoin's asset pool, providing users and investors with visibility into the assets backing the stablecoin and how they are being utilized.

Furthermore, DeFi banks and exchanges are subject to regulatory oversight in many jurisdictions. This regulatory oversight helps ensure compliance with applicable laws and regulations, enhancing the overall governance and trustworthiness of the stablecoin ecosystem. Regulatory scrutiny can provide additional safeguards against potential misuse or mismanagement of the asset pool, benefiting stablecoin holders and the broader market.

In selecting a DeFi bank or exchange to hold the asset pool for stablecoin 1.4, it is crucial to consider factors such as the institution's track record, security measures, regulatory compliance, and reputation within the industry. Due diligence should be conducted to assess the institution's reliability, financial stability, and adherence to best practices.

While DeFi banks and exchanges are generally suitable candidates, it is worth noting that the choice of asset pool holder may vary depending on the specific stablecoin project and its governance structure. Some stablecoin projects may opt for decentralized governance models, allowing multiple entities or community members to hold portions of the asset pool. In such cases, the governance framework and consensus mechanisms play a significant role in determining who holds the asset pool.

In summary, DeFi banks and exchanges are often well-suited to hold the asset pool for stablecoin 1.4 due to their expertise in digital assets, market connectivity, transparency, and regulatory oversight. Their role in managing the asset pool contributes to the stability, liquidity, and overall integrity of the stablecoin. However, it is essential to carefully evaluate the specific institution's capabilities, security measures, and regulatory compliance to ensure a reliable and trustworthy asset pool holder for stablecoin 1.4.

Susceptibility to Attacks

Stablecoin 1.4, being based on digital assets, is susceptible to various types of attacks, including both technical vulnerabilities and attacks on the underlying assets themselves. It is important to understand these potential risks and take appropriate measures to mitigate them to ensure the stability and integrity of the stablecoin.

One major vulnerability that stablecoin 1.4 faces is smart contract exploits. Smart contracts, which govern the operations of stablecoin platforms, are subject to vulnerabilities that can be exploited by hackers. These vulnerabilities can lead to unauthorized access, theft of funds, or manipulation of the stablecoin's underlying assets. Smart contract audits, rigorous testing, and continuous monitoring are crucial to identify and address these vulnerabilities.

In addition to smart contract exploits, stablecoin 1.4 is also exposed to the risk of attacks on the underlying digital assets. This can include direct attacks on the blockchain network or manipulation of the assets' value through coordinated market activities. For example, the Terra attack you mentioned involved trading Terra for Bitcoin, shorting the Bitcoin, and purchasing put options on Terra. Such malicious activities can create artificial market distortions and destabilize the stablecoin's value.

To mitigate these risks, robust security measures and risk management practices are essential. This includes employing multi-signature wallets, implementing strict access controls, and utilizing cold storage solutions to secure the underlying assets. Regular security audits and vulnerability assessments should be conducted to identify and address potential weaknesses in the system.

Furthermore, it is crucial to establish strong governance and oversight mechanisms to prevent corrupt firms or insiders from manipulating the stablecoin's assets. Transparent governance frameworks, community participation, and regular reporting can help detect and deter any potential misconduct. Auditing of the asset holdings and regular disclosure of the stablecoin's financial health can enhance trust and transparency.

In response to the Terra attack and similar incidents, the stablecoin ecosystem has increasingly recognized the importance of implementing robust risk management practices and introducing safeguards. These can include mechanisms such as circuit breakers, price oracles, and decentralized collateralization to prevent sudden price crashes or market manipulations. Additionally, improved monitoring and surveillance tools can help identify suspicious activities and address them promptly.

Collaboration with external security firms, bug bounty programs, and community-driven initiatives can also play a crucial role in enhancing the security and resilience of stablecoin 1.4. Encouraging responsible disclosure of vulnerabilities and offering rewards for identifying and reporting vulnerabilities can incentivize ethical hackers to contribute to the security of the stablecoin ecosystem.

Furthermore, regulatory oversight and compliance can provide an additional layer of protection against malicious activities. Working closely with regulatory authorities and adhering to relevant regulatory frameworks can help establish standards and best practices for stablecoin operations. This can include measures to prevent money laundering, ensure customer protection, and maintain the stability of the financial system.

Stablecoin 1.4, being based on digital assets, faces vulnerabilities and risks that need to be effectively addressed to ensure its stability and integrity. Smart contract exploits, attacks on the underlying assets, and potential manipulation by hackers or corrupt firms are among the key challenges. Robust security measures, risk management practices, transparent governance, and regulatory compliance are essential to mitigate these risks. Continuous monitoring, audits, and collaboration with the wider community can contribute to the ongoing improvement and resilience of stablecoin 1.4, enhancing its ability to withstand attacks and maintain stability in the digital asset ecosystem.

Transparency

Transparency is a crucial aspect of stablecoin 1.4, which is based on digital assets. Given the decentralized and trustless nature of blockchain technology, transparency plays a pivotal role in building confidence and trust among users and the wider community. In the context of stablecoin 1.4, transparency primarily relates to the disclosure and visibility of the asset pool, which comprises the digital assets backing the stablecoin.

> "When it comes to early adoption, invisible technologies are often the hardest for humans to relate with. It takes a lot of time and stories to gain a narrative that becomes viable. This is blockchain technology's problem too."
> — Olawale Daniel

One of the primary reasons for transparency in the asset pool is to ensure the stability and integrity of the stablecoin. By providing real-time access to information about the assets held in the pool, users can verify that the stablecoin is adequately collateralized. This transparency helps to mitigate counterparty risks and ensures that the stablecoin maintains its pegged value to the underlying assets.

Transparency also fosters accountability and helps to prevent fraudulent practices. By making the asset pool information publicly available, users and external auditors can monitor the stablecoin's compliance with stated asset ratios. This reduces the risk of mismanagement or manipulation of the asset pool, thereby enhancing the overall trustworthiness of the stablecoin.

Furthermore, transparency enables market participants to make informed decisions about using and trading the stablecoin. By having visibility into the underlying assets, users can assess the stability and quality of the asset pool. They can evaluate the risk profile, diversification, and overall health of the assets, which can be crucial factors in determining the stability and reliability of the stablecoin.

Transparency also helps in preventing conflicts of interest and ensuring fair treatment of users. When the asset pool composition is disclosed, it becomes easier to identify any potential biases or preferential treatment in the selection of assets. This promotes fairness and equal treatment of all stakeholders within the stablecoin ecosystem.

Moreover, transparency encourages innovation and collaboration within the stablecoin ecosystem. When the asset pool information is accessible to developers, researchers, and the broader community, it fosters the development of new tools, methodologies, and research to further enhance the stability and functionality of the stablecoin. Openness and collaboration can lead to improvements in risk management practices, security measures, and overall governance of the stablecoin.

It is worth noting that transparency does not necessarily mean full public disclosure of sensitive information. While the asset pool composition should be transparent, there may be legitimate reasons to protect certain details, such as specific trading strategies or proprietary information related to asset selection. Striking the right balance between transparency and the need for privacy and security is essential to protect the interests of users and maintain a competitive edge for stablecoin issuers.

To achieve transparency in stablecoin 1.4, various mechanisms can be employed. Regular and audited reports detailing the composition of the asset pool, including the types of digital assets held and their respective proportions, can be published. External audits conducted by reputable accounting firms can provide independent verification of the asset pool's integrity.

In addition, leveraging blockchain technology itself can enhance transparency. Public blockchains provide a decentralized and immutable ledger that allows anyone to verify transactions and holdings. By utilizing features such as on-chain asset verification and real-time reporting, stablecoin issuers can enhance transparency and facilitate trust among users.

Regulatory oversight can also play a role in ensuring transparency. Regulatory authorities can set guidelines and standards for disclosure and reporting requirements, promoting consistency and accountability across stablecoin issuers. Compliance with such regulations can provide an additional layer of transparency and strengthen market confidence in stablecoin 1.4.

Transparency is vital for stablecoin 1.4 based on digital assets. It enhances stability, fosters trust, and facilitates informed decision-making for users and market participants. By providing visibility into the asset pool, stablecoin issuers can demonstrate the collateralization of the stablecoin and mitigate risks. Transparency also promotes accountability, fairness, innovation, and collaboration within the stablecoin ecosystem. Striking the right balance between transparency, privacy, and security is crucial to protect user interests and maintain a competitive and trustworthy stablecoin ecosystem.

Viability in Major Markets

Stablecoin 1.4, based on digital assets, has experienced notable success in the international market, garnering widespread adoption and usage. However, its traction in the United States has been comparatively limited, facing certain challenges and regulatory constraints. Several factors contribute to this divergence in success.

1. Regulatory Environment: The regulatory landscape for cryptocurrencies and stablecoins in the United States is complex and evolving. The country has seen an increased focus on ensuring compliance with existing financial regulations, which has led to stricter oversight and licensing requirements for stablecoin issuers. The stringent regulatory environment has created challenges for stablecoin 1.4, requiring companies to navigate through various legal and compliance hurdles, resulting in slower adoption and limited market penetration.
2. Regulatory Clarity: Regulatory clarity plays a crucial role in fostering innovation and attracting investment in the stablecoin space. While some jurisdictions have provided clearer guidelines for the operation of stablecoins, the United States has experienced delays in providing comprehensive regulatory frameworks. The lack of clear and consistent regulations can create uncertainty for stablecoin issuers and deter potential market participants from entering the space.
3. Market Fragmentation: The United States comprises a large and diverse market, with multiple state-level regulatory bodies governing financial services. This fragmented regulatory landscape makes it challenging for stablecoin issuers to navigate

through varying compliance requirements across different states. The lack of a unified regulatory framework adds complexity and costs to operating stablecoin 1.4 within the country.

4. Trust and Consumer Confidence: Trust is a crucial factor for the widespread adoption of stablecoins. While stablecoin 1.4 has gained trust and recognition in international markets, it may face skepticism and concerns about security, transparency, and stability within the United States. The high-profile cases of fraud or mismanagement in the cryptocurrency industry have led to heightened caution among American consumers and regulators.
5. Market Competition: The stablecoin landscape in the United States is competitive, with established players like Tether (USDT) and USD Coin (USDC) dominating the market. These stablecoins have gained a significant foothold and enjoy wide acceptance among users and businesses. The strong market presence of existing stablecoins poses challenges for new entrants, including stablecoin 1.4, in establishing themselves and gaining market share.
6. Regulatory Focus on Investor Protection: Regulators in the United States have a strong focus on investor protection, especially in the context of cryptocurrencies and digital assets. This focus can result in stricter compliance requirements and heightened scrutiny for stablecoin issuers. While these measures aim to safeguard investors, they can also create barriers to entry and hinder the growth of stablecoin 1.4.

7. Financial Industry Resistance: The traditional financial industry in the United States has been cautious in embracing cryptocurrencies and stablecoins. Established financial institutions may perceive stablecoin 1.4 as a disruptive force that challenges their market dominance. Resistance from traditional players can impede the integration of stablecoin 1.4 into existing financial systems and limit its acceptance within the country.

Despite the challenges faced in the United States, stablecoin 1.4 has witnessed significant success in international markets. Factors such as regulatory clarity, supportive legal frameworks, and market demand in other jurisdictions have contributed to its adoption and acceptance. Moreover, in regions with unstable or inflationary fiat currencies, stablecoin 1.4 offers a stable alternative for individuals and businesses to store and transact value.

To improve the adoption of stablecoin 1.4 in the United States, regulatory clarity and harmonization across states are necessary. Clear guidelines and comprehensive frameworks would provide stability and instill confidence among market participants. Collaborative efforts between regulators, industry stakeholders, and stablecoin issuers can foster innovation, ensure investor protection, and establish a favorable environment for the growth of stablecoin 1.4 within the country.

It is also crucial for stablecoin 1.4 issuers to focus on building trust, enhancing transparency, and addressing concerns related to security and stability. Demonstrating robust security measures, providing regular audits, and offering transparent reporting of the asset pool can help alleviate doubts and foster trust among consumers and regulators.

Furthermore, partnerships with established financial institutions, compliance with existing financial regulations, and integration into existing payment systems can enhance the acceptance and usability of stablecoin 1.4 within the traditional financial ecosystem.

In summary, while stablecoin 1.4 has encountered challenges in gaining traction in the United States, regulatory clarity, market competition, trust-building efforts, and collaboration with traditional financial institutions are crucial for its success. By addressing these challenges and capitalizing on the unique advantages of stablecoin 1.4, such as global accessibility and stability, its adoption and usage in the United States can be significantly improved.

Conclusion

Stablecoin 1.4, which is based on digital assets, offers a unique approach to achieving stability in the volatile cryptocurrency market. While facing challenges in some jurisdictions, it has gained notable success and acceptance globally. Stablecoin 1.4 provides a valuable alternative to traditional fiat currencies and offers several advantages, including stability, efficiency, and accessibility.

One of the key strengths of stablecoin 1.4 is its ability to leverage digital assets as a basis for stability. While these assets may experience short-term volatility, they tend to exhibit positive stability over the long term. By utilizing digital assets as collateral, stablecoin 1.4 can maintain its value and provide a reliable store of wealth and medium of exchange.

The liquidity of stablecoin 1.4 is a significant factor contributing to its success. The use of digital assets allows for unprecedented liquidity compared to other stablecoin models. The digital nature of these assets enables quick and seamless transactions, facilitating the efficient movement of value within the ecosystem. This liquidity enhances the usability and accessibility of stablecoin 1.4 in various financial applications and transactions.

In terms of the required asset ratio, stablecoin 1.4 typically maintains an asset pool of around 50%. The digital asset-backed model allows for a lower asset ratio compared to traditional fiat-backed stablecoins. This flexibility is made possible by the continuous monitoring and rebalancing of the digital asset portfolio through algorithms, ensuring that the stablecoin remains within the desired stability range.

Regarding the management of the asset pool, it is advisable for stablecoin 1.4 to collaborate with trusted entities such as decentralized finance (DeFi) banks or reputable exchanges. These entities can effectively manage the asset pool, ensuring transparency, security, and compliance with regulatory requirements. By entrusting the asset pool to reliable partners, stablecoin 1.4 can enhance its credibility and instill confidence among users.

While stablecoin 1.4 offers numerous benefits, it is not without risks. The susceptibility to attacks, both in terms of the underlying digital assets and the stability mechanisms, poses a significant challenge. It is crucial to implement robust security measures, conduct regular audits, and maintain transparency to mitigate these risks effectively.

Transparency is paramount in the success of stablecoin 1.4. Users and market participants need clear and timely information about the asset pool, stability mechanisms, and governance processes. Transparent reporting, regular audits, and open communication channels build trust and confidence in stablecoin 1.4, fostering its acceptance and adoption in the market.

Although stablecoin 1.4 has achieved substantial success globally, its reception in the United States has been comparatively limited. Regulatory challenges, market fragmentation, and the cautious stance of traditional financial institutions have impeded its growth in the country. To overcome these obstacles, regulatory clarity, collaboration with regulatory bodies, and partnerships with established financial institutions are essential.

In conclusion, stablecoin 1.4 based on digital assets presents a promising solution for achieving stability in the cryptocurrency market. With its unique features, including liquidity, flexibility in asset ratios, and transparency, stablecoin 1.4 offers a reliable and efficient medium of exchange and store of value. By addressing challenges, enhancing transparency, and collaborating with key stakeholders, stablecoin 1.4 has the potential to play a significant role in the future of global financial systems.

Chapter 13: Stablecoin 1.5 (Collections of Fiat Currencies/Money Markets)

"War used to be something you could stand on a nearby hill and watch. How we have total war; everybody's in it. We have total economics as well. Everything affects everybody. The Malaysian currency shakes, and people around the world are seriously affected."[35]

In the first truly level two stablecoin, we see that the market drifted back where it is comfortable. In most cases, the market will move to a point where it is testing the boundaries of what society will accept, like the Budweiser/Target debacle of 2023, then the market will move back into an area where it knows it is safe and will recover. When Stablecoin 1.5 arose in 2019 ($bash), we saw the market crawling back into an area where it knew that it could be successful, money markets. Money markets have a positive stability that has been successful for years, it was only logical that the cryptocurrency market would adopt these. As cryptocurrency continues to evolve, this has become a stabilizing factor in the market having this comfort zone to return to.

In this chapter, we will follow the same pattern that we follow in all the chapters in this part of the book. First, we look at the stability of the asset class on which the token at this level is based. The asset class is very important because it is the 'pegged' underpinning of the system. Next, we look at the liquidity of the asset class, how fast can the asset be liquidated if there is a 'run' on the token. As a

---

[35] Salman Rusdie https://www.azquotes.com/quotes/topics/currency.html.

matter of course, we will then look at the asset ratio needed to maintain stability. The media tends to insist that there needs to be a 1:1 ratio, and letter agencies tend to echo this false statement; each class has a ratio that is needed for the token to be reliable in its function.

> "The stock market is a device for transferring money from the impatient to the patient."
>
> Warren Buffet

Also, as a matter of course, it is always important to look at the relationship between holders/creators of the currency and the asset pool. The more complex the system the more analysis needs to be done on how the assets are being managed. While a 1:1 ratio for stablecoin 1.0 can be managed by the local bank of the creator, more complex systems require funds, investment firms, and even multiple subsidiaries to ensure that the asset pool maintains value. The analysis of who holds the asset pool then leads to an analysis of susceptibility of to attack. The more complex the asset, the more stable it is- however, it does need to be protected from attacks. The discussions within these sections of each chapter will be quite similar. Transparency is also an issue that varies slightly as you look at different types of stablecoins, once again the more complex the tokenomics- the more transparent the company needs to be. Finally, there needs to be an analysis of viability in major markets. This results in a bimodal, cyclic model for the analysis of the system, viability seems to be highest for simple and highly complex models, with intermediate models being lukewarm in their reception.

Stablecoin 1.5 is one of the first true stablecoins that can point to a positive stability. While previous iterations have shown that they have a neutral to positive relationship with the market, stablecoin 1.5 truly has the ability to allow the system to adapt and overcome problems within the global economy. This allows for a small ability for the holding ratio to start low and grow over the course of the life of the token. The bigger the asset pool, the more diverse the fund can be. This leads to a strong level of positive stability which will define the level two stablecoins.

Stability of Asset Class

Stablecoin 1.5, which is based on a collection of different fiat currencies, represents a unique approach to achieving stability in the world of cryptocurrencies. Unlike single currency stablecoins, such as stablecoin 1.0, stablecoin 1.5 seeks to diversify its holdings across multiple fiat currencies to enhance stability and mitigate the risks associated with any individual currency. This diversification strategy aims to provide a more robust and resilient stablecoin model. In this discussion, we will explore the stability of stablecoin 1.5 in comparison to other stablecoin manifestations.

One of the primary advantages of stablecoin 1.5 is its inherent stability through diversification. By combining a variety of fiat currencies, stablecoin 1.5 aims to minimize the impact of individual currency fluctuations and macroeconomic factors that can affect any single currency. The diversification strategy helps to spread the risk and reduce the potential negative impact of inflation, monetary policy changes, or economic instability of a particular currency. As a result, stablecoin 1.5 may offer a more stable value proposition compared to stablecoin 1.0, which relies on a single fiat currency.

The stability of stablecoin 1.5 is further supported by the fact that it leverages established and widely recognized fiat currencies. These currencies are typically issued and regulated by central banks, which employ monetary policies aimed at maintaining price stability and managing inflation. By aligning with established currencies, stablecoin 1.5 benefits from the stability measures implemented by these central banks. This stability can instill confidence in users and market participants, as they perceive stablecoin 1.5 as a reliable store of value and medium of exchange.

> "To make money in the markets, you have to think independently and be humble."
>
> Ray Dalio

Moreover, stablecoin 1.5 can capitalize on the liquidity and depth of the foreign exchange (Forex) market. The Forex market is the largest and most liquid financial market globally, with significant daily trading volumes. By utilizing a collection of different fiat currencies, stablecoin 1.5 can tap into this deep market liquidity, enabling efficient conversion and hedging mechanisms. The access to diverse currency pairs and the ability to exchange between them can contribute to the overall stability and liquidity of stablecoin 1.5.

In terms of the required asset ratio, stablecoin 1.5 aims to maintain a certain level of backing from the different fiat currencies in its reserve. The exact asset ratio may vary depending on the design and governance framework of stablecoin 1.5. However, it is generally advisable for stablecoin 1.5 to maintain a sufficient reserve to back its outstanding supply. The asset ratio should be carefully determined to ensure that stablecoin 1.5 remains stable and can withstand potential shocks or fluctuations in the underlying fiat currencies.

It is worth noting that stablecoin 1.5 faces challenges related to the management and governance of its asset pool. The stability and value of the stablecoin heavily rely on the proper management of the reserve, including currency selection, rebalancing, and risk management. It is essential to establish robust governance mechanisms, compliance frameworks, and risk management protocols to ensure the stability and integrity of the asset pool. Collaboration with reputable financial institutions, asset managers, or custodians can provide the necessary expertise and oversight for effective management.

When compared to other stablecoin manifestations, stablecoin 1.5 offers unique advantages in terms of stability. While stablecoin 1.0 relies solely on a single fiat currency, stablecoin 1.5 diversifies its holdings, reducing the impact of any individual currency's fluctuations. This diversification strategy enhances stability and reduces the exposure to specific country risks. However, stablecoin 1.5 may face challenges related to managing multiple currencies, including the need for proper risk management and monitoring mechanisms.

In conclusion, stablecoin 1.5, based on a collection of different fiat currencies, represents a promising stablecoin model that offers enhanced stability through diversification. By spreading the risk across multiple currencies and leveraging established monetary policies, stablecoin 1.5 aims to provide a reliable store of value and medium of exchange. The diversification strategy helps mitigate the risks associated with individual currencies, while the access to deep liquidity in the Forex market contributes to overall stability. However, effective management and governance of the asset pool are crucial to ensure the stability and integrity of stablecoin 1.5. With proper risk management and governance frameworks in place, stablecoin 1.5 has the potential to offer stability and utility in the ever-evolving world of cryptocurrencies.

Liquidity of Asset Class

Liquidity is a critical factor for stablecoins as it determines their ability to be readily bought and sold in the market without significantly impacting their price. The level of liquidity influences the stability and usability of stablecoins in various financial transactions, including trading, remittances, and payment settlements.

Stablecoin 1.5 benefits from the liquidity of the underlying fiat currencies it holds in its reserve. By diversifying its holdings across different currencies, stablecoin 1.5 taps into the liquidity of each currency individually. This diversification offers an advantage over single currency stablecoins, such as stablecoin 1.0, which rely on the liquidity of a single fiat currency.

> "Markets are constantly in a state of uncertainty and flux and money is made by discounting the obvious and betting on the unexpected."
>
> George Soros

The liquidity of stablecoin 1.5 is closely tied to the liquidity of the foreign exchange (Forex) market. The Forex market is the largest and most liquid financial market globally, with trillions of dollars exchanged daily. As stablecoin 1.5 holds a collection of different fiat currencies, it can leverage the liquidity and depth of the Forex market to facilitate efficient conversions and transactions.

The liquidity of stablecoin 1.5 is further supported by the presence of reputable financial institutions and liquidity providers in the market. These entities play a crucial role in maintaining market depth and providing liquidity for stablecoin 1.5. As stablecoin 1.5 gains traction and adoption, it is likely to attract market makers, liquidity providers, and exchanges that offer seamless conversion between stablecoin 1.5 and other currencies, thereby enhancing liquidity.

Compared to stablecoin 1.0, which relies solely on the liquidity of a single fiat currency, stablecoin 1.5 has the advantage of diversification. This diversification can reduce liquidity risks associated with any specific currency, as market participants have the flexibility to exchange stablecoin 1.5 for multiple fiat currencies. The ability to tap into the liquidity of different currencies enhances stability and makes stablecoin 1.5 more attractive for various use cases.

However, it is important to note that stablecoin 1.5 may face certain liquidity challenges due to the nature of its asset pool. Managing liquidity across multiple fiat currencies requires careful monitoring and balancing to ensure that stablecoin 1.5 can meet the demands of its users. Adequate mechanisms should be in place to handle potential fluctuations in demand and supply and maintain an appropriate balance of liquidity across the different currencies.

Additionally, the liquidity of stablecoin 1.5 may also be influenced by the market's perception of its stability and reliability. Market participants may favor stablecoins with a proven track record and widespread adoption, as these tend to have more established liquidity pools. Building trust and establishing a strong user base are crucial for stablecoin 1.5 to attract liquidity and maintain its stability.

Comparing the liquidity of stablecoin 1.5 to other stablecoin models, we can observe varying degrees of liquidity depending on the underlying asset or mechanism. Stablecoins based on fiat currencies, such as stablecoin 1.0 and stablecoin 1.5, generally have access to the liquidity of the respective fiat currencies. However, stablecoin 1.5 has an advantage over stablecoin 1.0 due to its diversification, allowing for greater resilience to liquidity shocks specific to any single currency.

> "Investing requires quantities of temperament way more than quantities of intellect."
>
> Warren Buffet

On the other hand, stablecoins based on digital assets, such as stablecoin 1.4, may face different liquidity challenges. The liquidity of digital assets can be influenced by factors such as market sentiment, trading volume, and the underlying blockchain infrastructure. While the digital asset market has witnessed significant growth and liquidity improvements, it may still experience higher volatility and liquidity risks compared to fiat-based stablecoins like stablecoin 1.5.

In conclusion, stablecoin 1.5, based on a collection of different fiat currencies, has the potential to offer robust liquidity by leveraging the liquidity of each individual currency and the broader Forex market. Diversification across multiple currencies reduces the liquidity risks associated with any specific currency and enhances stability. However, managing liquidity across different currencies requires careful monitoring and balancing. Market perception and trust also play significant roles in attracting liquidity to stablecoin 1.5. Overall, stablecoin 1.5's liquidity profile presents a compelling option for users seeking stability and usability in the cryptocurrency ecosystem.

Required Asset Ratio for Stability

Stablecoin 1.5 operates with an initial asset pool that typically comprises around 40% of the total value of the pool. This means that when stablecoin 1.5 is first introduced, the asset pool backing the stablecoin is equivalent to approximately 40% of the total stablecoin supply. The remaining 60% is achieved through the growth potential of a money market and the smart burn system.

The growth potential of stablecoin 1.5 is primarily derived from the money market mechanism it employs. A money market is a financial market where short-term borrowing and lending of highly liquid assets occur. In the case of stablecoin 1.5, the money market allows for the generation of additional funds that contribute to the expansion of the stablecoin's asset pool.

Through the money market, stablecoin 1.5 can engage in various financial activities, such as lending and borrowing, investing in low-risk assets, or participating in yield-generating strategies. These activities generate returns on the funds held in the asset pool, leading to the growth of the pool over time.

One crucial aspect of stablecoin 1.5's growth potential is the implementation of a smart burn mechanism. The smart burn mechanism involves periodically reducing the supply of stablecoin 1.5 by removing a portion of the circulating tokens from circulation. This reduction in supply is achieved by burning stablecoins, effectively eliminating them from the market.

The smart burn mechanism serves two primary purposes. First, it creates a scarcity effect, increasing the value and stability of the remaining stablecoins in circulation. Second, it contributes to the growth potential of stablecoin 1.5 by reducing the supply and maintaining a balance between supply and demand dynamics.

> "If you aren't willing to own a stock for 10 years, don't think about owning it for 10 minutes."
>
> Warren Buffet

With the combination of the money market mechanism and the smart burn mechanism, stablecoin 1.5 has the potential to reach parity with the underlying asset pool within approximately 7 years. This timeline may vary depending on factors such as the growth rate of the money market, the effectiveness of the smart burn mechanism, and market dynamics.

The growth potential of stablecoin 1.5 offers several benefits. Firstly, it allows for the gradual expansion of the asset pool, ensuring that the stablecoin has sufficient backing to maintain its stability and peg to the underlying assets. As the asset pool grows, it provides increased confidence to users and market participants in the stability and value of the stablecoin.

Secondly, the growth potential provides opportunities for investors and holders of stablecoin 1.5 to earn additional returns on their holdings. The money market activities generate income or interest, which is shared with the holders of the stablecoin. This incentivizes users to hold stablecoin 1.5, contributing to its adoption and liquidity.

However, it's important to note that the growth potential of stablecoin 1.5 also introduces certain risks and considerations. The money market activities involve investments and lending, which inherently carry risks. Market fluctuations, defaults, or other unforeseen events can impact the returns generated by the money market and, consequently, the growth potential of stablecoin 1.5.

Furthermore, the smart burn mechanism should be carefully designed and implemented to ensure its effectiveness and avoid any unintended consequences. The timing, frequency, and magnitude of the burn events must be well-regulated to maintain stability and prevent disruptions in the market.

In conclusion, stablecoin 1.5 initiates with an asset pool comprising around 40% of the total value of the pool, with the remaining 60% being achieved through the growth potential of a money market. The money market mechanism allows for the expansion of the asset pool through various financial activities, while the smart burn mechanism enhances scarcity and contributes to long-term stability. This growth potential offers both benefits and risks, as it enables the stablecoin to reach parity with the asset pool within approximately 7 years but also introduces considerations related to market dynamics, risk management, and the effective implementation of the burn mechanism.

Relationship between Holders and access to backing.

When it comes to determining who should hold the asset pool in stablecoin 1.5, there are multiple factors to consider. Both the company issuing the stablecoin and third-party custodians can potentially fulfill this role, each with its own advantages and considerations.

Company Holding the Asset Pool: The company issuing stablecoin 1.5 may choose to hold the asset pool itself. This approach offers certain benefits:

- Control: The company has direct control over the management and security of the assets. They can implement strategies aligned with the stability and growth objectives of the stablecoin.
- Flexibility: The company can swiftly adapt to changing market conditions, implement necessary adjustments, and respond to user demands more efficiently.
- Integration: By holding the asset pool, the company can seamlessly integrate the stablecoin into its broader ecosystem, facilitating the use of stablecoin across its products and services.

However, there are considerations to keep in mind:

- Trust: Users and investors must trust the company's competence and integrity in managing the asset pool. The company needs to demonstrate transparency, robust security measures, and effective risk management practices.
- Concentration Risk: If the company encounters financial or operational difficulties, it could potentially impact the stability of the stablecoin.

Diversification of the asset pool and contingency plans are crucial to mitigate this risk.

- Regulatory Compliance: The company must ensure compliance with relevant financial regulations and adhere to industry standards to maintain trust and regulatory approval.

Third-Party Custodians: Alternatively, stablecoin 1.5's asset pool can be held by trusted third-party custodians. This approach offers certain advantages:

- Security and Expertise: Established custodians specialize in securely storing and managing digital assets. They employ robust security measures, multi-signature wallets, and best practices to safeguard the assets.
- Independence: A third-party custodian brings an additional layer of independence, reducing the potential for conflicts of interest and enhancing trust in the stablecoin ecosystem.
- Auditing and Transparency: External custodians often undergo regular audits by reputable firms, providing assurance regarding the integrity of the asset holdings. These audits enhance transparency and accountability.

However, there are considerations to keep in mind:

- Cost: Engaging third-party custodians involves additional costs, including custody fees and potentially higher transaction fees. These costs need to be factored into the stablecoin's economics and profitability.
- Regulatory Compliance: The third-party custodian must comply with relevant regulations and possess

the necessary licenses and credentials to operate legally and maintain trust.

In many cases, a hybrid approach may be adopted. The company issuing stablecoin 1.5 may partner with trusted custodians, leveraging their expertise while maintaining oversight and control over the asset pool. This arrangement can provide a balance between security, transparency, and operational flexibility.

> "Investing is simple. It's the financial industry that works hard to make it complex."
>
> Robert Rolih

Whichever approach is chosen, transparency is crucial. Users and investors should have access to regular reports and audits of the asset pool's holdings, demonstrating the stablecoin's backing and ensuring alignment with its stated objectives. The chosen entity, whether the company or a third-party custodian, should uphold the highest standards of transparency and demonstrate a commitment to the stablecoin's stability and integrity.

Overall, the decision of who should hold the asset pool in stablecoin 1.5 depends on various factors such as the company's capabilities, regulatory requirements, trust, security, and the desire for independence and transparency. It is essential to strike a balance between control, security, expertise, and adherence to regulatory standards to maintain the stability and trustworthiness of the stablecoin ecosystem.

Susceptibility to Attacks

Stablecoin 1.5, which is based on a collection of different fiat currencies, is designed to provide stability and security.

However, like any digital asset or financial instrument, it is not immune to potential attacks and vulnerabilities. Here are some potential susceptibility factors that stablecoin 1.5 may face:

Cybersecurity Attacks: Stablecoin 1.5, being a digital asset, is susceptible to various cybersecurity attacks, including:

- Hacking: Malicious actors may attempt to breach the systems holding the stablecoin's digital assets, compromising their security and potentially stealing funds.
- Phishing: Attackers might employ phishing techniques to trick users into revealing sensitive information like private keys or login credentials, which could lead to unauthorized access to the stablecoin's assets.
- Malware and Ransomware: Malicious software can infect users' devices, compromising the security of their wallets and potentially enabling attackers to gain control over the stablecoin's assets.

Smart Contract Vulnerabilities: Stablecoin 1.5 may rely on smart contracts to facilitate transactions and maintain the stability of the token. However, smart contracts are not immune to vulnerabilities, including:

- Exploitable Bugs: Coding errors or vulnerabilities in the smart contract's code may be exploited by attackers to manipulate or steal the stablecoin's assets.
- Oracle Manipulation: Smart contracts often rely on external data sources (oracles) for information. Manipulation or compromise of these oracles can lead to inaccurate or malicious data being used

within the smart contract, potentially impacting the stability of the stablecoin.

Regulatory and Compliance Risks: Stablecoin 1.5's operations may be subject to regulatory scrutiny and compliance requirements, which can pose potential risks, such as:

- Legal Challenges: Regulatory authorities may impose restrictions or regulations on stablecoin operations, impacting its ability to operate or limiting its market reach.
- Compliance Failures: Non-compliance with existing financial regulations, anti-money laundering (AML), or know-your-customer (KYC) requirements could lead to legal consequences and reputational damage.

Market Manipulation: Stablecoin 1.5's value may be influenced by market forces and external factors, making it vulnerable to manipulation, such as:

- Pump and Dump Schemes: Unscrupulous actors may artificially inflate or deflate the stablecoin's price, taking advantage of market sentiment and liquidity to profit at the expense of other users.
- Insider Trading: If individuals with privileged information manipulate the stablecoin's price or take advantage of their knowledge, it could harm other users' interests and undermine trust.

To mitigate these susceptibility factors, several measures can be implemented:

1. Robust Security Measures: Implementing robust cybersecurity measures, such as secure coding

practices, regular security audits, penetration testing, and multi-factor authentication, can enhance the resilience of stablecoin 1.5's infrastructure against hacking attempts and malicious activities.

> "Today the financial market is no good, but the money is there."
>
> Jack Ma

2. Smart Contract Audits: Conducting thorough audits of the smart contract's code by independent security experts can help identify and address vulnerabilities before deployment, reducing the risk of exploits and potential financial losses.
3. Regulatory Compliance: Collaborating with legal and compliance experts to ensure adherence to relevant regulations, AML, and KYC procedures can minimize the risk of legal challenges and reputational damage.
4. Transparency and Governance: Providing transparent and regular reporting on the stablecoin's asset holdings, audits, and risk management processes enhance trust and helps users make informed decisions. Implementing robust governance mechanisms can ensure the stablecoin's operations align with the best interests of its users.
5. Market Surveillance: Monitoring and surveillance mechanisms, such as detecting unusual trading patterns or suspicious activities, can help identify and prevent market manipulation, enhancing the stability and integrity of the stablecoin.
6. Collaboration with Security Experts: Engaging with reputable cybersecurity firms, blockchain security specialists, and digital asset custodians can leverage

their expertise to enhance the security posture of stablecoin 1.5.

It is essential for stablecoin 1.5's issuers and stakeholders to remain vigilant, adaptive, and proactive in addressing potential vulnerabilities and staying ahead of emerging threats. By implementing robust security practices, adhering to regulatory requirements, fostering transparency, and collaborating with experts, the susceptibility to attacks can be minimized, contributing to the stability and trustworthiness of the stablecoin.

Transparency

Transparency plays a crucial role in the success and adoption of stablecoin 1.5. It not only builds trust among users but also provides the necessary accountability and assurance regarding the stablecoin's operations and asset holdings. Here are several reasons why transparency is essential for stablecoin 1.5:

1. Trust and Credibility: Transparency fosters trust and credibility in stablecoin 1.5. When users have access to comprehensive information about the stablecoin's asset holdings, issuance, and operations, they can make informed decisions and have confidence in the stability and value of the token. Trust is vital for widespread adoption and long-term sustainability.
2. Proof of Reserves: Transparency enables stablecoin 1.5 to provide proof of reserves, validating that the token is backed by the claimed assets. By regularly publishing audited reports or real-time data on the asset pool, users can verify the stablecoin's backing and ensure it is adequately collateralized. This helps

prevent instances of fractional reserve or insolvency risks.

3. Risk Assessment and Mitigation: Transparent disclosure of information about the stablecoin's asset composition and risk management strategies allows users to assess the level of risk associated with holding the token. It enables them to make informed decisions based on their risk tolerance and understanding of the underlying assets, reducing the potential for surprises or unexpected losses.
4. Regulatory Compliance: Transparency is essential for maintaining compliance with regulatory frameworks. By providing clear and accurate information about the stablecoin's operations, asset holdings, and adherence to regulatory requirements, stablecoin 1.5 can demonstrate compliance and minimize legal and regulatory risks. Transparent reporting also helps in the enforcement of anti-money laundering (AML) and know-your-customer (KYC) regulations.
5. Market Confidence: Transparent stablecoin projects are more likely to gain confidence from market participants, including investors, businesses, and financial institutions. When stakeholders can access reliable and up-to-date information about the stablecoin's operations, they can make informed decisions about its integration into their business processes or investment strategies. This, in turn, promotes market liquidity and stability.
6. Investor Protection: Transparency protects the interests of stablecoin 1.5's investors. Clear and accessible information about the stablecoin's assets, governance, and risk management practices helps

investors assess the stability and reliability of the investment. It enables them to evaluate the credibility of the stablecoin issuer and make informed choices to protect their investments.

7. Auditability and Accountability: Transparent stablecoin projects are subject to external audits and scrutiny, ensuring accountability and reducing the potential for fraudulent activities or mismanagement. Regular audits by reputable auditing firms provide an additional layer of assurance and help maintain the stablecoin's integrity.
8. Market Differentiation: In a competitive landscape, transparency can differentiate stablecoin 1.5 from other similar offerings. By proactively providing detailed information about its asset composition, operations, and compliance practices, stablecoin 1.5 can establish itself as a trustworthy and reliable option. This can attract users and businesses looking for stablecoins with a high level of transparency.

To achieve transparency, stablecoin 1.5 should consider implementing the following practices:

- Regular and comprehensive reporting of asset holdings and their valuation.
- Audits by reputable third-party firms to verify the stablecoin's reserves.
- Real-time or periodic disclosure of transaction data and addresses holding the stablecoin.
- Clear governance frameworks and decision-making processes that involve community participation.
- Disclosure of risk management strategies and contingency plans for different scenarios.

- Open communication channels to address user inquiries, concerns, and feedback.

Overall, transparency is vital for stablecoin 1.5 to build trust, attract users and investors, comply with regulations, and differentiate itself in the market. By being transparent in its operations and asset holdings, stablecoin 1.5 can enhance its credibility and foster long-term success.

Viability in Major Markets

> "The key to making money in stocks in not to get scared out of them."
>
> Peter Lynch

Stablecoin 1.5 has shown promising performance in major markets, gaining significant traction and acceptance among users, businesses, and investors. Its unique composition, backed by a collection of different fiat currencies, provides stability and diversification, making it an attractive option for market participants. Here are some key factors that contribute to stablecoin 1.5's performance in major markets:

Stability and Value Preservation:

Stablecoin 1.5's asset pool consisting of a basket of fiat currencies helps maintain stability and preserves the value of the token. By combining different fiat currencies, stablecoin 1.5 mitigates the risk associated with the individual currencies, providing a more stable value proposition compared to single currency stablecoins. This stability makes it an ideal medium of exchange, store of value, and unit of account, enabling seamless transactions and reducing currency-related risks.

Market Acceptance:

Stablecoin 1.5 has gained widespread acceptance in major markets due to its ability to address the volatility and uncertainties associated with cryptocurrencies. The inclusion of multiple fiat currencies makes it more familiar and comfortable for users who are accustomed to traditional currencies. This broader acceptance allows stablecoin 1.5 to be readily integrated into various financial systems and transactions, expanding its use cases and market reach.

Enhanced Liquidity:

Stablecoin 1.5 benefits from increased liquidity in major markets. The broad base of users and businesses utilizing stablecoin 1.5 creates a robust ecosystem, facilitating seamless and efficient transactions. The liquidity of stablecoin 1.5 allows for easy conversion into other cryptocurrencies or fiat currencies, enabling users to access and exit the stablecoin quickly. This liquidity feature enhances its appeal to traders, investors, and businesses.

Market Stability:

Stablecoin 1.5's presence in major markets contributes to overall market stability. As a reliable and widely adopted stablecoin, it helps reduce the volatility and price fluctuations often associated with other cryptocurrencies. Stablecoin 1.5 provides a benchmark for value stability, offering a reference point for market participants and fostering confidence in the broader cryptocurrency ecosystem.

Regulatory Compliance:

Stablecoin 1.5's compliance with regulatory frameworks in major markets enhances its market performance. By adhering to applicable laws and regulations, stablecoin 1.5

establishes itself as a trusted and compliant digital asset. This compliance ensures that stablecoin 1.5 can operate within established regulatory boundaries, offering a secure and reliable option for users and businesses seeking stability and regulatory certainty.

Market Integration:
Stablecoin 1.5's integration into existing financial systems and platforms further contributes to its performance in major markets. As stablecoin 1.5 gains adoption, it becomes more widely supported by cryptocurrency exchanges, wallets, and payment processors. This integration enhances accessibility and usability, enabling users to seamlessly transact with stablecoin 1.5 across various platforms and services. The ability to integrate with existing financial infrastructure facilitates its widespread usage and market penetration.

User and Business Confidence: Stablecoin 1.5's performance and reputation in major markets have built confidence among users and businesses. The stability and reliability of the stablecoin, combined with its market acceptance, create a positive perception of its value and usability. Users and businesses are more likely to engage with stablecoin 1.5, considering it as a trusted medium of exchange, a safe store of value, and an efficient unit of account. This confidence fosters continued adoption and growth in major markets.

Despite its success in major markets, stablecoin 1.5 still faces challenges. These challenges include regulatory uncertainties, market competition, and potential risks associated with the underlying fiat currencies. However, by addressing these challenges and maintaining its stability, transparency, and market integration, stablecoin 1.5 can continue to thrive and play a significant role in the evolving digital economy.

Conclusion

In conclusion, stablecoin 1.5 has emerged as a compelling solution in major markets, offering stability, liquidity, and transparency. Its unique composition, based on a collection of different fiat currencies, provides an effective hedge against volatility and risks associated with individual currencies. Stablecoin 1.5 has demonstrated strong performance, gaining acceptance among users, businesses, and investors in major markets.

One of the key strengths of stablecoin 1.5 lies in its stability and value preservation. By combining multiple fiat currencies, it offers a reliable store of value, making it an attractive option for those seeking stability in the digital asset space. The broad market acceptance of stablecoin 1.5 further enhances its performance, as it becomes widely integrated into financial systems and transactions.

The liquidity of stablecoin 1.5 is another crucial factor contributing to its success. Its presence in major markets fosters a robust ecosystem, enabling seamless and efficient transactions. The enhanced liquidity allows for easy conversion into other cryptocurrencies or fiat currencies, providing users with flexibility and accessibility.

Transparency plays a vital role in the success of stablecoin 1.5. By providing clear visibility into the asset pool and its management, stablecoin 1.5 builds trust among users and businesses. Transparency also contributes to regulatory compliance, ensuring that stablecoin 1.5 operates within established frameworks and maintains the necessary safeguards.

While stablecoin 1.5 has shown significant success in major markets, challenges remain. Regulatory uncertainties, market competition, and risks associated with the underlying fiat currencies are factors that require continuous attention and adaptation. However, by addressing these challenges and maintaining stability, transparency, and market integration, stablecoin 1.5 can continue to thrive and expand its market presence.

As the digital economy evolves, stablecoin 1.5 provides a bridge between traditional financial systems and the world of cryptocurrencies. Its stability, liquidity, and market acceptance position it as a valuable tool for individuals, businesses, and investors seeking a reliable and efficient medium of exchange, store of value, and unit of account. Stablecoin 1.5 has demonstrated its ability to offer stability, liquidity, and transparency in major markets, setting the stage for its continued growth and adoption in the global financial landscape.

Chapter 14: Stablecoin 1.6 (Stocks and Derivatives)

"Mandatory allocation of capital between bonds and stocks by mutual funds creates tremendous short-term opportunities for investors."[36]

Growth of a fund is one of the most important elements of any stablecoin after stablecoin 1.0. Without growth, the inflation rate slowly dissolves the value of the fund until it is just a fraction of its previous worth. In Chapter 13, we looked at the evolution of stablecoin to the point where there was a money market backing the product. It only stands to reason that the next evolutionary step for the currency system is to have a mutual fund style system. While mutual funds do not provide the returns that direct investment and day trading funds do, they also do not bear the massive risks associated with those types of funds. Therefore, the mutual fund backed stablecoins, if run properly, have the ability to sit back and wait for the fund to mature and for the token to grow to an 'S' class stablecoin.

In this chapter, we will follow the same pattern that we follow in all the chapters in this part of the book. First, we look at the stability of the asset class on which the token at this level is based. The asset class is very important because it is the 'pegged' underpinning of the system. Next, we look at the liquidity of the asset class, how fast can the asset be liquidated if there is a 'run' on the token. As a matter of course, we will then look at the asset ratio needed

---

[36] Naved Abdali at https://www.goodreads.com/quotes/tag/mutual-funds.

to maintain stability. The media tends to insist that there needs to be a 1:1 ratio, and letter agencies tend to echo this false statement; each class has a ratio that is needed for the token to be reliable in its function.

> "The world of derivatives is full of holes that very few people are really aware of. Its like hydrogen and oxygen sitting on the corner waiting for a little flame."
>
> Charlie Munger

Also, as a matter of course, it is always important to look at the relationship between holders/creators of the currency and the asset pool. The more complex the system the more analysis needs to be done on how the assets are being managed. While a 1:1 ratio for stablecoin 1.0 can be managed by the local bank of the creator, more complex systems require funds, investment firms, and even multiple subsidiaries to ensure that the asset pool maintains value. The analysis of who holds the asset pool then leads to an analysis of susceptibility of to attack. The more complex the asset, the more stable it is- however, it does need to be protected from attacks. The discussions within these sections of each chapter will be quite similar. Transparency is also an issue that varies slightly as you look at different types of stablecoins, once again the more complex the tokenomics- the more transparent the company needs to be. Finally, there needs to be an analysis of viability in major markets. This results in a bimodal, cyclic model for the analysis of the system, viability seems to be highest for simple and highly complex models, with intermediate models being lukewarm in their reception.

Mutual fund backed stablecoins are some of the finest level two stablecoins on the market. Because they have the ability to 'wait it out,' they are able to mature at a more reasonable level. This allows for early investors to make more money, mid-term investors to still make money, and late term investors to hedge against the market. Stablecoin 1.6 provides some of the best opportunity for investors to beat inflation (prior to stablecoin 2.0). This means this is one of the best evolutionary elements of the stablecoin system.

Stability of Asset Class

Stablecoin 1.6, backed by a well-diversified mutual fund, offers a unique approach to stability in the realm of stablecoins. By leveraging the inherent stability and diversification of a mutual fund, stablecoin 1.6 aims to provide a reliable and secure digital asset that can serve as a medium of exchange and store of value.

The stability of stablecoin 1.6 stems from the underlying mutual fund's investment portfolio. Mutual funds are designed to pool funds from various investors and invest in a diversified range of assets, such as stocks, bonds, and other financial instruments. This diversification helps to mitigate risk and protect against market volatility. As a result, stablecoin 1.6 derives its stability from the balanced and well-managed nature of the mutual fund's holdings.

One of the key advantages of a stablecoin backed by a mutual fund is the potential for consistent returns. Mutual funds are managed by experienced professionals who follow a disciplined investment strategy. The fund managers aim to generate stable and predictable returns over time, aligning with the objective of a stablecoin to maintain a steady value. By diversifying across different asset classes and sectors, the mutual fund can mitigate the impact of market fluctuations and deliver more stable performance.

Furthermore, the use of a mutual fund as the asset pool for stablecoin 1.6 provides an additional layer of risk management. The fund's portfolio is regularly monitored and adjusted to adapt to changing market conditions and investment opportunities. This active management helps to ensure that the stablecoin remains stable and resilient in the face of market uncertainties. It also allows for the inclusion of assets that have the potential to generate attractive returns, contributing to the long-term stability and growth of the stablecoin.

> "A Derivative is a bet on whether a stock, or a bond, or a real estate asset, is going to go up or down. There's a winner and a loser. It's like betting on a horserace."
>
> Michael Hudson

In terms of liquidity, stablecoin 1.6 benefits from the underlying mutual fund's liquidity profile. Mutual funds are designed to provide investors with easy access to their funds through redemption mechanisms. This means that stablecoin holders can potentially convert their stablecoins back into the underlying assets of the mutual fund if they wish to exit their position. The liquidity of the mutual fund translates into the liquidity of stablecoin 1.6, allowing users to efficiently convert their stablecoins into other assets or fiat currencies.

While stablecoin 1.6 offers stability through a diversified mutual fund, it is essential to consider potential risks and challenges. The performance of the stablecoin is still subject to the performance of the underlying mutual fund's assets. Economic downturns, market volatility, or mismanagement of the mutual fund can impact the stability of stablecoin 1.6. Therefore, it is crucial for investors to carefully evaluate the reputation and track record of the mutual fund and its managers before investing in stablecoin 1.6.

Transparency is another critical aspect of stablecoin 1.6, as investors need visibility into the composition and performance of the mutual fund's portfolio. Regular reporting and disclosure of the fund's holdings, investment strategies, and performance metrics provide transparency to stablecoin holders. This transparency fosters trust and enables investors to make informed decisions regarding their investments.

In conclusion, stablecoin 1.6 backed by a diversified mutual fund offers stability through the well-managed and balanced nature of the fund's portfolio. The diversification across different asset classes and sectors helps to mitigate risks and protect against market volatility. The potential for consistent returns, combined with the liquidity provided by the underlying mutual fund, makes stablecoin 1.6 an attractive option for individuals seeking a stable digital asset. However, it is crucial for investors to carefully evaluate the mutual fund's reputation, performance, and transparency to ensure the long-term stability and reliability of stablecoin 1.6.

Liquidity of Asset Class

When it comes to the liquidity of stablecoin 1.6, which is backed by a mutual fund, it is important to consider its unique characteristics and compare them to other types of stablecoins that we have discussed. The liquidity of stablecoin 1.6 can be seen in two dimensions: the liquidity of the stablecoin itself and the liquidity of the underlying mutual fund. Let's examine each aspect in more detail.

Firstly, stablecoin 1.6 offers liquidity as a digital asset. It can be easily traded on cryptocurrency exchanges, allowing users to buy, sell, and transfer stablecoins with relative ease. This provides liquidity and flexibility for individuals and businesses that utilize stablecoin 1.6 for transactions or as a store of value. The digital nature of stablecoin 1.6 enables fast and seamless transactions, making it a convenient means of exchange.

Secondly, the liquidity of stablecoin 1.6 is closely tied to the liquidity of the underlying mutual fund. Mutual funds generally offer high liquidity to investors, allowing them to redeem their shares and access their funds when needed. The mutual fund managers strive to maintain a level of liquidity that meets investor demands and regulatory requirements. This means that stablecoin holders have the potential to convert their stablecoins into the underlying assets of the mutual fund if they wish to exit their position. This redemption mechanism provides a certain level of liquidity to stablecoin 1.6, enabling holders to access the value of their stablecoins.

Comparing the liquidity of stablecoin 1.6 backed by a mutual fund to other types of stablecoins, it is important to consider the underlying assets and the redemption mechanisms.

Stablecoins backed by fiat currency, such as stablecoin 1.0, generally offer a high level of liquidity. Fiat currencies are widely accepted and liquid, making it relatively easy for stablecoin holders to convert their stablecoins back into fiat currency. However, the liquidity of stablecoin 1.0 is limited to the availability of the fiat currency in the asset pool. In contrast, stablecoin 1.6 backed by a mutual fund can potentially offer higher liquidity due to the diversified nature of the underlying assets in the fund. The mutual fund can include a variety of liquid assets such as stocks, bonds, and other financial instruments, which may provide a more robust and flexible liquidity profile.

Stablecoins backed by digital assets, like stablecoin 1.4, can have varying levels of liquidity depending on the market conditions and the specific digital assets in the asset pool. Digital assets, such as cryptocurrencies, can be subject to high price volatility and liquidity fluctuations. This can affect the liquidity of stablecoin 1.4, especially during periods of market stress or extreme volatility. Stablecoin 1.6 backed by a mutual fund may provide a more stable liquidity profile due to the diversified nature of the mutual fund's assets.

> "I view derivatives as time bombs, both for the parties that deal in them and the economic system."
>
> Warren Buffet

It is important to note that while stablecoin 1.6 backed by a mutual fund may offer a certain level of liquidity, it is still subject to market conditions and the liquidity of the underlying assets. Economic downturns, market volatility, or illiquid assets in the mutual fund can potentially impact the liquidity of stablecoin 1.6. Additionally, the redemption process of the mutual fund may have certain limitations, such as redemption fees or redemption periods, which could affect the speed and ease of accessing the funds.

In conclusion, stablecoin 1.6 backed by a mutual fund can provide a reasonable level of liquidity due to its digital nature and the liquidity of the underlying mutual fund. The digital aspect of stablecoin 1.6 allows for easy and efficient trading on cryptocurrency exchanges, while the redemption mechanisms of the mutual fund provide potential access to the underlying assets. However, the liquidity of stablecoin 1.6 should be evaluated in comparison to other stablecoins, taking into account the specific characteristics of each type of stablecoin and the liquidity of their respective asset pools. It is essential for investors and users of stablecoin 1.6 to understand the potential risks and limitations related to liquidity and to consider their individual liquidity needs and preferences when engaging with stablecoin 1.6.

Required Asset Ratio for Stability

When it comes to the asset pool ratio of stablecoin 1.6, it is important to ensure a sufficient level of backing to maintain stability and provide confidence to the holders of the stablecoin. While the specific asset pool ratio may vary depending on the design and implementation of stablecoin 1.6, a commonly suggested ratio is to start with and never go below a 50% ratio.

The asset pool ratio represents the proportion of the stablecoin's total supply that is backed by the underlying assets held in the mutual fund. In the case of stablecoin 1.6, the asset pool consists of a diversified portfolio of assets within the mutual fund.

Starting with a 50% asset pool ratio means that at the launch of stablecoin 1.6, at least half of the stablecoin's total supply is backed by the assets in the mutual fund. This ensures that there is a substantial backing for the stablecoin, instilling confidence among users that their stablecoin holdings are adequately supported by tangible assets.

Maintaining the asset pool ratio at or above 50% is crucial for the stability of stablecoin 1.6. If the asset pool ratio were to fall below 50%, it could raise concerns about the stability and solvency of the stablecoin, potentially leading to a loss of trust and value erosion. By setting a minimum threshold of 50%, stablecoin 1.6 aims to maintain a robust level of backing, mitigating the risk of insolvency and protecting the value of the stablecoin for its holders.

The 50% asset pool ratio serves as a safeguard against market fluctuations and potential losses in the value of the underlying assets. It provides a buffer, ensuring that even if the value of the assets in the mutual fund experiences a decline, there is still sufficient backing to support the stablecoin at a predetermined ratio.

This conservative approach to the asset pool ratio aims to strike a balance between stability and flexibility. It allows for the potential growth and diversification of the mutual fund while ensuring a strong foundation of backing for stablecoin 1.6. The 50% asset pool ratio provides stability by limiting the potential impact of market volatility on the stablecoin's value and enhances the confidence of users in the stability and reliability of stablecoin 1.6 as a medium of exchange and store of value.

Furthermore, by setting a minimum threshold for the asset pool ratio, stablecoin 1.6 aims to comply with regulatory requirements and address concerns related to systemic risk. Regulators often emphasize the need for stablecoins to maintain sufficient backing to protect users and the broader financial system from potential disruptions. A 50% asset pool ratio demonstrates a commitment to robust risk management and compliance with regulatory standards.

The 50% asset pool ratio for stablecoin 1.6 is designed to provide a solid foundation of backing, ensuring stability, and instilling confidence among users. By maintaining a minimum ratio of 50%, stablecoin 1.6 aims to mitigate risks associated with market fluctuations, safeguard the value of the stablecoin, and comply with regulatory requirements. The asset pool ratio serves as an important parameter for the stability and reliability of stablecoin 1.6, contributing to its usability as a trusted medium of exchange and store of value.

Relationship between Holders and access to backing.

> "To say that derivative accounting is a sewer is an insult to sewage."
>
> Charlie Munger

When it comes to managing the asset fund of stablecoin 1.6, there are considerations regarding the expertise and experience required to ensure the effective management of the fund. While some stablecoin models involve the company managing its own fund, stablecoin 1.6 could benefit from being managed by a well-staffed mutual fund with years of experience in the industry. This approach brings several advantages, including professional expertise, regulatory compliance, and a focus on investor protection.

Managing an asset fund requires a deep understanding of financial markets, asset allocation strategies, risk management, and regulatory compliance. A well-staffed mutual fund typically comprises a team of investment professionals with diverse expertise and experience in managing funds. These professionals are dedicated to conducting thorough research, analysis, and due diligence to make informed investment decisions.

By entrusting the management of stablecoin 1.6's asset fund to an established mutual fund, investors can benefit from the professional expertise and knowledge of seasoned investment professionals. These professionals have experience navigating various market conditions, identifying attractive investment opportunities, and implementing sound investment strategies. Their expertise can contribute to maximizing the performance of the asset fund and enhancing the stability of stablecoin 1.6.

Furthermore, a well-staffed mutual fund is likely to have a robust framework in place for regulatory compliance. The mutual fund industry is subject to strict regulatory oversight to ensure investor protection, transparency, and fair practices. Mutual funds are required to comply with regulatory guidelines, disclosure requirements, and reporting standards. They must adhere to specific investment restrictions and risk management protocols.

By utilizing the expertise of a mutual fund with years of experience in regulatory compliance, stablecoin 1.6 can benefit from a well-defined framework that promotes transparency, accountability, and regulatory adherence. This can help instill trust among investors and ensure compliance with regulatory requirements in different jurisdictions where stablecoin 1.6 operates.

Moreover, the involvement of a well-staffed mutual fund can provide additional layers of protection for investors. Mutual funds are subject to governance structures, including independent boards of directors and custodians, designed to safeguard investors' interests. The presence of independent oversight helps ensure that the fund's operations are conducted in the best interest of investors and minimizes the risk of conflicts of interest.

In contrast, if the company managing its own asset fund, there may be concerns regarding potential conflicts of interest, lack of expertise, or insufficient resources dedicated solely to the management of the fund. The company's primary focus may be on the development and promotion of the stablecoin itself, which could divert attention and resources away from effective fund management.

Additionally, by engaging a well-staffed mutual fund, stablecoin 1.6 can benefit from the mutual fund's existing infrastructure, operational capabilities, and risk management systems. Mutual funds typically have established processes for asset valuation, portfolio monitoring, compliance, and reporting. These systems contribute to efficient and effective management of the fund, enhancing the stability and reliability of stablecoin 1.6.

In conclusion, entrusting the management of stablecoin 1.6's asset fund to a well-staffed mutual fund with years of experience brings significant advantages. It ensures access to professional expertise, regulatory compliance, and a focus on investor protection. By leveraging the knowledge and experience of investment professionals within the mutual fund industry, stablecoin 1.6 can benefit from sound investment strategies, robust risk management, and compliance with regulatory standards. This approach promotes stability, transparency, and investor confidence, contributing to the long-term success of stablecoin 1.6 as a reliable and trusted digital asset.

Susceptibility to Attacks

Stablecoin 1.6, being backed by a mutual fund, may face certain vulnerabilities and potential attacks that can impact its stability and security. While the use of a mutual fund provides diversification and stability, it is still essential to address the potential risks and develop appropriate safeguards to protect against attacks. Here are some key considerations regarding the susceptibility of stablecoin 1.6 to attacks:

1. Cybersecurity Risks: Like any digital asset, stablecoin 1.6 is exposed to cybersecurity risks. Hackers may attempt to exploit vulnerabilities in the technology infrastructure supporting the stablecoin or target the wallets and exchanges used for transactions. To mitigate these risks, it is crucial to implement robust security measures, such as multi-factor authentication, encryption protocols, and regular security audits. Ongoing monitoring and response mechanisms should be in place to detect and address any potential security breaches promptly.
2. Smart Contract Vulnerabilities: Stablecoin 1.6 may utilize smart contracts for various functions, such as asset tracking, fund management, and transaction processing. However, smart contracts are not immune to vulnerabilities. Flaws in the code or improper implementation can be exploited by malicious actors, leading to financial losses or manipulation of the stablecoin's operations. Rigorous code review, third-party audits, and regular updates to address vulnerabilities are crucial to mitigate these risks.
3. Market Manipulation: Stablecoin 1.6's value can be influenced by market forces, including supply and demand dynamics. Malicious actors may attempt to manipulate the market to artificially inflate or

deflate the stablecoin's value for their own gain. This can disrupt the stability of the stablecoin and undermine its credibility. Robust monitoring mechanisms, including market surveillance tools and regulatory oversight, are essential to detect and prevent market manipulation activities.

> "If you have trouble imagining a 20% loss in the stock market, then you should not be in stocks."
>
> John C. Boyle

4. Regulatory Compliance Risks: Stablecoin 1.6's operations may be subject to regulatory scrutiny and compliance obligations in various jurisdictions. Non-compliance with applicable laws and regulations can result in legal and reputational risks. It is important to engage legal experts and ensure that the stablecoin's operations adhere to relevant regulatory frameworks, including anti-money laundering (AML) and know-your-customer (KYC) requirements. Transparency and reporting mechanisms should be established to provide regulators with necessary information and demonstrate compliance.
5. Insider Threats: The involvement of individuals within the mutual fund managing stablecoin 1.6's asset pool introduces the risk of insider threats. Employees or individuals with access to sensitive information may misuse their privileges for personal gain or engage in fraudulent activities. Proper internal controls, segregation of duties, and stringent background checks can help mitigate these

risks. Regular audits and independent oversight can provide additional checks and balances to detect and prevent insider threats.

6. Legal and Regulatory Challenges: Stablecoin 1.6 may face legal and regulatory challenges, particularly as the regulatory landscape for digital assets continues to evolve. Uncertainty regarding the classification, taxation, and licensing requirements of stablecoins can create compliance challenges and legal risks. Engaging with regulators, staying abreast of regulatory developments, and actively participating in industry discussions can help navigate these challenges and ensure adherence to applicable laws.

To enhance the security and resilience of stablecoin 1.6, it is crucial to adopt a multi-layered approach to risk management. This includes implementing robust cybersecurity measures, conducting regular security audits, establishing strong governance structures, engaging in transparent reporting, and auditing practices, and actively collaborating with regulatory authorities. Additionally, fostering collaboration and knowledge-sharing within the industry can help identify and address emerging threats and vulnerabilities collectively.

By addressing these susceptibility factors and implementing comprehensive risk management strategies, stablecoin 1.6 can mitigate potential attacks and enhance its overall stability and security. This will instill confidence in users, promote wider adoption, and contribute to the long-term success of stablecoin 1.6 as a reliable and secure digital asset in the financial ecosystem.

Transparency

Transparency is a crucial aspect of stablecoin 1.6 to ensure trust, integrity, and accountability within the ecosystem. Given the increased adoption of stablecoins in various financial applications, it becomes paramount to establish transparency mechanisms that provide stakeholders with visibility into the underlying assets, operations, and decision-making processes. Here are key reasons why transparency is essential for stablecoin 1.6:

1. Asset Verification and Trust: Transparent reporting of the asset pool held by stablecoin 1.6 helps verify the adequacy and authenticity of the underlying assets. By disclosing information about the assets and their valuation, users can have confidence that the stablecoin is indeed backed by reliable and sufficient collateral. This transparency builds trust and mitigates the risk of fraudulent or misleading claims about the asset pool, enhancing the stability and credibility of stablecoin 1.6.
2. Investor Protection and Risk Management: Transparency enables investors and users to make informed decisions by understanding the composition, quality, and risk profile of the underlying assets. Clear disclosure of the investment strategy, risk management practices, and performance metrics empowers users to assess the stability and suitability of stablecoin 1.6 for their needs. Transparent reporting also helps regulators and auditors evaluate compliance with regulatory requirements and identify potential risks or misconduct.
3. Market Confidence and Stability: Stablecoin 1.6 operates within a broader market ecosystem that

includes participants, investors, and service providers. Transparency fosters market confidence by providing visibility into the stablecoin's operations, reserves, and stability mechanisms. When stakeholders have access to accurate and up-to-date information, they can assess the stability and reliability of stablecoin 1.6, reducing uncertainty and promoting market stability.

4. Auditing and Accountability: Transparent reporting facilitates independent audits of stablecoin 1.6's operations, financials, and compliance with regulatory standards. Independent audits provide an external validation of the stablecoin's claims, verifying the accuracy of asset holdings, valuation methodologies, and operational procedures. This auditing process ensures accountability and helps identify any discrepancies or mismanagement, protecting the interests of users and investors.
5. Regulatory Compliance: Transparency is vital for complying with existing and evolving regulatory frameworks. By providing detailed information about the stablecoin's operations, asset pool, and risk management practices, stablecoin 1.6 can demonstrate compliance with relevant regulations, such as anti-money laundering (AML) and know-your-customer (KYC) requirements. Transparent reporting helps regulators assess the stablecoin's adherence to regulatory standards, contributing to a more compliant and regulated financial ecosystem.
6. Market Integrity and Avoiding Fraud: Transparent reporting mechanisms can help identify and prevent fraudulent activities within stablecoin 1.6. By disclosing information about the underlying assets,

transactions, and operational processes, suspicious or anomalous activities can be identified and investigated promptly. This proactive approach to fraud detection and prevention enhances market integrity, protects users from potential scams, and fosters a safer financial environment.

7. Community Governance and Participation: Transparency promotes community involvement and governance in stablecoin 1.6. Open communication channels, public disclosure of decision-making processes, and community feedback mechanisms allow stakeholders to actively participate in the stablecoin's development and evolution. This inclusivity strengthens community trust, facilitates the resolution of conflicts or concerns, and encourages a sense of ownership among stakeholders.

To achieve transparency in stablecoin 1.6, various measures can be implemented. Regular and audited financial reporting, disclosure of asset composition and valuation methodologies, publication of governance structures and policies, and open communication channels with stakeholders are key elements. Leveraging blockchain technology can enhance transparency by providing immutable and publicly accessible transaction records. By embracing transparency, stablecoin 1.6 can establish itself as a reliable and trustworthy digital asset in the market. Transparency not only benefits users and investors but also contributes to the overall stability, integrity, and sustainability of the stablecoin ecosystem.

Viability in Major Markets

Stablecoin 1.6, based on its unique characteristics and value proposition, has experienced significant success in the markets. Here are several key factors that have contributed to its popularity and adoption:

1. Stability and Reliability: Stablecoin 1.6 offers a stable value proposition, which is a crucial attribute for users in volatile cryptocurrency markets. The use of a well-diversified asset pool, managed by a reputable mutual fund, provides stability, and reduces the risk of value fluctuations. This stability attracts investors and users who seek a reliable medium of exchange, store of value, and unit of account.
2. Diversification: Stablecoin 1.6 stands out from other stablecoins by leveraging the diversification benefits of a mutual fund. The asset pool comprises a wide range of assets, such as stocks, bonds, commodities, and currencies. This diversification helps to spread risk and minimize the impact of market fluctuations. By offering exposure to multiple asset classes, stablecoin 1.6 appeals to investors who value a balanced and diversified portfolio.

> "In reality the origin of the notion of derivatives is in the vague feeling of the mobility of things, and the greater or less speed with which phenomena take place."
>
> Emile Picard

3. Professional Fund Management: The involvement of well-established and experienced mutual fund managers in managing the asset pool of stablecoin 1.6 adds credibility and expertise to the stablecoin. Professional fund managers have extensive knowledge of investment strategies, risk management, and asset allocation. Their active management approach ensures that the asset pool is optimized for stability and long-term growth, instilling confidence in investors and users.
4. Market Confidence: Stablecoin 1.6 has gained market confidence through its transparent and regulated operations. The involvement of regulated mutual fund managers provides oversight and compliance with industry standards and regulatory requirements. This regulatory compliance and transparency foster trust among users, investors, and regulators, resulting in increased market adoption and acceptance.
5. Enhanced Liquidity: Stablecoin 1.6 benefits from the liquidity of the underlying assets held within the mutual fund. The ability to quickly convert the assets into cash or other liquid instruments allows stablecoin 1.6 to maintain a high level of liquidity. This liquidity provides users with the confidence

that they can easily transact with the stablecoin and access their funds when needed.

6. Market Demand: The success of stablecoin 1.6 can also be attributed to the increasing demand for stablecoins in general. As the cryptocurrency market expands and matures, stablecoins have emerged as a crucial bridge between traditional financial systems and digital assets. Stablecoin 1.6, with its stability, diversification, and professional management, meets the growing demand for a reliable and secure digital asset.
7. Global Reach: Stablecoin 1.6's success is not limited to a specific geographic region but extends to a global market. The ability to transact and hold stablecoin 1.6 across borders with minimal barriers facilitates international trade, remittances, and cross-border transactions. The global accessibility and utility of stablecoin 1.6 have contributed to its widespread adoption and success.
8. Partnerships and Integrations: Stablecoin 1.6 has formed strategic partnerships and integrations with various financial institutions, payment processors, and digital platforms. These collaborations enhance the usability and acceptance of stablecoin 1.6 in different sectors, such as e-commerce, remittances, and decentralized finance (DeFi). By expanding its ecosystem and integrations, stablecoin 1.6 gains access to a broader user base and market opportunities.

In summary, stablecoin 1.6 has experienced success in the markets due to its stability, diversification, professional fund management, regulatory compliance, enhanced liquidity, global reach, market demand, and strategic partnerships. These factors have collectively positioned stablecoin 1.6 as a reliable, transparent, and widely accepted digital asset that caters to the needs of investors, users, and businesses in the evolving cryptocurrency landscape.

Conclusion

In conclusion, Stablecoin 1.6 has emerged as a prominent and successful digital asset in the cryptocurrency market. Its unique value proposition, based on a mutual fund-backed model, has proven to be highly attractive to investors, users, and businesses. The stability, diversification, and professional management of the asset pool contribute to its reputation as a reliable and secure digital currency.

Stablecoin 1.6's success can be attributed to several key factors. Firstly, the stability it offers addresses the volatility concerns that are inherent in the cryptocurrency market. By leveraging a well-diversified asset pool, Stablecoin 1.6 minimizes the risk of value fluctuations and provides users with a reliable medium of exchange, store of value, and unit of account.

The involvement of experienced mutual fund managers in managing the asset pool ensures that it is optimized for stability and long-term growth. Their expertise in investment strategies, risk management, and asset allocation instills confidence in investors and users, making Stablecoin 1.6 a trusted digital asset.

Transparency and regulatory compliance are crucial aspects of Stablecoin 1.6's success. The involvement of regulated mutual fund managers and adherence to industry standards and regulations enhance market confidence and foster trust among users, investors, and regulators. This transparency enables users to have a clear understanding of the underlying assets and their management.

Another key factor is the enhanced liquidity of Stablecoin 1.6. The ability to quickly convert the underlying assets into cash or other liquid instruments ensures that the stablecoin remains easily tradable and accessible to users. This liquidity feature enhances its usability and acceptance in various sectors, including international trade, remittances, and decentralized finance.

Stablecoin 1.6's success is not limited to a specific region but has gained global reach. Its ability to facilitate cross-border transactions with minimal barriers has positioned it as a valuable digital asset in the international market. Strategic partnerships and integrations with financial institutions, payment processors, and digital platforms have further expanded its ecosystem and user base.

Overall, Stablecoin 1.6's stability, diversification, professional fund management, transparency, liquidity, global reach, and strategic partnerships have propelled its success in the markets. It offers a reliable and secure digital asset solution that addresses the needs of investors, users, and businesses in the evolving cryptocurrency landscape.

As the cryptocurrency market continues to evolve, Stablecoin 1.6's position as a trusted and widely accepted digital asset is likely to strengthen. Its value proposition as a stable and well-diversified digital currency sets it apart from other stablecoins and positions it for continued success in the future.

Chapter 15: Stablecoin 1.7 (Collections of Properties)

"The Wise young man or wage earner of today invests his money in real estate."[37]

Or

"He is not a full man who does not own a piece of land."[38]

Stablecoin 1.7 is an analogy within the pantheon of stablecoins. The concept that property has an inherent value is one that everyone is taught, but because the governments of the world are buying up more and more of the property, they really are not highlighting property as one of the things that everyone should have. There is a constant truth about property that was eloquently expressed by Samuel Clement (Mark Twain), "Buy land, they're not making it anymore."[39] While some nations such as Australia and China are trying to make new land (successfully), largely the land that is out there in the world today is all the land that there is going to be, which makes it an interesting investment opportunity, especially for digital investors.

In this chapter, we will follow the same pattern that we follow in all the chapters in this part of the book. First, we look at the stability of the asset class on which the token at this level is based. The asset class is very important because it is the 'pegged' underpinning of the system. Next, we look at the liquidity of the asset class, how fast can the

---

[37] Andrew Carnegie at https://garrettsrealty.com/blog/real-estate-investing-quotes.html.
[38] Hebrew Proverb at https://garrettsrealty.com/blog/real-estate-investing-quotes.html.
[39] Id.

asset be liquidated if there is a 'run' on the token. As a matter of course, we will then look at the asset ratio needed to maintain stability. The media tends to insist that there needs to be a 1:1 ratio, and letter agencies tend to echo this false statement; each class has a ratio that is needed for the token to be reliable in its function.

> "Real estate cannot be lost or stolen, nor can it be carried away. Purchases with common sese, paid for in full, and managed with reasonable care, it is about the safest investment in the world."
>
> Franklin D. Roosevelt

Also, as a matter of course, it is always important to look at the relationship between holders/creators of the currency and the asset pool. The more complex the system the more analysis needs to be done on how the assets are being managed. While a 1:1 ratio for stablecoin 1.0 can be managed by the local bank of the creator, more complex systems require funds, investment firms, and even multiple subsidiaries to ensure that the asset pool maintains value. The analysis of who holds the asset pool then leads to an analysis of susceptibility of to attack. The more complex the asset, the more stable it is- however, it does need to be protected from attacks. The discussions within these sections of each chapter will be quite similar. Transparency is also an issue that varies slightly as you look at different types of stablecoins, once again the more complex the tokenomics- the more transparent the company needs to be. Finally, there needs to be an analysis of viability in major markets. This results in a bimodal, cyclic model for the analysis of the system, viability seems to be highest for simple and highly complex models, with intermediate models being lukewarm in their reception.

Land is a powerful investment tool, with a return of about 16% in many areas of the world. Collaboratively owning land in a large cryptocurrency fund can be a viable way to see increases across the board. While this is a "good investment" it is not very liquid, which means that when there is a run on stablecoin 1.7, the run is hard to cut off. However, as more and more people flee the poorly run cities of the world, we are seeing more people who want that little slice of heaven, and thus we are seeing an increase in the value of land around the world.

Stability of Asset Class

Stablecoin 1.7, based on a portfolio of land, offers a unique approach to achieving stability in the realm of digital assets. Unlike traditional stablecoins backed by fiat currencies or other financial instruments, Stablecoin 1.7 derives its stability from the underlying value of land assets. This alternative model presents both advantages and challenges when compared to other stablecoin variations.

One of the key aspects of Stablecoin 1.7 is its stability through tangible assets. Land, as a physical asset, tends to have a relatively stable value over time. While it may experience fluctuations based on market dynamics and economic conditions, the long-term value of land tends to appreciate or remain relatively constant. This inherent stability provides a solid foundation for Stablecoin 1.7, making it an attractive option for individuals and businesses seeking a reliable digital currency.

Furthermore, the use of land as a backing asset offers diversification benefits. Unlike fiat-based stablecoins that are subject to the economic and monetary policies of a single country, Stablecoin 1.7's land portfolio can span multiple geographic regions. This diversification reduces the concentration risk associated with a specific economy or jurisdiction, enhancing the stability and resilience of the stablecoin.

> "The major fortunes in America have been made in land."
>
> John D. Rockefeller

However, it is important to acknowledge that stablecoin 1.7's stability based on land assets may also introduce certain challenges. The valuation of land can vary significantly depending on the location, market conditions, and other factors. This means that Stablecoin 1.7 may experience differing levels of stability across different land portfolios. It requires careful management and ongoing monitoring to ensure that the land assets backing the stablecoin are adequately diversified and managed to mitigate any potential risks.

Liquidity is another crucial factor to consider when evaluating stablecoin 1.7's stability. Land, being an illiquid asset, may present challenges in terms of quickly converting it into cash or other liquid forms. This could potentially impact the liquidity and tradability of the stablecoin, particularly during times of increased demand or in situations where immediate access to funds is required. It is essential to strike a balance between the stability derived from land assets and the need for sufficient liquidity to support the usability and acceptance of stablecoin 1.7.

Additionally, the stability of stablecoin 1.7 may be influenced by external factors such as regulatory frameworks, property rights, and land ownership laws in different jurisdictions. The legal and regulatory landscape surrounding land ownership and transactions can vary significantly across countries, which may introduce additional complexities and potential risks. It is crucial to navigate these legal and regulatory challenges to ensure the stability and viability of stablecoin 1.7 in various markets.

In terms of the required asset ratio for stablecoin 1.7, it would depend on the specific land portfolio and its valuation. While it may be challenging to provide an exact ratio, it is essential to ensure that the asset pool is adequately capitalized to withstand market fluctuations and maintain stability. A comprehensive and professional valuation of the land assets is necessary to determine the appropriate asset ratio and ensure the stablecoin's stability in the long run.

Overall, stablecoin 1.7 based on a portfolio of land assets offers a unique approach to achieving stability in the digital asset space. Its stability derived from tangible assets provides an alternative to traditional stablecoins backed by fiat currencies or financial instruments. However, it also introduces challenges related to land valuation, liquidity, and legal and regulatory considerations. By addressing these challenges and implementing robust risk management practices, stablecoin 1.7 can provide a stable and reliable digital currency option backed by real-world land assets.

Liquidity of Asset Class

Stablecoin 1.7, which is backed by a portfolio of land assets, presents a unique challenge when it comes to liquidity. Unlike other types of assets, such as fiat currencies or highly liquid financial instruments, land investments are known for their inherent illiquidity. This illiquidity can create certain challenges for Stablecoin 1.7, affecting its ability to provide immediate access to funds and potentially impacting the stability of the stablecoin.

Land is a long-term investment that typically requires time to convert into cash or other liquid forms. The process of selling land and realizing its value can be complex and time-consuming. Factors such as market conditions, property location, and the legal and regulatory framework surrounding land transactions can all influence the speed at which land can be sold and converted into cash.

> "Where justice is denied, where poverty is enforced, where ignorance prevails, and where any one class is made to feel that society is an organized conspiracy to oppress, rob, and degrade them, neither persons nor property will be safe."
>
> Frederick Douglass

The illiquidity of land assets poses challenges for Stablecoin 1.7 in terms of meeting immediate redemption demands or accommodating sudden increases in demand for the stablecoin. In situations where stablecoin holders seek to convert their stablecoin holdings into cash quickly, the illiquidity of the land portfolio backing the stablecoin can limit the speed at which these transactions can be executed.

This lack of liquidity can impact the stability of Stablecoin 1.7 in several ways. Firstly, if there is a sudden surge in demand for the stablecoin and holders wish to redeem their stablecoins for cash, the limited liquidity of the land assets may make it difficult for the stablecoin issuer to meet these redemption requests in a timely manner. This could potentially erode trust in the stablecoin and lead to a loss of confidence among users.

Furthermore, the illiquidity of land investments can result in challenges for price discovery and valuation of the stablecoin. Unlike liquid assets that have readily available market prices, land assets often require independent appraisals and valuations to determine their current market value. The infrequent and time-consuming nature of land transactions can make it challenging to establish real-time and accurate valuations for the land portfolio backing Stablecoin 1.7. This lack of transparent and up-to-date pricing information can undermine the stability and trustworthiness of the stablecoin.

To address the liquidity challenges associated with land investments, several measures can be implemented. Firstly, the stablecoin issuer can establish a reserve fund or liquidity pool to ensure that there are sufficient liquid assets available to meet potential redemption requests. This fund can be composed of highly liquid financial instruments or short-term investments that can be easily converted into cash.

Additionally, the stablecoin issuer can explore partnerships with financial institutions or liquidity providers that have access to capital markets and can provide liquidity in times of high demand. These partners can facilitate the conversion of land assets into cash or provide lines of credit to meet redemption requests promptly.

Moreover, the stablecoin issuer can implement mechanisms to incentivize long-term holding of the stablecoin and discourage short-term trading or frequent redemption. By encouraging stability and discouraging excessive liquidity demands, the stablecoin can better manage the inherent illiquidity of the land asset portfolio.

It is important to note that while land investments may present liquidity challenges, they offer other advantages such as long-term stability and potential appreciation in value. The illiquidity of land assets can be viewed as a trade-off for the stability and resilience they provide to the stablecoin. The key is to strike a balance between the liquidity needs of stablecoin holders and the long-term investment nature of land assets.

In conclusion, Stablecoin 1.7 backed by a portfolio of land assets faces liquidity challenges due to the inherent illiquidity of land investments. The illiquid nature of land can affect the stability of the stablecoin by limiting immediate access to funds and impacting price discovery and valuation. However, by implementing measures such as reserve funds, strategic partnerships, and incentives for long-term holding, the stablecoin issuer can mitigate these challenges and ensure the stability and viability of Stablecoin 1.7 in the market. It is crucial to strike a balance between liquidity needs and the long-term stability benefits that land investments provide to create a robust and resilient stablecoin.

Required Asset Ratio for Stability

Stablecoin 1.7, which is backed by a portfolio of land assets, requires careful management of the asset-to-token ratio to ensure its stability. The asset-to-token ratio represents the proportion of the underlying asset value relative to the total number of stablecoin tokens in circulation. In the case of Stablecoin 1.7, maintaining an asset-to-token ratio of approximately 60-62% is crucial for maintaining stability and confidence in the stablecoin.

The asset-to-token ratio is a critical factor in the stability of any stablecoin, as it determines the level of collateralization and ensures that there are sufficient assets backing the value of the tokens in circulation. A higher asset-to-token ratio provides a stronger buffer against market fluctuations and enhances the stability of the stablecoin.

> "Nobody spends somebody else's money as carefully as he spends his own. Nobody uses somebody else's resources as carefully as he uses his own. So, if you want efficiency and effectiveness, if you want knowledge to be properly utilized, you have to do it through the means of private property."
>
> Milton Freidman

For Stablecoin 1.7, the required asset-to-token ratio of 60-62% is relatively higher compared to other stablecoin models. This is due to the nature of the underlying land assets, which are typically considered less liquid and may require more conservative collateralization. The higher ratio serves to safeguard against potential price fluctuations and illiquidity associated with land investments.

The stability of Stablecoin 1.7 is further enhanced by the relatively quick maturity of land assets compared to other forms of stablecoin backing. While land investments are generally considered long-term assets, they tend to have a more predictable growth pattern and appreciation in value over time. This characteristic allows the land asset portfolio backing Stablecoin 1.7 to reach maturity more quickly, ensuring stability and reducing the risk of value depreciation.

By setting a slightly higher asset-to-token ratio, Stablecoin 1.7 can achieve a balance between stability and liquidity. The higher ratio provides a cushion to absorb market fluctuations and potential liquidity demands, ensuring that the stablecoin remains well-collateralized and maintaining confidence among token holders.

It is worth noting that the asset-to-token ratio of Stablecoin 1.7 can be influenced by factors such as the type and location of the land assets in the portfolio. Different types of land, such as residential, commercial, or agricultural, may have varying liquidity profiles and appreciation potential. Additionally, the location of the land assets in different markets can impact their value and liquidity.

To manage the asset-to-token ratio effectively, the stablecoin issuer of Stablecoin 1.7 needs to carefully monitor the performance and valuation of the land assets in the portfolio. Regular assessments and appraisals of the land assets will help ensure that the asset-to-token ratio remains within the desired range.

Furthermore, ongoing due diligence and risk management practices are crucial for maintaining the stability of Stablecoin 1.7. The stablecoin issuer should conduct thorough analysis of the land assets, including factors such as title deeds, legal and regulatory considerations, and environmental assessments. This comprehensive approach mitigates potential risks and strengthens the stability and resilience of the stablecoin.

In summary, Stablecoin 1.7, backed by a portfolio of land assets, requires a target asset-to-token ratio of approximately 60-62% to maintain stability. The higher ratio accounts for the unique characteristics of land investments, including their relatively lower liquidity and longer-term growth potential. By setting a slightly higher ratio, Stablecoin 1.7 can ensure robust collateralization, absorb market fluctuations, and enhance investor confidence. Continuous monitoring, due diligence, and risk management practices are vital to maintaining the desired asset-to-token ratio and ensuring the stability of Stablecoin 1.7 in the market.

Relationship between Holders and access to backing.

In the case of Stablecoin 1.7, which is backed by a portfolio of land assets, it is prudent for the assets to be held in a trust rather than by the company that created the token. The use of a trust structure offers several advantages in terms of transparency, security, and independence, which ultimately contribute to the stability and credibility of the stablecoin.

> "Don't want to buy real estate. Buy real estate and wait."
>
> Will Rodgers

One of the primary reasons for holding the assets in a trust is to ensure transparency and mitigate potential conflicts of interest. By entrusting the assets to an independent trust, the ownership and management of the land portfolio are separated from the stablecoin issuer. This separation minimizes the risk of misappropriation or misuse of assets and provides clear accountability and oversight.

A trust structure also enhances the security of the assets backing Stablecoin 1.7. Land assets are tangible and valuable holdings that require protection against theft, fraud, or other threats. By placing the assets in a trust, specific safeguards and legal protections can be implemented to safeguard the assets from unauthorized access or misuse. Trust structures often involve robust legal mechanisms and governance frameworks that enhance the security and integrity of the asset holdings.

Moreover, holding the assets in a trust provides a layer of insulation against the potential bankruptcy or financial distress of the stablecoin issuer. In the event that the issuing company faces financial difficulties, the assets held in a trust are separate and protected from the company's liabilities. This separation safeguards the value of the assets and ensures that they remain available to support the stability and redeemability of the stablecoin.

A trust structure also contributes to the long-term stability of Stablecoin 1.7 by providing a clear and dedicated custodian for the assets. The trust is managed by professional trustees who have fiduciary responsibilities to act in the best interest of the beneficiaries, in this case, the stablecoin holders. These trustees are experienced in managing and safeguarding assets, ensuring proper valuation, and making informed decisions regarding the asset portfolio.

Additionally, holding the assets in a trust enhances investor confidence and credibility in Stablecoin 1.7. Trusts are subject to strict regulatory oversight and reporting requirements, ensuring that the asset holdings and their management are transparent and accountable. This transparency builds trust among stablecoin holders and potential investors, as they can have visibility into the underlying asset portfolio and the governance mechanisms in place.

The use of a trust structure also provides legal clarity and stability in the event of regulatory changes or potential disputes. Trusts are established under specific legal frameworks that provide clear guidelines for the management and distribution of assets. This legal structure enhances stability, reduces regulatory risks, and ensures compliance with relevant laws and regulations.

Furthermore, holding the assets in a trust facilitates effective risk management. The trustees of the trust can diversify the land asset portfolio to minimize concentration risk and maximize overall stability. They can assess the performance of the assets, monitor market conditions, and make informed decisions regarding the composition and management of the asset pool. This active risk management approach ensures that the assets backing Stablecoin 1.7 remain robust and resilient.

In summary, holding the assets of Stablecoin 1.7 in a trust rather than by the issuing company offers numerous benefits. A trust structure enhances transparency, security, and independence, mitigates conflicts of interest, and insulates the assets from the financial risks faced by the stablecoin issuer. Trusts provide legal clarity, regulatory oversight, and governance mechanisms that enhance investor confidence and contribute to the long-term stability of the stablecoin. By employing a trust structure, Stablecoin 1.7 can ensure the integrity and credibility of the asset holdings, ultimately supporting the stability and trustworthiness of the stablecoin in the market.

Susceptibility to Attacks

Stablecoin 1.7, which is backed by a portfolio of land assets, exhibits limited susceptibility to attacks due to its unique characteristics and the inherent nature of land as an investment. While no system is completely immune to attacks, the specific properties of land as an asset class make Stablecoin 1.7 relatively resilient to certain types of attacks commonly seen in other stablecoin models.

One of the primary reasons for the limited susceptibility to attacks is the physical nature of land assets. Unlike digital assets or financial instruments, land is tangible and cannot be easily manipulated or altered. Land ownership is governed by well-established property rights and legal frameworks, which provide a solid foundation for stability and protection against attacks.

One common attack vector in the stablecoin space is the manipulation of the underlying assets to artificially inflate or deflate the value of the stablecoin. In the case of Stablecoin 1.7, the land assets held in the portfolio are relatively stable and less susceptible to price manipulation compared to digital assets or volatile financial instruments. Land values tend to exhibit more gradual and stable growth or decline over time, reducing the likelihood of sudden and drastic value fluctuations that could be exploited by attackers.

> "Or houses are such unwieldy property that we are often imprisoned rather than housed by them."
>
> Henry David Thoreau

Another advantage of land-backed stablecoins like Stablecoin 1.7 is the limited exposure to technological vulnerabilities. Many attacks on stablecoins involve exploiting vulnerabilities in digital platforms or smart contracts. However, with land assets, the primary risks lie in physical threats such as theft, encroachment, or natural disasters. While these risks exist, they are typically managed through established insurance mechanisms and physical security measures, which contribute to the overall stability and resilience of the land asset portfolio.

Additionally, the use of a diversified land portfolio in Stablecoin 1.7 further enhances its resistance to attacks. By holding a variety of land assets across different geographic locations and property types, the stablecoin mitigates concentration risk. Attackers would need to target multiple properties simultaneously, which significantly increases the complexity and reduces the likelihood of a successful attack.

Moreover, the limited susceptibility to attacks is reinforced by the stringent due diligence and risk assessment processes associated with land investments. Land acquisition involves thorough research, title verification, and legal assessments to ensure the integrity and authenticity of the assets. This rigorous process minimizes the potential for fraudulent or compromised assets to be included in the portfolio, reducing the vulnerability to attacks.

Furthermore, stablecoin 1.7 can leverage established property rights systems and legal frameworks to protect against attacks. Land ownership is typically governed by national or regional laws, ensuring clear ownership records and dispute resolution mechanisms. These legal protections provide a strong defense against fraudulent claims or attempts to manipulate ownership rights, bolstering the stability and security of the land-backed stablecoin.

While stablecoin 1.7 exhibits limited susceptibility to attacks, it is essential to acknowledge that no system is entirely immune. It is crucial for the stablecoin issuer to implement robust security measures, including physical security for land assets, secure custody practices, and comprehensive risk management protocols. Ongoing monitoring and auditing of the land portfolio's integrity and compliance with regulatory requirements are also vital to identify and address potential vulnerabilities.

In summary, stablecoin 1.7, backed by a portfolio of land assets, demonstrates limited susceptibility to attacks due to the physical nature of land, the stability of land values, and the established legal frameworks governing land ownership. The diversified nature of the land portfolio, rigorous due diligence processes, and adherence to risk management protocols further enhance the stability and resilience of the stablecoin. While no system is impervious to attacks, stablecoin 1.7 leverages the unique characteristics of land investments to mitigate specific attack vectors commonly observed in other stablecoin models.

Transparency

Transparency is a critical factor in the stability and credibility of Stablecoin 1.7, which is backed by a portfolio of land assets. The nature of land investments necessitates a high level of transparency to instill trust among token holders and stakeholders in the stablecoin ecosystem.

First and foremost, land is a tangible and physical asset with clear ownership rights and legal frameworks. Transparency in the form of clear documentation and public records of land ownership is essential to ensure the integrity of the land-based assets backing Stablecoin 1.7. Token holders and investors need access to information regarding the land properties, including their location, boundaries, legal titles, and any encumbrances or restrictions that may impact the value or use of the assets.

Transparent land ownership records play a crucial role in preventing fraud, double-counting, or misrepresentation of assets. By making land ownership information publicly available, stakeholders can verify the authenticity and legal status of the underlying assets, ensuring that the stablecoin is backed by genuine and properly registered land properties.

Furthermore, transparency in the management of the land portfolio is vital to address potential conflicts of interest and ensure proper governance. Land investments often involve multiple stakeholders, such as investors, land managers, and custodians. Transparent reporting on the performance of the land assets, financial statements, and audits helps ensure accountability and mitigate the risk of mismanagement or improper handling of the assets.

In addition to legal and financial transparency, environmental and social transparency is also crucial for a land-backed stablecoin like Stablecoin 1.7. Land holdings can have significant environmental and social impacts, such as deforestation, habitat destruction, displacement of communities, or violation of indigenous rights. Transparent reporting and disclosure mechanisms allow stakeholders to assess and monitor the environmental and social sustainability of the land assets.

Transparency also extends to the valuation of land assets. Appropriate and reliable valuation methodologies should be used to determine the fair value of the land properties. Transparent valuation processes provide clarity on how the stablecoin's value is derived from the underlying land assets, reducing the risk of overvaluation or undervaluation, and ensuring an accurate representation of the stablecoin's stability.

> "Everyone wants a piece of land. It's the only sure investment. It can never depreciate like a car or washing machine, Land will only double in 10 years."
>
> Sam Shepard

Moreover, transparency in the land-based stablecoin ecosystem enhances market confidence and facilitates regulatory compliance. Regulators, financial institutions, and other participants in the ecosystem require transparency to assess the stability and compliance of Stablecoin 1.7. Open and transparent communication with regulators can help ensure that the stablecoin operates within legal frameworks and regulatory requirements.

Furthermore, transparency can attract institutional investors who prioritize disclosure and visibility into the underlying assets. Institutional investors often have stringent due diligence processes and risk management frameworks that require access to detailed information and transparency in the investment structure. By providing transparency, Stablecoin 1.7 can increase its appeal to these investors and potentially enhance liquidity and market adoption.

It is worth noting that achieving transparency in a land-based stablecoin does come with challenges. Land ownership records and information may vary across jurisdictions, and establishing standardized and accessible data sources can be complex. However, advancements in technology, such as blockchain-based land registries or digital land platforms, can facilitate transparency by providing immutable and transparent records of land ownership and transactions.

In summary, transparency is paramount for Stablecoin 1.7 due to its reliance on land-based assets. Transparent land ownership records, financial reporting, environmental and social disclosures, and valuation methodologies are necessary to build trust, ensure accountability, and facilitate regulatory compliance. By embracing transparency, Stablecoin 1.7 can enhance market confidence, attract institutional investors, and demonstrate its commitment to maintaining the stability and integrity of the stablecoin ecosystem.

Viability in Major Markets

Stablecoin 1.7, backed by a portfolio of land assets, has experienced notable success in the market, garnering attention, and interest from various participants in the financial ecosystem. Its unique value proposition and characteristics have contributed to its success and widespread adoption.

One key factor driving the success of Stablecoin 1.7 is the perceived stability and security offered by its underlying land assets. Land, as a tangible and finite resource, has long been considered a safe investment option. It provides intrinsic value, serving as a store of wealth and a hedge against inflation. By utilizing land as the backing for Stablecoin 1.7, it provides reassurance to investors and token holders who seek stability in their digital assets.

The market success of Stablecoin 1.7 can also be attributed to the diversification and potential for appreciation inherent in a portfolio of land assets. The stablecoin benefits from the value appreciation of the land holdings over time, which can act as a positive catalyst for its market performance. As land assets tend to have long-term value appreciation, Stablecoin 1.7 holders can potentially enjoy capital gains, making it an attractive investment option.

Moreover, Stablecoin 1.7's success can be attributed to its compatibility and integration within the existing financial infrastructure. It offers seamless integration with decentralized finance (DeFi) platforms, allowing users to leverage the stability of land-backed stablecoins while benefiting from the efficiency and flexibility of blockchain technology. The interoperability of Stablecoin 1.7 with various blockchain networks and smart contract platforms enables its use in a wide range of applications and financial services, further enhancing its market acceptance.

> "Freedom and Property rights are inseparable. You cannot have one without the other."
>
> George Washington

Additionally, the transparency and trust provided by Stablecoin 1.7 have played a significant role in its success. Land investments, being tangible and legally regulated assets, require a high level of transparency and due diligence. Stablecoin 1.7 has implemented robust reporting mechanisms, providing investors and stakeholders with detailed information about the land portfolio, its performance, and valuation methodologies. This transparency fosters trust and confidence, attracting investors who prioritize transparency and accountability.

Furthermore, Stablecoin 1.7's success is fueled by the increasing interest and demand for digital assets and stablecoins in the market. The rise of cryptocurrencies and the growing acceptance of blockchain technology have created a favorable environment for stablecoins. Investors and users are seeking stable and reliable digital assets that can bridge the gap between traditional finance and the digital economy. Stablecoin 1.7 fills this need by combining the stability of land assets with the benefits of blockchain technology, positioning itself as a trusted and efficient medium of exchange and store of value.

The success of Stablecoin 1.7 can also be attributed to its ability to address specific market niches and cater to the needs of different user segments. For example, it appeals to individuals and institutions who seek exposure to real estate and land assets without the high entry barriers or liquidity constraints associated with direct property investments. Stablecoin 1.7 provides a liquid and easily accessible avenue for investors to participate in the land market, allowing for diversification and risk mitigation.

Furthermore, the success of Stablecoin 1.7 is amplified by partnerships and collaborations with established financial institutions, real estate firms, and land developers. These partnerships provide credibility, market access, and expertise, enabling Stablecoin 1.7 to tap into existing networks and leverage established relationships. Collaborations with reputable entities enhance market confidence and facilitate the adoption of Stablecoin 1.7 as a reliable and trusted stablecoin in the industry.

Lastly, regulatory compliance and adherence to legal frameworks have contributed to the success of Stablecoin 1.7. By operating within regulatory guidelines, Stablecoin 1.7 has gained the trust of regulators, financial institutions, and institutional investors. Its commitment to compliance ensures that it meets regulatory requirements, mitigates risks associated with money laundering and illicit activities, and provides a secure and compliant stablecoin solution.

Stablecoin 1.7 has experienced substantial success in the market due to several key factors. The stability and security offered by land assets, the diversification and potential for appreciation, compatibility with existing financial infrastructure, transparency, and trust, increasing demand for stablecoins, targeted market niches, strategic partnerships, and regulatory compliance have all contributed to its success. As the digital asset landscape continues to evolve, Stablecoin 1.7's innovative approach combining land assets and blockchain technology positions it as a promising and sought-after stablecoin solution in the market.

Conclusion

In conclusion, Stablecoin 1.7 has proven to be a compelling and successful stablecoin in the market. Its unique proposition of being backed by a portfolio of land assets has resonated with investors, offering stability, diversification, and long-term value appreciation. The liquidity challenges associated with land-based assets have been effectively managed through careful portfolio management and innovative mechanisms.

Stablecoin 1.7 has addressed the need for stability in the volatile cryptocurrency market, providing a reliable digital asset that combines the benefits of blockchain technology with the tangible value of land. By utilizing land as a backing asset, Stablecoin 1.7 offers intrinsic value and serves as a hedge against inflation, attracting investors looking for a secure store of wealth.

The success of Stablecoin 1.7 can be attributed to its seamless integration within the existing financial infrastructure. Its compatibility with decentralized finance platforms and blockchain networks allows for efficient transactions, interoperability, and a wide range of use cases. This has facilitated its adoption and utilization in various financial services and applications, further enhancing its market acceptance.

Transparency has been a key element in the success of Stablecoin 1.7. The trust and confidence of investors and stakeholders are fostered through robust reporting mechanisms and detailed information about the land portfolio. By providing transparency in the valuation methodologies, performance, and compliance, Stablecoin 1.7 ensures accountability and builds credibility in the market.

Furthermore, the success of Stablecoin 1.7 is driven by the increasing demand for stablecoins and digital assets. Investors are seeking reliable and trustworthy digital assets that bridge the gap between traditional finance and the digital economy. Stablecoin 1.7 caters to this demand by offering stability, liquidity, and exposure to the real estate market without the constraints of direct property investments.

Partnerships and collaborations have played a vital role in the success of Stablecoin 1.7. Collaborations with established financial institutions, real estate firms, and land developers have provided credibility, market access, and expertise. These partnerships have expanded the reach of Stablecoin 1.7, facilitated adoption, and enhanced its reputation as a reliable stablecoin solution.

Lastly, regulatory compliance has been a cornerstone of Stablecoin 1.7's success. By operating within regulatory frameworks and ensuring adherence to legal requirements, Stablecoin 1.7 has gained the trust of regulators, financial institutions, and institutional investors. Its commitment to compliance has minimized risks associated with money laundering and illicit activities, making it a secure and compliant stablecoin solution.

Overall, Stablecoin 1.7 has demonstrated its ability to provide stability, security, and value appreciation through its unique land-based asset portfolio. Its success in the market can be attributed to factors such as stability, compatibility, transparency, increasing demand for stablecoins, strategic partnerships, and regulatory compliance. As the digital asset landscape continues to evolve, Stablecoin 1.7 is well-positioned to continue its growth and solidify its position as a trusted and sought-after stablecoin in the industry.

Chapter 16: Stablecoin 1.8 (Based on Derivatives)

"Derivatives are financial weapons of mass destruction."[40]

Or,

"Derivatives in and of themselves are not evil. There's nothing evil about how they're traded, how they're accounted for, and how they're financed, like any other financial instrument, if done properly."[41]

People in the investing world either love or hate derivatives. For those that do not know, derivatives are simply additional ways to trade more traditional assets. If you think that a value is going to go up, you buy a call option. You think a value is going to go down, you buy a put option. You think that the income streams are going to be better for something you do not have than something you do have, find somebody that believes the converse and create a swap with them. There are many ways to trade assets and the more complex they are the riskier they are. This is why some hate derivatives, and some investors live and die by how they work.

In this chapter, we will follow the same pattern that we follow in all the chapters in this part of the book. First, we look at the stability of the asset class on which the token at this level is based. The asset class is very important because it is the 'pegged' underpinning of the system. Next, we look at the liquidity of the asset class, how fast can the

---

[40] Warren Buffet at https://www.brainyquote.com/topics/derivatives-quotes.
[41] James Chanos at https://www.brainyquote.com/topics/derivatives-quotes.

asset be liquidated if there is a ‘run’ on the token. As a matter of course, we will then look at the asset ratio needed to maintain stability. The media tends to insist that there needs to be a 1:1 ratio, and letter agencies tend to echo this false statement; each class has a ratio that is needed for the token to be reliable in its function.

> “It’s hard to speculate on what tomorrow will bring.”
>
> Bob Dylan

Also, as a matter of course, it is always important to look at the relationship between holders/creators of the currency and the asset pool. The more complex the system the more analysis needs to be done on how the assets are being managed. While a 1:1 ratio for stablecoin 1.0 can be managed by the local bank of the creator, more complex systems require funds, investment firms, and even multiple subsidiaries to ensure that the asset pool maintains value. The analysis of who holds the asset pool then leads to an analysis of susceptibility of to attack. The more complex the asset, the more stable it is- however, it does need to be protected from attacks. The discussions within these sections of each chapter will be quite similar. Transparency is also an issue that varies slightly as you look at different types of stablecoins, once again the more complex the tokenomics- the more transparent the company needs to be. Finally, there needs to be an analysis of viability in major markets. This results in a bimodal, cyclic model for the analysis of the system, viability seems to be highest for simple and highly complex models, with intermediate models being lukewarm in their reception.

Derivative backed stablecoins are among the most challenging stablecoins to 'clock' on the market. If they are managed correctly, then they can have a very quick level of maturity and can be expanded very quickly. However, if they are done poorly, they can be a disaster that is difficult to even watch. It all comes down to management, and it is the management of stablecoins that separates them from other assets.

Stability of Asset Class

Stablecoin 1.8 represents a new breed of stablecoins backed exclusively by derivatives. Unlike traditional stablecoins that rely on fiat currencies or other underlying assets, stablecoin 1.8 leverages derivatives contracts to provide stability and value preservation. This unique approach introduces a different level of stability and offers a range of advantages and challenges compared to other stablecoin models.

The stability of stablecoin 1.8 stems from the nature of derivatives contracts. Derivatives are financial instruments whose value is derived from an underlying asset. They allow market participants to speculate on the future price movements of assets without actually owning them. By using derivatives, stablecoin 1.8 achieves stability by maintaining a balanced exposure to multiple asset classes, such as commodities, currencies, or indices.

The use of derivatives in stablecoin 1.8 provides several benefits. First, it offers a high degree of diversification, mitigating the risks associated with holding a single asset as collateral. The underlying derivatives contracts cover a wide range of assets, which helps stabilize the value of the stablecoin by spreading risk across different markets. This diversification contributes to the stability and resilience of stablecoin 1.8, making it less susceptible to sudden price fluctuations in any particular asset class.

Furthermore, stablecoin 1.8's reliance on derivatives introduces dynamic flexibility in managing its asset portfolio. The derivative contracts can be adjusted or rebalanced to adapt to changing market conditions and optimize the stability of the stablecoin. This adaptability allows stablecoin 1.8 to respond to market trends, adjust the composition of its derivatives holdings, and potentially seize opportunities for enhanced performance.

> "Speculation is not knowledge."
>
> Robert M. Price

However, stablecoin 1.8's reliance on derivatives also presents unique challenges, particularly in terms of liquidity and counterparty risk. Derivatives markets can be less liquid than traditional asset markets, meaning it may be more challenging to execute trades quickly without impacting market prices. This illiquidity can pose challenges for stablecoin 1.8, as it may hinder the ability to maintain stable and real-time redemption mechanisms.

Additionally, the use of derivatives contracts introduces counterparty risk. Derivatives contracts rely on agreements between multiple parties, and the stability of stablecoin 1.8 is inherently tied to the performance and creditworthiness of these counterparties. Any default or failure to honor contractual obligations by the counterparties could potentially impact the stability and value of stablecoin 1.8.

To mitigate these challenges and ensure stability, rigorous risk management practices and due diligence are crucial. Stablecoin 1.8 should carefully select and monitor its derivative counterparties, ensuring they have a strong financial standing and robust risk management frameworks. Regular assessments of counterparties' creditworthiness, collateral requirements, and monitoring of market conditions are essential to maintain the stability and integrity of stablecoin 1.8.

Transparency is another critical aspect for stablecoin 1.8. Given the complex nature of derivatives and their role in maintaining stability, transparent reporting and disclosure of the underlying derivative positions, valuation methodologies, and risk exposures are essential. Investors and stakeholders must have access to accurate and timely information to assess the stability and risks associated with stablecoin 1.8. This transparency builds trust and confidence, enhancing market adoption and mitigating concerns related to potential manipulation or hidden risks.

In conclusion, stablecoin 1.8, backed exclusively by derivatives, offers a unique approach to achieving stability in the cryptocurrency market. By leveraging diversified derivatives contracts, it introduces a new level of stability and flexibility compared to traditional stablecoin models. However, the liquidity challenges and counterparty risks associated with derivatives require careful risk management and transparency. The success of stablecoin 1.8 will depend on its ability to navigate these challenges effectively, maintain robust risk management practices, and provide transparent reporting to ensure the stability and trustworthiness of the stablecoin in the market.

Liquidity of Asset Class

Stablecoin 1.8, which is backed exclusively by derivatives, presents a unique stability profile compared to other types of stablecoins we have discussed. To assess whether stablecoin 1.8 is more or less stable, we need to consider its characteristics and compare them with other stablecoin models.

One key factor that influences stability is the underlying asset or collateral backing the stablecoin. Traditional stablecoins, such as fiat-backed or asset-backed stablecoins, rely on relatively stable assets like fiat currencies or physical commodities. These assets provide a sense of stability and value preservation, as their prices tend to be less volatile compared to derivatives.

In contrast, stablecoin 1.8's reliance on derivatives introduces a different level of stability. Derivatives derive their value from underlying assets, which can include commodities, currencies, or indices. The stability of stablecoin 1.8 is linked to the performance of these underlying assets and the effectiveness of the derivatives contracts in replicating their value.

The stability of stablecoin 1.8 depends on the diversification and risk management strategies employed within its derivative portfolio. By maintaining a balanced exposure to multiple asset classes, stablecoin 1.8 can potentially mitigate the risks associated with a single asset and achieve a more stable value. This diversification factor contributes to the stability of stablecoin 1.8, as it spreads the risk across different markets and reduces vulnerability to sudden price fluctuations in any specific asset class.

However, the stability of stablecoin 1.8 may be influenced by the liquidity and counterparty risks inherent in derivatives markets. Derivatives contracts can be less liquid than traditional asset markets, which means it may be more challenging to execute trades quickly without impacting market prices. This illiquidity could potentially affect stablecoin 1.8's ability to maintain stable and real-time redemption mechanisms, which could impact its stability compared to other stablecoin models.

Counterparty risk is another factor that may impact the stability of stablecoin 1.8. Derivatives contracts rely on agreements between multiple parties, and the stability of the stablecoin is closely tied to the performance and creditworthiness of these counterparties. Any default or failure to honor contractual obligations by the counterparties could potentially impact the stability and value of stablecoin 1.8.

Comparing stablecoin 1.8's stability to other stablecoin models, it is important to consider the specific characteristics and underlying assets of each model. Fiat-backed stablecoins, for instance, benefit from the stability of fiat currencies, which are typically less volatile in the short term. These stablecoins aim to maintain a 1:1 ratio with the underlying fiat currency, providing a high level of stability in terms of value preservation.

Asset-backed stablecoins, on the other hand, rely on physical commodities or other assets to maintain stability. These stablecoins can be subject to price fluctuations in the underlying assets, but their value is typically backed by tangible assets that provide a level of stability over the long term.

When comparing stablecoin 1.8 to these models, stablecoin 1.8's stability may differ due to its reliance on derivatives. While derivatives introduce a dynamic and potentially more diverse range of underlying assets, they also bring additional complexities and risks. The stability of stablecoin 1.8 is closely tied to the performance of these derivatives contracts, which may introduce more volatility and uncertainty compared to traditional stablecoins.

In conclusion, stablecoin 1.8, backed exclusively by derivatives, offers a unique stability profile compared to other stablecoin models. Its stability depends on the effectiveness of diversification and risk management strategies within its derivative portfolio. The liquidity and counterparty risks associated with derivatives may impact the stability of stablecoin 1.8 compared to other stablecoin models backed by more traditional assets. Assessing the relative stability of stablecoin 1.8 requires considering the specific characteristics and underlying assets of each stablecoin model and evaluating their ability to maintain a stable value over time.

> "It is remarkable, that person who speculate the most boldly often conform with the most perfect quietude of the external regulations of society."
>
> Nathaniel Hawthorne

Required Asset Ratio for Stability

Stablecoin 1.8, which is backed exclusively by derivatives, presents a unique challenge when it comes to determining an asset ratio. Unlike other stablecoin models that have a fixed or target asset ratio, stablecoin 1.8's asset ratio cannot be predetermined due to the inherent volatility of derivative valuation. Instead, investors need to assess their risk tolerance and find a ratio that they are comfortable with.

Derivatives are financial instruments whose value is derived from an underlying asset or benchmark. They include options, futures, swaps, and other contracts that allow investors to speculate on or hedge against price movements of the underlying assets. The valuation of derivatives is complex and relies on factors such as market conditions, interest rates, and volatility.

The volatility of derivative valuation introduces challenges when determining an asset ratio for stablecoin 1.8. The value of the underlying assets can experience significant fluctuations, especially in times of market turbulence or economic uncertainty. As a result, it becomes challenging to assign a fixed asset ratio for stablecoin 1.8 because the value of the derivatives can deviate significantly from their initial valuation.

In this context, investors in stablecoin 1.8 need to carefully assess their risk appetite and determine the asset ratio they are comfortable with. A higher asset ratio may provide a greater level of stability, as it indicates a larger portion of the stablecoin's value is backed by the underlying assets. However, a higher asset ratio also means less flexibility and potential for growth, as a larger portion of the stablecoin's value is tied up in the underlying assets.

Conversely, a lower asset ratio allows for more flexibility and growth potential, as a smaller portion of the stablecoin's value is backed by the underlying assets. This may be suitable for investors with a higher risk tolerance who are willing to accept a higher level of volatility in exchange for potential returns.

It is important to note that stablecoin 1.8's asset ratio should not be viewed as a static parameter but rather as a dynamic range within which investors can adjust their positions based on market conditions and their risk preferences. The asset ratio can be seen as a guiding principle that investors use to determine their exposure to the underlying assets, but it should be subject to ongoing evaluation and adjustment as market conditions evolve.

> "Speculation is a parasite feeding upon value, creating none."
>
> Andrew Carnegie

Additionally, the management of stablecoin 1.8 should adopt risk management strategies to mitigate the inherent volatility of derivative valuation. This includes implementing robust risk monitoring systems, stress testing the derivative portfolio, and closely monitoring market conditions and counterparties' performance. By actively managing and adjusting the portfolio, the stablecoin issuer can aim to maintain stability and protect the value of the stablecoin over time.

In conclusion, stablecoin 1.8, backed exclusively by derivatives, poses challenges when it comes to determining an asset ratio due to the inherent volatility of derivative valuation. Investors need to assess their risk tolerance and find a ratio that aligns with their investment objectives. The asset ratio should be viewed as a dynamic parameter subject to ongoing evaluation and adjustment based on market conditions. Risk management strategies play a crucial role in ensuring the stability of stablecoin 1.8 and protecting the value of the stablecoin over time.

Relationship between Holders and access to backing.

The question of who should hold the asset fund for a stablecoin, whether independent firms or the creating firm, is an important consideration in ensuring the stability and security of the stablecoin ecosystem. Both options have their advantages and potential drawbacks, and the decision ultimately depends on various factors, including regulatory requirements, risk management practices, and investor preferences.

One argument in favor of independent firms holding the asset fund is the principle of separation of duties and increased transparency. By entrusting the asset fund to a specialized custodial or trust company, the stablecoin issuer can demonstrate a commitment to impartiality and avoid conflicts of interest. Independent firms are typically subject to stringent regulatory oversight, audited financial reporting, and compliance with robust security measures. This can enhance transparency, mitigate the risk of misappropriation or mismanagement, and provide investors with assurance that their funds are held securely.

Moreover, independent firms often have extensive experience and expertise in asset custody and fund management. They possess robust operational infrastructure, advanced security protocols, and dedicated teams that specialize in safeguarding digital assets. By outsourcing the asset custody function to these entities, stablecoin issuers can leverage their knowledge and capabilities, reducing operational risks and ensuring the proper implementation of best practices in asset security.

On the other hand, some stablecoin creators argue in favor of holding the asset fund within the creating firm itself. This approach allows for more direct control over the assets, enabling the stablecoin issuer to respond quickly to market dynamics and optimize the management of the fund. It can provide flexibility in adjusting the asset allocation, rebalancing the portfolio, and implementing risk management strategies tailored to the stablecoin's specific objectives.

Keeping the asset fund in-house may also be seen as a means of maintaining a competitive advantage and preserving the stablecoin issuer's proprietary knowledge. By maintaining control over the asset fund, the issuer can leverage its expertise, research capabilities, and insights into the market to make informed decisions that enhance the stability and value of the stablecoin. This approach may be particularly relevant in cases where the stablecoin issuer possesses unique insights or has access to specific investment opportunities that can contribute to the stability and growth of the fund.

However, it is important to note that if the creating firm holds the asset fund, there is a potential for conflicts of interest and concerns about transparency. Critics argue that self-custody may create an information asymmetry between the stablecoin issuer and its investors. Investors may have limited visibility into the actual holdings and risk exposure of the stablecoin, potentially undermining trust and confidence in the stability of the token. This lack of transparency can hinder broader market adoption and regulatory acceptance.

To strike a balance between the benefits of independence and issuer control, some stablecoin projects have pursued a hybrid approach. They may opt for a consortium-based custody model, where independent custodians or qualified third parties hold the asset fund in a distributed and decentralized manner. This model allows for increased security, diversification of risk, and improved transparency through the involvement of multiple trusted entities.

In conclusion, the decision of who should hold the asset fund for a stablecoin depends on various factors, including regulatory requirements, risk management practices, and investor preferences. Independent firms offer impartiality, transparency, and expertise in asset custody, while creating firms may value the control, flexibility, and competitive advantages of holding the fund in-house. A hybrid approach that involves independent custodians or a consortium-based custody model can provide a balanced solution that combines the benefits of independence, security, and transparency. Ultimately, the choice should prioritize the stability, security, and trustworthiness of the stablecoin ecosystem to foster broader market adoption and regulatory acceptance.

Susceptibility to Attacks

Stablecoin 1.8, which is backed exclusively by derivatives, presents a unique set of vulnerabilities and susceptibility to attacks compared to other stablecoin models. While it offers advantages in terms of liquidity and flexibility, its reliance on derivatives introduces additional risks that need to be carefully managed to ensure the stability and security of the stablecoin ecosystem.

One of the primary vulnerabilities of stablecoin 1.8 is the risk of counterparty failure. Derivatives are financial contracts that derive their value from an underlying asset or benchmark, such as commodities, currencies, or interest rates. The stablecoin's value is linked to the performance of these derivatives. If the counterparty fails to fulfill their obligations or defaults on the derivative contracts, it can have a severe impact on the stability and value of the stablecoin. This risk is heightened in times of market stress or systemic financial crises when counterparty risk increases.

Another vulnerability is the potential for market manipulation or fraudulent practices related to the derivatives used to back the stablecoin. Derivatives markets can be subject to various forms of manipulation, such as spoofing, wash trading, or insider trading. Malicious actors can exploit these vulnerabilities to artificially influence the value of the derivatives and, consequently, the stablecoin. This can lead to price manipulation, volatility, and potential losses for stablecoin holders.

Furthermore, stablecoin 1.8 is exposed to the risk of regulatory scrutiny and legal challenges. Derivatives markets are subject to complex regulatory frameworks aimed at ensuring fair and transparent trading practices, investor protection, and systemic stability. Non-compliance with these regulations can result in legal consequences, fines, or even the suspension of derivative contracts. Any regulatory actions or unfavorable legal outcomes can significantly impact the stability and viability of the stablecoin.

The interconnectedness of stablecoin 1.8 with the broader financial system also poses risks. The stablecoin's value depends on the performance of derivatives, which are often interconnected with other financial instruments and institutions. A systemic event affecting the broader financial system, such as a market crash or a bank failure, can propagate through derivative markets and impact the stability of stablecoin 1.8. This interconnectedness highlights the importance of risk management, stress testing, and contingency plans to mitigate systemic risks and ensure the stability of the stablecoin.

Cybersecurity threats are also significant concerns for stablecoin 1.8. As a digital asset, the stablecoin is susceptible to various cyber-attacks, including hacking, phishing, and malware. Cybercriminals may target the stablecoin platform, derivative exchanges, or the infrastructure supporting the stablecoin ecosystem. Successful attacks can result in the theft of funds, manipulation of derivative prices, or disruption of stablecoin operations. Robust security measures, including encryption, multi-factor authentication, and regular security audits, are essential to mitigate these risks.

> "One of the peculiarities of recent speculation, especially in America, is that ideas are abandoned in virtue of a mere change of feeling, without any new evidence or new arguments. We do not nowadays refute our predecessors, we pleasantly bid them good-bye."
>
> George Santayana

To address these vulnerabilities and enhance the stability of stablecoin 1.8, comprehensive risk management practices are crucial. Stablecoin issuers need to implement rigorous due diligence processes to select reliable and reputable counterparties for derivative contracts. Ongoing monitoring and evaluation of counterparty risk should be performed to ensure their continued creditworthiness. Additionally, effective risk management frameworks should be in place to identify, assess, and mitigate the various risks associated with derivative trading.

Regulatory compliance is also paramount. Stablecoin issuers must adhere to relevant regulations, obtain necessary licenses, and maintain transparency in their operations. This includes robust anti-money laundering (AML) and know-your-customer (KYC) procedures to prevent illicit activities and ensure the integrity of the stablecoin ecosystem.

Education and awareness are critical components in minimizing vulnerabilities. Stablecoin issuers should educate users and investors about the risks associated with stablecoin 1.8 and provide clear disclosures on the nature of the derivative-backed assets. Transparent reporting, regular audits, and third-party verification can enhance trust and confidence in the stablecoin's stability.

In conclusion, stablecoin 1.8, backed exclusively by derivatives, faces unique susceptibility to attacks compared to other stablecoin models. Counterparty risk, market manipulation, regulatory challenges, systemic risks, and cybersecurity threats are among the key vulnerabilities. Effective risk management practices, compliance with regulations, transparency, and robust security measures are crucial to mitigate these risks and enhance the stability and security of stablecoin 1.8. Continuous monitoring, adaptation to evolving market conditions, and collaboration with regulators and industry participants are essential for the long-term success of stablecoin 1.8 and the broader adoption of derivative-backed stablecoins.

Transparency

Transparency plays a critical role in ensuring the stability and credibility of stablecoin 1.8, which is backed exclusively by derivatives. Given the complexity and potential risks associated with derivative markets, transparency is essential to provide stakeholders with the necessary information and confidence in the stablecoin's underlying assets and operations. By promoting transparency, stablecoin 1.8 can enhance trust, facilitate risk assessment, and enable market participants to make informed decisions.

One of the primary reasons for transparency in stablecoin 1.8 is to ensure clarity regarding the composition and valuation of the derivative assets that back the stablecoin. Derivatives are complex financial instruments whose value is derived from an underlying asset or benchmark. The stablecoin holders and investors need to have visibility into the types of derivatives held in the portfolio, their respective counterparties, and the methodology used to value these assets. Transparent reporting of these details allows stakeholders to assess the risk profile and understand the stability of the stablecoin.

Transparency also plays a crucial role in addressing counterparty risk in stablecoin 1.8. Counterparties involved in derivative contracts can have varying levels of creditworthiness and reliability. Transparent disclosure of counterparties and their credit ratings enables investors to evaluate the stability and trustworthiness of the counterparties. This transparency ensures that the stablecoin issuer is engaging in contracts with reputable and reliable counterparties, reducing the likelihood of counterparty defaults and associated risks.

Furthermore, transparency in stablecoin 1.8 helps address concerns related to market manipulation and fraudulent practices. Derivatives markets can be susceptible to various forms of manipulation, including price manipulation, insider trading, and false reporting. By providing transparency in the stablecoin's operations, including the trading and valuation of derivatives, the stablecoin issuer can demonstrate a commitment to fair and transparent practices. This helps to deter potential manipulative activities and enhances the stability and integrity of the stablecoin ecosystem.

Transparency also contributes to regulatory compliance and addresses regulatory concerns. Stablecoin 1.8, being backed exclusively by derivatives, operates within a regulatory framework that governs derivative markets. Transparent reporting and disclosure of relevant information regarding the stablecoin's operations, derivative holdings, and risk management practices help ensure compliance with applicable regulations. By proactively sharing information with regulators, stablecoin issuers can foster a collaborative relationship and address any regulatory concerns, promoting a stable and compliant ecosystem.

> "Speculation, I retorted, is never a waste of time. It clears the deadwood in the thickets of deduction."
>
> Elizabeth Peters.

From an investor perspective, transparency is crucial for risk assessment and making informed investment decisions. Transparent disclosure of the stablecoin's underlying assets, risk factors, and performance metrics allows investors to evaluate the risk-return profile of stablecoin 1.8. It enables them to understand the potential volatility, liquidity, and valuation risks associated with the derivative holdings. Transparent reporting also assists investors in assessing the stability and sustainability of the stablecoin's value, providing them with the necessary information to align their investment strategies with their risk appetite.

Transparency also helps build trust and confidence among stablecoin users and the wider market participants. By providing timely and accurate information, stablecoin issuers demonstrate a commitment to accountability and openness. This transparency fosters trust, which is crucial for the adoption and acceptance of stablecoin 1.8. It allows market participants to assess the stability, credibility, and operational soundness of the stablecoin, creating a favorable environment for its usage and growth.

To achieve transparency in stablecoin 1.8, stablecoin issuers should adopt robust reporting and disclosure practices. This includes regular publication of audited financial statements, periodic reports on the stablecoin's derivative holdings, valuation methodologies, risk management practices, and compliance with relevant regulations. External audits and third-party verification can further enhance transparency by providing an independent assessment of the stablecoin's operations and adherence to best practices.

In conclusion, transparency is a fundamental requirement for stablecoin 1.8, backed exclusively by derivatives, to ensure the stability, credibility, and trustworthiness of the stablecoin. Transparent reporting and disclosure of derivative holdings, valuation methodologies, counterparties, and risk management practices empower stakeholders to assess the stability, risk profile, and regulatory compliance of the stablecoin. It promotes investor confidence, deters market manipulation, and facilitates informed decision-making. By prioritizing transparency, stablecoin issuers can contribute to the long-term success and widespread adoption of stablecoin 1.8 in the digital asset ecosystem.

Viability in Major Markets

Stablecoin 1.8, which is backed exclusively by derivatives, has experienced significant success in the market. The innovative approach of leveraging derivatives as the underlying assets of the stablecoin has presented unique opportunities and advantages that have contributed to its success.

One key aspect that has contributed to the success of stablecoin 1.8 is its potential for high liquidity. Derivatives are financial instruments that derive their value from an underlying asset or benchmark. They are highly liquid instruments traded on established exchanges, providing stablecoin 1.8 with a deep and liquid market. This liquidity allows for efficient trading and conversion of stablecoin tokens, ensuring that users can easily buy, sell, and transfer stablecoin 1.8 without significant price impact or slippage.

Another factor driving the success of stablecoin 1.8 is its ability to provide stability while offering exposure to a diversified range of underlying assets. Derivatives allow stablecoin 1.8 to be backed by a portfolio of different assets, such as commodities, currencies, or indices. This diversification helps to mitigate risks associated with any single asset's volatility, providing a stable value proposition for the stablecoin. By accessing a broad range of derivatives, stablecoin 1.8 can maintain stability even in the face of market fluctuations, making it an attractive option for investors seeking stability and exposure to various asset classes.

Stablecoin 1.8's success can also be attributed to its flexibility and adaptability. Derivatives markets offer a wide array of products, including futures, options, swaps, and forwards, among others. This flexibility allows stablecoin 1.8 to adjust its underlying assets to align with market conditions or investor preferences. For example, if certain assets become more volatile or less desirable, the stablecoin can rebalance its derivatives portfolio to maintain stability and optimize risk-return characteristics. This adaptability makes stablecoin 1.8 an appealing choice for users who value flexibility and want their stablecoin to adapt to changing market dynamics.

Moreover, stablecoin 1.8 has gained recognition and acceptance from both retail and institutional investors. The use of derivatives provides access to a sophisticated financial instrument widely utilized by institutional investors and professional traders. This integration with traditional financial markets has helped stablecoin 1.8 bridge the gap between traditional finance and the digital asset space, attracting institutional investors who seek exposure to stable digital assets. The increasing institutional adoption of stablecoin 1.8 has provided credibility and stability to the token, fostering further market growth and acceptance.

Furthermore, stablecoin 1.8's success can be attributed to the transparent and accountable nature of its operations. As discussed earlier, transparency is essential for stablecoins backed by derivatives. Stablecoin 1.8 ensures transparency by providing regular and comprehensive reporting on the underlying derivative assets, valuation methodologies, risk management practices, and compliance with regulations. This transparency builds trust among users and market participants, as they can assess the stability, reliability, and compliance of stablecoin 1.8. It also enables stakeholders to make informed investment decisions, aligning their risk appetite with the stablecoin's risk profile.

In terms of market acceptance, stablecoin 1.8 has witnessed growing adoption across various sectors. It has become a preferred stablecoin for traders and investors engaged in derivatives markets, allowing them to easily move funds between stable assets and exposure to derivative contracts. Furthermore, stablecoin 1.8 has found utility in decentralized finance (DeFi) protocols, where it serves as a stable asset for lending, borrowing, and liquidity provision. The integration of stablecoin 1.8 into DeFi platforms has opened up new avenues for its usage and contributed to its liquidity and market acceptance.

> "War and threat of war stimulate speculation on the conditions of peace."
>
> Kenneth N. Waltz

Additionally, stablecoin 1.8 has gained international recognition and acceptance, with its availability on multiple global exchanges and widespread use in cross-border transactions. Its stability, liquidity, and compatibility with existing financial infrastructure have facilitated its integration into global markets, making it a reliable and widely accepted stablecoin option.

Stablecoin 1.8, backed exclusively by derivatives, has achieved significant success in the market due to its high liquidity, stability, flexibility, institutional adoption, transparency, and international acceptance. The unique combination of derivative-backed assets and the advantages they offer have positioned stablecoin 1.8 as a viable and attractive option for users seeking stability and exposure to diverse asset classes. As the market continues to evolve, stablecoin 1.8 is likely to play a significant role in the digital asset ecosystem and contribute to the broader adoption of stablecoins worldwide.

Conclusion

In conclusion, Stablecoin 1.8, backed exclusively by derivatives, has emerged as a successful and innovative solution in the stablecoin landscape. The use of derivatives as the underlying assets provides unique advantages, contributing to its stability, liquidity, flexibility, and market acceptance.

Stablecoin 1.8 has demonstrated its stability by leveraging a diversified portfolio of derivatives. While derivatives may exhibit short-term volatility, the overall portfolio is designed to mitigate risks and provide long-term stability. This stability is particularly appealing to users who seek a reliable digital asset that can maintain its value even during market fluctuations.

One of the key strengths of stablecoin 1.8 lies in its high liquidity. Derivatives are traded on well-established exchanges, offering deep and liquid markets. This ensures that stablecoin 1.8 holders can easily buy, sell, and transfer their tokens without significant price impact or difficulty in executing trades. The robust liquidity of stablecoin 1.8 enhances its usability and accessibility in various financial transactions and investment strategies.

The flexibility of stablecoin 1.8 is another factor contributing to its success. Derivatives markets provide a wide range of products and instruments, allowing stablecoin 1.8 to adapt its underlying assets based on market conditions and investor preferences. This adaptability enables stablecoin 1.8 to optimize its risk-return characteristics and maintain stability even in changing market dynamics. The ability to adjust the portfolio composition ensures that stablecoin 1.8 remains resilient and aligned with the evolving needs of its users.

Stablecoin 1.8 has garnered substantial attention and adoption from both retail and institutional investors. Its integration with traditional financial markets has allowed it to bridge the gap between traditional finance and the digital asset space. Institutional investors, in particular, are attracted to the use of derivatives and the stability they provide. The increasing institutional adoption of stablecoin 1.8 has bolstered its credibility and trustworthiness, further solidifying its position in the market.

Transparency is a crucial aspect of stablecoin 1.8. Given its derivative-backed nature, it is paramount to provide transparent reporting on the underlying derivative assets, valuation methodologies, risk management practices, and compliance with regulations. Transparent operations inspire trust among users and market participants, enabling them to assess the stability and reliability of stablecoin 1.8. The commitment to transparency ensures that stakeholders can make informed decisions and align their risk appetite with the risk profile of stablecoin 1.8.

Stablecoin 1.8 has also witnessed success in major markets, both domestically and internationally. Its availability on global exchanges and widespread usage in cross-border transactions have contributed to its acceptance and recognition on a global scale. The compatibility of stablecoin 1.8 with existing financial infrastructure has facilitated its integration into various sectors, including trading, decentralized finance (DeFi), and international remittances. The growing market acceptance of stablecoin

1.8 is a testament to its value proposition and the benefits it offers to users.

In summary, stablecoin 1.8, backed exclusively by derivatives, has demonstrated its stability, liquidity, flexibility, and market acceptance. The innovative approach of using derivatives as underlying assets has provided unique advantages that have contributed to its success. As the stablecoin ecosystem continues to evolve, stablecoin 1.8 is poised to play a significant role in enabling stable and efficient digital transactions and fostering broader adoption of stablecoins globally.

Chapter 17: Stablecoin 1.9 (Single Asset Classes)

"Managing a portfolio is like managing a garden. You don't just want different kinds of plants in your garden. You want those different plants to have synergy and to work together harmoniously to maximize productivity. In the same way, when different elements in the portfolio have synergy and work together to help each other maximize individual productivity, their collective yields can then be reinvested to maximize the productivity of the whole portfolio. There's a compounding effect and a multiplicative value effect that takes place with the permaculture investing approach."[42]

Evolution refined is evolution based on the needs of the population. In the case of cryptocurrency, the needs of the population are somewhere stable to hold their assets when the market is moving in a direction that the holder does not like. Single asset class stablecoins were the ideal area until stablecoin 2.0 came along. With the ability to diversify across an entire asset class, this give investors the ability to have a stability feature that was based as broadly as possible, pre-2.0.

In this chapter, we will follow the same pattern that we follow in all the chapters in this part of the book. First, we look at the stability of the asset class on which the token at this level is based. The asset class is very important because it is the 'pegged' underpinning of the system. Next, we look at the liquidity of the asset class, how fast can the

---

[42] Hendrith Vanlon Smith at https://www.goodreads.com/quotes/tag/diversification.

asset be liquidated if there is a ‘run’ on the token. As a matter of course, we will then look at the asset ratio needed to maintain stability. The media tends to insist that there needs to be a 1:1 ratio, and letter agencies tend to echo this false statement; each class has a ratio that is needed for the token to be reliable in its function.

Also, as a matter of course, it is always important to look at the relationship between holders/creators of the currency and the asset pool. The more complex the system the more analysis needs to be done on how the assets are being managed. While a 1:1 ratio for stablecoin 1.0 can be managed by the local bank of the creator, more complex systems require funds, investment firms, and even multiple subsidiaries to ensure that the asset pool maintains value. The analysis of who holds the asset pool then leads to an analysis of susceptibility of to attack. The more complex the asset, the more stable it is- however, it does need to be protected from attacks. The discussions within these sections of each chapter will be quite similar. Transparency is also an issue that varies slightly as you look at different types of stablecoins, once again the more complex the tokenomics- the more transparent the company needs to be. Finally, there needs to be an analysis of viability in major markets. This results in a bimodal, cyclic model for the analysis of the system, viability seems to be highest for simple and highly complex models, with intermediate models being lukewarm in their reception.

As stablecoin evolved, 1.9 was the principal method for good hedging until KIROS came along. Even in the post stablecoin 2.0 world, stablecoin 1.9 is one of the best places to put money during a down crypto market. Some stablecoin 1.9s are starting to integrate smart burn technology into their system to ensure that they keep up with stablecoin 2.0, but when in a pinch an investor cannot go wrong with stablecoin 1.9.

Stability of Asset Class

> If you can count all your assets, you always show a profit."
>
> Wilson Mizner

Stablecoin 1.9, which is based on a full family of a single asset class, aims to provide stability in the cryptocurrency market by leveraging the characteristics of a specific asset class. Unlike other stablecoins that rely on diversified asset portfolios, stablecoin 1.9 focuses on a single asset class, such as real estate, commodities, or equities. This unique approach to stability offers both advantages and challenges in the context of stablecoin design and market performance.

The stability of stablecoin 1.9 is heavily influenced by the performance and behavior of the chosen asset class. For example, if the stablecoin is based on a family of real estate assets, its stability will depend on factors such as property values, rental income, and market demand. The value of the stablecoin will closely track the performance of the underlying real estate assets. This can provide a relatively stable and reliable value proposition for users, especially in comparison to more volatile cryptocurrencies.

However, stability in stablecoin 1.9 is not immune to the risks and fluctuations inherent in the chosen asset class. For instance, if the stablecoin is based on equities, it will be subject to market volatility, corporate performance, and regulatory factors that impact stock prices. While stablecoin 1.9 may offer stability in the long run due to the underlying asset class's historical performance, short-term fluctuations and market events can still introduce variability and potential risks.

The design of stablecoin 1.9 necessitates careful consideration of the asset class's characteristics and the implementation of robust risk management strategies. For example, if the stablecoin is backed by a family of commodities, the issuer must navigate challenges such as price volatility, supply chain disruptions, and geopolitical factors that impact commodity markets. Implementing risk mitigation measures such as hedging strategies, diversification within the asset class, and prudent asset selection can help maintain stability and protect the stablecoin's value.

The stability of stablecoin 1.9 is also influenced by the management and governance of the underlying asset class. Effective management ensures proper oversight, risk assessment, and alignment with market dynamics. The stability of the asset class relies on factors such as professional expertise, regulatory compliance, and transparent reporting. For example, if the stablecoin is based on a family of real estate assets, professional property management, diligent market analysis, and compliance with local regulations are essential to maintain stability.

One potential advantage of stablecoin 1.9 is its ability to attract investors and users who have confidence in the underlying asset class. Investors seeking exposure to a specific asset class can find stability in stablecoin 1.9, allowing them to participate in the market while minimizing some of the risks associated with direct ownership or investment. Stablecoin 1.9 can also offer diversification benefits within the asset class itself, allowing users to access a broad range of assets within a single digital token.

However, it is crucial to note that stability in stablecoin 1.9 may vary depending on the performance and resilience of the chosen asset class. Economic downturns, regulatory changes, or disruptions specific to the asset class can impact the stability of the stablecoin. Therefore, investors and users must conduct thorough due diligence, evaluate the historical performance of the asset class, and consider potential risks before engaging with stablecoin 1.9.

In conclusion, stablecoin 1.9, which is based on a full family of a single asset class, offers stability by closely tracking the performance of the chosen asset class. While it provides advantages in terms of long-term stability and exposure to specific asset classes, it is not immune to the risks and fluctuations associated with the selected asset class. Careful management, risk mitigation strategies, and transparent governance are crucial to maintaining stability in stablecoin 1.9. Users and investors should thoroughly evaluate the underlying asset class's characteristics, historical performance, and potential risks before engaging with stablecoin 1.9.

Liquidity of Asset Class

Stablecoin 1.9, which is based on a full family of a single asset class, presents unique considerations in terms of liquidity compared to other stablecoin classes. Liquidity refers to the ease with which an asset can be bought or sold without causing significant price movements or impacting its market value. In the case of stablecoin 1.9, the liquidity profile is influenced by the characteristics of the underlying asset class and the mechanisms in place for maintaining liquidity within the stablecoin ecosystem.

The liquidity of stablecoin 1.9 can vary depending on the specific asset class it represents. Different asset classes have varying degrees of liquidity, which affects the stability and tradability of the stablecoin. For example, if stablecoin 1.9 is based on a family of equities, the liquidity of the stablecoin will be closely tied to the liquidity of the underlying stocks. Equities generally have a higher level of liquidity compared to other asset classes, as they are frequently traded on regulated exchanges. This can contribute to the liquidity and ease of trading for stablecoin 1.9.

On the other hand, if stablecoin 1.9 is based on a less liquid asset class, such as real estate or private equity, the liquidity profile may be different. Real estate transactions typically involve longer settlement periods and fewer market participants compared to publicly traded securities. This can potentially impact the liquidity of the stablecoin, as it may take more time and effort to buy or sell the underlying assets to support the stablecoin's value. However, advancements in blockchain technology and the emergence of real estate tokenization platforms are working to increase the liquidity and tradability of real estate-based stablecoins.

The liquidity of stablecoin 1.9 is also influenced by the mechanisms in place to ensure liquidity within the stablecoin ecosystem. These mechanisms can include market makers, liquidity pools, or decentralized exchanges (DEXs). Market makers play a crucial role in providing liquidity by continuously offering to buy or sell stablecoin 1.9 at a specified price. Liquidity pools aggregate funds from multiple participants, enabling efficient trading and minimizing price slippage. DEXs, which operate on blockchain networks, facilitate peer-to-peer trading, and provide liquidity through automated smart contracts. These mechanisms can enhance the liquidity of stablecoin 1.9, allowing for easier and more efficient trading.

> "Rich people acquire assets, poor and middle-class people acquire liabilities that they think are assets."
>
> Robert Kiyosaki

Compared to other stablecoin classes, stablecoin 1.9 may face unique liquidity considerations. Stablecoins backed by diversified asset portfolios, such as stablecoin 1.2 or stablecoin 1.5, may benefit from the combined liquidity of multiple asset classes. Diversification can provide stability and liquidity across different markets, reducing the reliance on a single asset class. In contrast, stablecoin 1.9's liquidity is tied to the liquidity of the chosen asset class, which can have implications for the stablecoin's market depth and trading volume.

However, it is essential to note that the liquidity of stablecoin 1.9 is not solely determined by the asset class itself but also by the adoption and usage of the stablecoin in the market. The demand for stablecoin 1.9, the number of participants using the stablecoin, and the availability of trading platforms all contribute to its liquidity. A stablecoin with a robust user base and a wide range of trading platforms can enhance liquidity and market accessibility.

To enhance liquidity, stablecoin 1.9 issuers and ecosystem participants can employ various strategies. These may include partnering with reputable market makers or liquidity providers, integrating with popular decentralized exchanges, or actively promoting the stablecoin's adoption among users and traders. Furthermore, transparent reporting of asset holdings and regular audits can enhance investor confidence and attract liquidity providers to engage with stablecoin 1.9.

In conclusion, the liquidity of stablecoin 1.9, which is based on a full family of a single asset class, is influenced by the liquidity of the underlying assets and the mechanisms in place to maintain liquidity within the stablecoin ecosystem. The specific asset class, such as equities or real estate, plays a significant role in determining the liquidity profile. Additionally, the adoption and usage of stablecoin 1.9 in the market, along with strategies to enhance liquidity, are essential factors. By considering these aspects and implementing measures to improve liquidity, stablecoin 1.9 can strive to provide a reliable and accessible trading experience for users.

Required Asset Ratio for Stability

Stablecoin 1.9, which is based on a full family of a single asset class, operates under a unique asset ratio framework. The asset ratio refers to the proportion of the stablecoin's total value that is backed by the underlying assets. This ratio is crucial in maintaining the stability and value proposition of the stablecoin.

In the case of stablecoin 1.9, an asset ratio of around 50% is recommended to start with. This means that at the launch of the stablecoin, approximately half of its value is backed by the underlying assets. The choice of the 50% ratio is influenced by various factors, including risk management, market conditions, and investor expectations.

A 50% asset ratio provides a reasonable level of backing for stablecoin 1.9, striking a balance between stability and growth potential. It offers a solid foundation of asset support, reassuring investors that their stablecoin holdings have tangible value. At the same time, it allows for some flexibility and potential for growth in the stablecoin's ecosystem.

The maturity period of stablecoin 1.9, estimated to be around 7 years, refers to the time it takes for the stablecoin's asset pool to reach parity with the stablecoin's total value. This means that within 7 years, the asset pool will have grown to match the stablecoin's overall value, assuming the stability of the underlying asset class and consistent growth in the stablecoin's usage.

The choice of the 7-year maturity period is influenced by several considerations. First, it aligns with the typical investment horizon for long-term investors, allowing them to benefit from the stablecoin's growth potential over time. Second, it takes into account the nature of the chosen asset class and its expected performance over the long term.

However, it is important to note that the actual maturity period of stablecoin 1.9 may vary depending on various factors, including market conditions, asset performance, and the stability of the stablecoin's ecosystem. The 7-year estimate is a general guideline and should be adjusted based on specific market dynamics and risk factors.

During the maturity period, it is crucial to manage the stablecoin's asset pool effectively. This involves active asset allocation, risk management, and monitoring of the asset class's performance. Regular assessments and adjustments may be necessary to ensure that the asset pool remains aligned with the stablecoin's total value and maintains stability.

As stablecoin 1.9 progresses towards maturity, the asset pool's growth should reflect the stablecoin's increasing adoption and usage. As more users and investors engage with the stablecoin, the asset pool can expand, providing further support and enhancing the stability of the stablecoin's value.

It is worth noting that while the 50% asset ratio is recommended to start with, the stability and maturity of stablecoin 1.9 are not solely dependent on the asset ratio alone. Other factors, such as the quality and diversification of the underlying assets, risk management strategies, and the overall health of the stablecoin's ecosystem, also contribute to stability and long-term value.

In conclusion, stablecoin 1.9 adopts an asset ratio of around 50% at launch, with an estimated maturity period of approximately 7 years. This ratio provides a solid foundation for stability and value while allowing for potential growth. However, the actual maturity period may vary, and effective management of the asset pool is essential throughout the stablecoin's lifecycle. By striking the right balance between asset backing, risk management, and growth potential, stablecoin 1.9 aims to provide a reliable and sustainable stablecoin solution for users and investors.

Relationship between Holders and access to backing.

Stablecoin 1.9, backed by a full family of a single asset class, requires careful consideration when determining who should hold the asset pool. While the creating company may seem like a logical choice, there are compelling reasons why an independent firm should be entrusted with managing the asset pool for stablecoin 1.9. This essay will explore the advantages and rationale behind relying on an external firm for holding the asset pool.

> "Investment- is an asset or item that is purchased with the hope that it will generate value or appreciate in the future."
>
> Investment Brothers

First and foremost, the decision to entrust an independent firm with the asset pool aligns with the principles of transparency and decentralization, which are fundamental to many cryptocurrencies and blockchain-based systems. By assigning the responsibility to an external entity, the process becomes more transparent and less prone to potential conflicts of interest. Users and investors can have greater confidence knowing that the asset pool is held by a third-party firm that operates independently and upholds stringent fiduciary obligations.

Additionally, an independent firm specializing in managing assets brings a wealth of expertise and experience to the table. Asset management firms have the necessary infrastructure, resources, and knowledge to effectively monitor, evaluate, and make strategic decisions regarding the underlying assets. They possess a deep understanding of market dynamics, risk management techniques, and regulatory compliance, which are essential for maintaining the stability and long-term growth potential of the asset pool.

Moreover, engaging an external firm can mitigate certain risks associated with the creating company directly holding the asset pool. Companies may face operational risks, conflicts of interest, or financial challenges that could impact their ability to safeguard the asset pool effectively. By outsourcing this responsibility to a specialized firm, the stability and integrity of the asset pool can be enhanced, providing greater assurance to users and investors.

Another crucial aspect is the potential for enhanced governance and accountability when an independent firm manages the asset pool. Such firms typically adhere to strict regulatory frameworks and reporting requirements. They are subject to audits and oversight by regulatory bodies, which ensures that they operate within legal boundaries and follow best practices in asset management. This level of accountability instills confidence and transparency in the stablecoin ecosystem, fostering trust among stakeholders.

Furthermore, entrusting the asset pool to an independent firm allows for the segregation of duties. Separating the creating company's role from that of the asset holder reduces the risk of internal fraud, mismanagement, or misuse of funds. It establishes a clear delineation of responsibilities, reducing the potential for conflicts of interest and enhancing overall governance.

Additionally, an independent asset management firm brings a broader network and access to diverse investment opportunities. They have established relationships with financial institutions, access to market data, and the ability to identify and seize potential investment avenues that align with the stablecoin's asset class. This can optimize the performance of the asset pool and potentially enhance returns for stablecoin holders.

Moreover, by leveraging the expertise and networks of external firms, stablecoin 1.9 can benefit from continuous monitoring and adaptation to changing market conditions.

Asset management firms actively track market trends, perform risk assessments, and adjust investment strategies accordingly. This proactive approach ensures that the asset pool remains resilient and well-positioned to navigate challenges or seize opportunities as they arise.

Lastly, engaging an external firm for holding the asset pool introduces an additional layer of governance and oversight. This can contribute to the overall stability and integrity of the stablecoin ecosystem. External firms are subject to regulatory scrutiny, internal control requirements, and compliance obligations, which act as safeguards against potential malfeasance or negligence. This framework of checks and balances serves to protect the interests of stablecoin holders and promotes long-term sustainability.

In conclusion, the decision to have an independent firm hold the asset pool for stablecoin 1.9 is driven by several key considerations. It aligns with the principles of transparency and decentralization, enhances expertise and experience in asset management, mitigates risks, improves governance and accountability, provides access to diverse investment opportunities, enables continuous monitoring and adaptation, and adds an extra layer of oversight. By leveraging the capabilities of an independent firm, stablecoin 1.9 can offer users and investors a robust, secure, and stable platform for value transfer, while maintaining the integrity of the asset pool and upholding investor trust.

Susceptibility to Attacks

Stablecoin 1.9, which is based on a full family of a single asset class, offers inherent advantages that make it less susceptible to attacks compared to other types of stablecoins. The asset structure of stablecoin 1.9 plays a crucial role in enhancing its security and reducing vulnerability to attacks. This essay will delve into the reasons why stablecoin 1.9 is less susceptible to attacks due to its asset structure.

One of the key factors that contribute to the reduced susceptibility of stablecoin 1.9 to attacks is the diversified nature of the asset class. By relying on a full family of a single asset class, the stablecoin is backed by a broad portfolio of assets within that particular class. This diversification mitigates the impact of any single asset's performance or vulnerability to attacks. In other words, even if one or a few assets within the asset class face challenges or come under attack, the overall stability of the stablecoin is not significantly affected, thanks to the presence of other assets that can compensate for potential losses.

Furthermore, the asset structure of stablecoin 1.9 ensures that the underlying assets have intrinsic value and are backed by tangible assets or productive enterprises. This reduces the risk of the stablecoin being subjected to speculative attacks or manipulation. Since the value of the stablecoin is directly linked to the value of the underlying assets, any attack aimed at destabilizing the stablecoin would require undermining the fundamental value of the asset class itself. This adds an extra layer of protection and makes it more challenging for malicious actors to manipulate the stablecoin's value.

Additionally, the asset structure of stablecoin 1.9 can incorporate mechanisms that enhance security and resilience. For instance, smart contract technology can be utilized to enforce strict rules and protocols governing the management and utilization of the underlying assets. Smart contracts can help automate and secure the process of asset allocation, redemption, and tracking, minimizing the potential for human error or fraudulent activities. This technology provides transparency and immutability, making it more difficult for attackers to manipulate or exploit the underlying assets.

Moreover, the asset structure of stablecoin 1.9 can include robust risk management strategies. The diversified asset portfolio allows for risk mitigation through effective portfolio management techniques, such as asset allocation, hedging, and risk diversification. These strategies help protect the stablecoin from sudden market fluctuations, systemic risks, or unexpected events that could impact the stability of the asset class. By proactively managing risks, stablecoin 1.9 reduces its vulnerability to attacks that exploit market vulnerabilities or systemic weaknesses.

> "Never walk away from failure. On the contrary, study it carefully and imaginatively for its hidden assets."
>
> Michael Korda

Another aspect that strengthens the security of stablecoin 1.9 is the potential for incorporating regulatory oversight and compliance measures. The underlying assets within the asset class may be subject to regulatory frameworks, ensuring that they meet certain standards, undergo audits, and comply with legal requirements. This regulatory oversight adds an extra layer of scrutiny and protection, making it more challenging for attackers to exploit vulnerabilities within the asset class. Regulatory compliance helps maintain the integrity of the stablecoin and instills confidence in users and investors.

Furthermore, the asset structure of stablecoin 1.9 can enable a higher level of transparency. Since the stablecoin is backed by a full family of assets within a single asset class, investors can have visibility into the composition and performance of the underlying assets. This transparency fosters trust and confidence among users and investors, as they can assess the stability and value of the stablecoin based on the performance and quality of the asset class. Increased transparency reduces the risk of attacks based on misinformation or lack of information, as users can make informed decisions based on accurate and comprehensive data.

In conclusion, stablecoin 1.9's asset structure provides inherent advantages that make it less susceptible to attacks. The diversified nature of the asset class, the intrinsic value of the underlying assets, the incorporation of risk management strategies, and the potential for regulatory oversight and transparency all contribute to enhancing the security and stability of stablecoin 1.9. By leveraging these attributes, stablecoin 1.9 can offer users and investors a reliable and resilient platform for value transfer while minimizing the risk of attacks and manipulation.

Transparency

Transparency is a fundamental aspect of stablecoin 1.9 that plays a crucial role in ensuring trust, credibility, and stability within the ecosystem. By providing clear and accessible information about the underlying assets, operations, and governance, transparency enhances the overall integrity of stablecoin 1.9. In this essay, we will delve into the reasons why transparency is vitally important for stablecoin 1.9 and how it contributes to its success.

First and foremost, transparency fosters trust among users and investors. Stablecoin 1.9 relies on the confidence and trust of its participants to function effectively. By providing transparent information about the underlying assets, users can have a clear understanding of what backs the stablecoin and evaluate its stability and value. This transparency reduces uncertainty and helps to build credibility, ensuring that users have faith in the stablecoin and its ability to maintain its pegged value.

Moreover, transparency in stablecoin 1.9 allows for effective risk assessment and management. By providing comprehensive and accurate information about the underlying assets, users and investors can evaluate the risks associated with the stablecoin. This includes assessing the quality, diversification, and performance of the asset class that backs the stablecoin. Armed with this information, users can make informed decisions about their involvement with stablecoin 1.9, understanding the potential risks and rewards involved.

Transparency also facilitates accountability and oversight. By making the relevant information accessible, users and regulators can monitor the activities and operations of stablecoin 1.9. This accountability reduces the potential for misuse or fraudulent activities within the ecosystem. Users can hold the stablecoin issuer accountable for the proper management of the underlying assets, ensuring that they are being handled in accordance with the stated guidelines and objectives. Regulators can also enforce compliance and regulatory standards, promoting a healthy and transparent stablecoin ecosystem.

Furthermore, transparency in stablecoin 1.9 enables market participants to conduct due diligence. Users and investors can examine the composition of the asset class, including the individual assets within it. This allows for an assessment of the asset quality, liquidity, and risk characteristics. Market participants can evaluate the stability and resilience of the asset class, assessing its ability to weather market fluctuations and external shocks. This due diligence empowers investors to make informed investment decisions and mitigates the risk of potential fraudulent schemes or misleading representations.

In addition, transparency enhances the stability and credibility of stablecoin 1.9 in the face of external scrutiny. The cryptocurrency industry is subject to regulatory scrutiny and public skepticism. By embracing transparency, stablecoin 1.9 can differentiate itself from other opaque or suspicious projects, positioning itself as a trustworthy and compliant option. Transparency helps to address concerns related to money laundering, fraud, and market manipulation, which are prevalent in the crypto industry. By adhering to transparency standards, stablecoin 1.9 can showcase its commitment to integrity and regulatory compliance, attracting both users and institutional investors seeking a reliable and transparent stablecoin solution.

Moreover, transparency promotes innovation and collaboration within the stablecoin ecosystem. By openly sharing information about the underlying assets and operations, stablecoin 1.9 encourages developers, researchers, and stakeholders to contribute their expertise and insights. This collaborative approach fosters innovation in areas such as risk management, governance mechanisms, and asset diversification strategies. Transparency can also enable peer reviews and audits, which provide an additional layer of scrutiny and verification of the stablecoin's operations. By welcoming external contributions and feedback, stablecoin 1.9 can continuously improve and adapt to changing market dynamics, enhancing its stability and value proposition.

In conclusion, transparency is of utmost importance for stablecoin 1.9. It builds trust, facilitates effective risk management, promotes accountability and oversight, enables due diligence, enhances stability in the face of external scrutiny, and fosters innovation and collaboration. By embracing transparency and providing comprehensive and accessible information about the underlying assets, operations, and governance, stablecoin 1.9 can establish itself as a reliable and credible stablecoin solution. Transparency ensures that users and investors have the necessary information to make informed decisions, mitigates risks, and safeguards the stability and integrity of stablecoin 1.9 in the broader cryptocurrency ecosystem.

Viability in Major Markets

Stablecoin 1.9 has experienced significant success in the markets due to several key factors that have contributed to its appeal and widespread adoption. In this essay, we will explore the reasons behind the strong performance of stablecoin 1.9 and how it has gained traction in the cryptocurrency ecosystem.

> "Healthy citizens are the greatest asset any country can have."
>
> Winston Churchill

One of the primary reasons for the success of stablecoin 1.9 is its stability. Unlike other cryptocurrencies that are prone to high volatility, stablecoin 1.9 is designed to maintain a stable value, typically pegged to a specific asset or basket of assets. This stability makes stablecoin 1.9 an attractive option for users who seek a reliable medium of exchange and store of value within the cryptocurrency space. The ability to transact and hold assets without the fear of significant value fluctuations provides a sense of security and trust for users, leading to increased adoption and usage.

Another factor contributing to the success of stablecoin 1.9 is its versatility and usability. Stablecoin 1.9 can be seamlessly integrated into various decentralized finance (DeFi) applications, enabling users to engage in lending, borrowing, yield farming, and other financial activities. The stability of stablecoin 1.9 makes it a suitable medium of exchange for these DeFi protocols, providing users with a reliable base currency to interact with other digital assets and financial instruments. This versatility and compatibility with DeFi platforms have expanded the use cases and utility of stablecoin 1.9, attracting both retail and institutional investors.

Furthermore, stablecoin 1.9 has gained market success due to its regulatory compliance and adherence to industry standards. The cryptocurrency industry has faced significant regulatory challenges, with concerns around anti-money laundering (AML), know-your-customer (KYC) regulations, and investor protection. Stablecoin 1.9 has sought to address these concerns by implementing robust compliance measures, conducting regular audits, and ensuring transparency in its operations. This commitment to regulatory compliance has instilled confidence among institutional investors and market participants, leading to increased adoption and support for stablecoin 1.9.

Additionally, the strong market performance of stablecoin 1.9 can be attributed to the trust and confidence it has garnered from investors. The transparency and accountability of stablecoin 1.9 have helped build trust among users and investors. By providing clear information about the underlying assets, asset management practices, and regular audits, stablecoin 1.9 offers transparency that instills confidence in its stability and value proposition. This trust has attracted a wide range of investors, including retail users, institutional investors, and even traditional financial institutions, who recognize the potential of stablecoin 1.9 as a reliable and regulated digital asset.

Moreover, the success of stablecoin 1.9 can be attributed to its ability to bridge the gap between traditional finance and the crypto ecosystem. Stablecoin 1.9 has positioned itself as a viable alternative to traditional fiat currencies, offering the benefits of blockchain technology, such as faster and cheaper transactions, while maintaining stability. This has made stablecoin 1.9 an attractive option for individuals and businesses seeking to leverage the advantages of cryptocurrencies without exposing themselves to the high volatility typically associated with other digital assets.

Furthermore, the global nature of stablecoin 1.9 has contributed to its success in the markets. Stablecoin 1.9 operates on a decentralized infrastructure, allowing users to transact and hold assets across borders without the need for intermediaries or traditional banking systems. This accessibility and borderless nature have made stablecoin 1.9 particularly appealing for individuals in regions with limited access to stable banking systems or unstable local currencies. The ability to transact in stablecoin 1.9 provides a reliable and efficient financial solution for individuals in underserved markets, contributing to its widespread adoption and success.

Stablecoin 1.9 has achieved significant success in the markets due to its stability, versatility, regulatory compliance, transparency, trustworthiness, ability to bridge traditional finance with the crypto ecosystem, and global accessibility. These factors have collectively contributed to the strong market performance of stablecoin 1.9, attracting a diverse range of users and investors. As the cryptocurrency industry continues to evolve, stablecoin 1.9 stands as a robust and reliable digital asset, offering stability, utility, and trust in an ever-changing financial landscape.

Conclusion

Stablecoin 1.9 has emerged as a significant player in the cryptocurrency ecosystem, offering stability, versatility, and trust to users and investors. Throughout this chapter, we have explored various aspects of stablecoin 1.9, including its stability, tokenomics, asset pools, susceptibility to attacks, need for transparency, and its performance in major markets. With its unique characteristics and value proposition, stablecoin 1.9 has gained substantial adoption and market success. In this concluding section, we will summarize the key findings and discuss the future prospects of stablecoin 1.9.

Stability has been a defining feature of stablecoin 1.9, setting it apart from other cryptocurrencies that are notorious for their high volatility. By pegging its value to a specific asset or a basket of assets, stablecoin 1.9 provides users with a reliable store of value and medium of exchange. This stability has made stablecoin 1.9 an attractive option for individuals and businesses seeking a more predictable and less risky digital asset. Furthermore, stablecoin 1.9's stability has enabled its integration into various decentralized finance (DeFi) applications, expanding its utility and use cases within the rapidly growing DeFi ecosystem.

Tokenomics has played a crucial role in the success of stablecoin 1.9. The unique token economics of stablecoin 1.9 have differentiated it from other stablecoin models. Whether it is backed by a specific asset class, a mutual fund, land, or derivatives, stablecoin 1.9 has presented innovative approaches to maintaining its value stability. The careful balance between asset holdings, liquidity management, and growth potential has allowed stablecoin 1.9 to cater to the diverse needs of users and investors, offering a range of options depending on risk tolerance and investment preferences.

Asset pools have been a critical aspect of stablecoin 1.9's value proposition. The composition and management of the asset pool have played a significant role in ensuring the stability and reliability of stablecoin 1.9. Whether the asset pool is held by an independent firm or a third party, the transparent and responsible management of the assets has been crucial in building trust among users and investors. Additionally, the asset ratio for stablecoin 1.9 has varied depending on the underlying assets and risk management strategies. A careful balance between asset allocation, diversification, and risk mitigation has been necessary to maintain stability and meet the long-term objectives of stablecoin 1.9.

The susceptibility to attacks has been an ongoing concern in the cryptocurrency industry, and stablecoin 1.9 has not been immune to these risks. However, the specific asset structures, asset classification, and management practices of stablecoin 1.9 have contributed to its resilience and reduced vulnerability. By implementing robust security measures, conducting regular audits, and ensuring transparency, stablecoin 1.9 has mitigated the potential risks associated with attacks and fraud. However, continuous vigilance and improvement in security protocols are essential to maintain the trust and confidence of users and investors.

Transparency has been a core principle for stablecoin 1.9. The need for transparency is particularly crucial when dealing with assets such as land, derivatives, or a portfolio of assets. Clear information about the underlying assets, asset management practices, audits, and regulatory compliance have been vital in building trust and ensuring the stability of stablecoin 1.9. Transparency has allowed users and investors to make informed decisions, understand the risks associated with stablecoin 1.9, and verify the integrity of the underlying assets and operations.

Stablecoin 1.9 has experienced significant success in the markets, attracting a broad range of users and investors. Its stability, versatility, regulatory compliance, transparency, and trustworthiness have been key drivers of its market performance. By offering a bridge between traditional finance and the crypto ecosystem, stablecoin 1.9 has appealed to both crypto enthusiasts and traditional investors looking to explore the world of digital assets. The global accessibility and ease of use have further contributed to the adoption of stablecoin 1.9 across various jurisdictions, fostering its growth and acceptance.

Looking ahead, stablecoin 1.9 is poised to continue its growth and play an integral role in the evolving financial landscape. As regulations surrounding stablecoins become more defined, stablecoin 1.9's compliance with regulatory frameworks will be crucial in maintaining its market position and expanding its user base. The ongoing advancements in blockchain technology and the integration of stablecoin 1.9 into DeFi applications offer exciting opportunities for further innovation and adoption. Continued efforts to enhance security, transparency, and stability will be vital to address emerging challenges and ensure the long-term success of stablecoin 1.9.

In conclusion, stablecoin 1.9 has emerged as a prominent and successful stablecoin model, offering stability, versatility, and trust to users and investors. Its unique characteristics, including stability, tokenomics, asset pools, transparency, and resilience to attacks, have set it apart from other stablecoin manifestations. Stablecoin 1.9's success in the markets is a testament to its ability to meet the diverse needs of users and navigate the evolving regulatory landscape. With its strong foundation and future prospects, stablecoin 1.9 is well-positioned to continue its growth and contribute to the broader adoption of digital assets in the global financial ecosystem.

Chapter 18: Stablecoin 2.0 (Multi-Asset Independent Funds with a Smart Burn System)

"We as a contingent are barely scratching the surface on what blockchain technology can do for us. Imagin if government could harness the benefits of traceability the technology offers. This could increase accountability in all sectors."[43]

I suppose with success comes challenges. As we begin to look at the next evolution in cryptocurrency, it is important to note that not all cryptocurrencies are a tool for the government to spy on you. True, CBDCs, which are the government's 'wish' cryptocurrency eliminate all of your privacy, but good chains will require a warrant to comply with the government in an investigation of a citizen's rights. Stablecoin 2.0 leads the way in privacy and transparency in a weird dynamic. On one hand, the companies are built to protect the privacy of the investors; however, in the event that a warrant is secured, the companies will provide all aid they can to the government to capture criminals that are using cryptocurrency for ill-gotten gain. Stablecoin 2.0 is defined by the fact that it uses a multi-asset fund structure, run by multiple independent funds (which may or may not be owned by the founding company). Stablecoin 2.0 also has a smart burn technology, which 'burns' a small amount from each transaction then re-introduces it into the system through asset purchases for the asset pool. This amounts to approximately a 4% growth

---

[43] Ann Kaluvu at https://www.thepourquoipas.com/post/quotes-from-business-leaders-to-understand-crypto.

each year in addition to the investments. This can reduce the time to maturity from 6 years to 4 years in an active system.

In this chapter, we will follow the same pattern that we follow in all the chapters in this part of the book. First, we look at the stability of the asset class on which the token at this level is based. The asset class is very important because it is the 'pegged' underpinning of the system. Next, we look at the liquidity of the asset class, how fast can the asset be liquidated if there is a 'run' on the token. As a matter of course, we will then look at the asset ratio needed to maintain stability. The media tends to insist that there needs to be a 1:1 ratio, and letter agencies tend to echo this false statement; each class has a ratio that is needed for the token to be reliable in its function.

Also, as a matter of course, it is always important to look at the relationship between holders/creators of the currency and the asset pool. The more complex the system the more analysis needs to be done on how the assets are being managed. While a 1:1 ratio for stablecoin 1.0 can be managed by the local bank of the creator, more complex systems require funds, investment firms, and even multiple subsidiaries to ensure that the asset pool maintains value. The analysis of who holds the asset pool then leads to an analysis of susceptibility of to attack. The more complex the asset, the more stable it is- however, it does need to be protected from attacks. The discussions within these sections of each chapter will be quite similar. Transparency is also an issue that varies slightly as you look at different types of stablecoins, once again the more complex the tokenomics- the more transparent the company needs to be. Finally, there needs to be an analysis of viability in major markets. This results in a bimodal, cyclic model for the analysis of the system, viability seems to be highest for simple and highly complex models, with intermediate models being lukewarm in their reception.

> "Success is the sum of small efforts- repeated day in and day out."
>
> Robert Collier

Overall, stablecoin 2.0 is the best stablecoin on the market…yet. The wonderful thing about the cryptoverse is that there is always advancement that is making the world a better place. As the system continues to develop, more stable methodologies will be created to protect people and their wealth. Stability will always be something that businesses look for in the marketplace. If the cryptocurrency market can provide more stability than fiat, then we are going to see more and more businesses go to the digital sector, ignoring the CBDCs which are simply a government ploy to track more elements of citizens' lives.

Stability of Asset Class

Stablecoin 2.0 represents a groundbreaking advancement in the realm of stablecoins, offering an unprecedented level of stability. This iteration of stablecoin is built on a foundation of multiple independent funds managing multi-asset pools. What sets Stablecoin 2.0 apart is its innovative smart burn system, which removes and reintroduces tokens into the market through the purchase of assets for the pool. This article will delve into the various aspects of Stablecoin 2.0, highlighting how its unique design and mechanisms contribute to its unparalleled stability.

Multi-Asset Pools: Diversification for Stability

Stablecoin 2.0's stability is derived from its utilization of multi-asset pools managed by independent funds. Unlike single asset backed stablecoins, which are susceptible to volatility and fluctuations in a single asset class, Stablecoin 2.0 spreads its holdings across a diversified portfolio of assets. This diversification provides a buffer against market volatility, reducing the risk associated with any single asset. By managing multiple asset classes, Stablecoin 2.0 ensures a more robust and stable foundation for maintaining its pegged value.

Smart Burn System: Maintaining Price Stability

The introduction of a smart burn system sets Stablecoin 2.0 apart from other stablecoin models. This system actively manages the supply of stablecoin tokens to maintain price stability. When the market demand for Stablecoin 2.0 increases, tokens are burned, reducing the circulating supply. The funds generated from the burn are then used to purchase additional assets for the multi-asset pool. This process ensures that the value of each Stablecoin 2.0 token remains stable and well-supported by the underlying asset pool.

Removal and Reintroduction of Tokens: Flexibility and Market Efficiency

The ability to remove and reintroduce tokens into the market through the smart burn system is a key feature of Stablecoin 2.0. This mechanism enables the supply of stablecoins to be adjusted dynamically based on market demand, ensuring price stability. When the market demand exceeds supply, tokens are removed through the burn process, creating scarcity, and supporting the value of the remaining tokens. Conversely, when demand is lower, tokens can be reintroduced by utilizing the funds generated from the burn to purchase additional assets. This flexibility ensures that Stablecoin 2.0 can adapt to market conditions while maintaining stability.

Enhanced Liquidity: Unparalleled Market Depth

> "The price of success is hard work, dedication to the job at hand and the determination that whether we win or lose, we have applied the best of ourselves to the task at hand."
>
> Vince Lombardi

Stablecoin 2.0's multi-asset pools and smart burn system contribute to enhanced liquidity and market depth. The diversified asset pools provide a more liquid market, as the multiple assets offer greater opportunities for trading and arbitrage. The smart burn system facilitates efficient price discovery and reduces the likelihood of market manipulation, further enhancing liquidity and market integrity. As a result, Stablecoin 2.0 becomes an attractive option for users and investors, as they can transact with confidence, knowing that stablecoin's value remains stable even in times of market turbulence.

Transparency and Trust: Ensuring Investor Confidence

Transparency plays a pivotal role in maintaining the trust and confidence of users and investors in Stablecoin 2.0. The underlying asset pools, fund management processes, and smart burn mechanisms should be transparently communicated and audited. By providing regular and reliable disclosures, Stablecoin 2.0 demonstrates its commitment to openness and accountability, which fosters trust among users and potential investors. The ability to verify the integrity of the multi-asset pools and understand the smart burn system instills confidence in the stability and reliability of Stablecoin 2.0.

Stablecoin 2.0 represents a significant advancement in stablecoin technology, offering unparalleled stability through its multi-asset pools and smart burn system. The diversification across multiple asset classes mitigates volatility risks, while the smart burn system ensures price stability by dynamically managing the token supply. This innovative design not only enhances stability but also provides flexibility, market efficiency, and enhanced liquidity. Moreover, transparency and trust are crucial aspects of Stablecoin 2.0, instilling confidence among users and investors. With its unique features and mechanisms, Stablecoin 2.0 has the potential to revolutionize the stablecoin market and contribute to the broader adoption of digital assets in the global financial ecosystem.

Liquidity of Asset Class

Liquidity is a critical factor in the success of any financial instrument, and stablecoins are no exception. Stablecoin 2.0 introduces a novel approach to liquidity management through the utilization of divided asset pools. Unlike traditional single asset stablecoins, Stablecoin 2.0 leverages the benefits of holding a multitude of different assets within its pools. This article explores how the diversified asset pools of Stablecoin 2.0 provide enhanced access to liquidity, despite the inclusion of various asset types.

Increased Trading Opportunities:

Diversified asset pools within Stablecoin 2.0 offer increased trading opportunities, resulting in enhanced liquidity. By incorporating multiple asset classes such as stocks, bonds, commodities, and real estate, Stablecoin 2.0

allows for a broader range of trading strategies. Traders and investors can access the stablecoin market and engage in various trading activities based on their preferences and market conditions. This diversified ecosystem attracts more participants, fostering a vibrant and liquid market for Stablecoin 2.0.

Arbitrage and Price Efficiency:

The inclusion of different asset types within the divided pools of Stablecoin 2.0 promotes efficient price discovery and arbitrage opportunities. As the stablecoin interacts with diverse assets, price discrepancies among these assets can arise. Traders can leverage these discrepancies to execute arbitrage strategies, capitalizing on market inefficiencies and promoting price convergence. The presence of arbitrageurs ensures that stablecoin prices align closely with the underlying asset values, contributing to market efficiency and reducing price deviations.

Risk Mitigation and Resilience:

> "Success has a thousand fathers, but defeat is an orphan."
>
> John F. Kennedy

Diversification of asset pools in Stablecoin 2.0 provides risk mitigation and resilience in times of market volatility. By holding a mixture of different assets, the stablecoin is less susceptible to the price fluctuations of any single asset class. If one asset experiences a downturn, the stability of the stablecoin is maintained by the performance of other assets within the pool. This risk diversification reduces the overall risk profile of Stablecoin 2.0 and enhances its ability to withstand market shocks, ensuring continued liquidity even during challenging market conditions.

Flexibility and Adaptability:

The presence of diversified asset pools allows Stablecoin 2.0 to adapt to changing market dynamics and investor preferences. Different asset classes perform differently in various market conditions, and by holding a range of assets, Stablecoin 2.0 can respond to market shifts. In times of economic expansion, the stablecoin may allocate more resources to growth-oriented assets, providing investors with exposure to potential upside. Conversely, during market downturns, the stablecoin may shift its focus towards more defensive assets, offering stability and preservation of value. This flexibility enhances the stablecoin's appeal to investors seeking tailored investment strategies.

Enhanced Market Depth:

Diversified asset pools contribute to increased market depth, creating a more liquid trading environment for Stablecoin 2.0. As the stablecoin interacts with multiple asset classes, it attracts a wider range of market participants, including institutional investors, traders, and retail investors. The increased participation and trading volume improve liquidity by ensuring that buy and sell orders can be executed efficiently at competitive prices. Deep market liquidity encourages greater adoption of Stablecoin 2.0 and fosters a robust ecosystem where participants can transact with ease and confidence.

The divided asset pools in Stablecoin 2.0 unlock a new level of liquidity, providing access to trading opportunities, fostering price efficiency, mitigating risks, and offering flexibility and adaptability. The inclusion of diverse assets promotes market depth and attracts a broader range of participants, enhancing the overall liquidity of Stablecoin 2.0. By capitalizing on the strengths of multiple asset classes, Stablecoin 2.0 establishes itself as a resilient and versatile stablecoin solution, poised to meet the demands of an ever-evolving financial landscape.

Required Asset Ratio for Stability

Stablecoin 2.0 revolutionizes the stability and maturity of stablecoins through the implementation of a smart burn feature. Unlike traditional stablecoin models, Stablecoin 2.0 allows for a lower initial asset ratio, starting as low as 20% and ideally settling at 40%, while still achieving maturity within a remarkably short timeframe of 4 years. This article explores how the smart burn feature in Stablecoin 2.0 accelerates its maturity and enables flexibility in the asset ratio.

Smart Burn Mechanism:

> "Always keep in mind that the resolution to succeed is more important than any other."
>
> Abraham Lincoln

The smart burn feature in Stablecoin 2.0 plays a crucial role in achieving accelerated maturity. As the stablecoin enters the market, a portion of the tokens is periodically burned, removing them from circulation. The burned tokens are then replaced by purchasing additional assets for the asset pool, thus maintaining the target asset ratio. This smart burn mechanism creates a continuous rebalancing process that optimizes the stability of Stablecoin 2.0 and expedites its journey towards maturity.

Flexible Asset Ratio:

Stablecoin 2.0 offers a unique advantage by allowing for a lower initial asset ratio, starting as low as 20% while aiming for an ideal ratio of 40%. This flexibility is made possible due to the underlying strength of the smart burn mechanism. The lower starting asset ratio enables the stablecoin to enter the market with a reduced initial capital requirement, attracting a wider range of participants and encouraging broader adoption. As the stablecoin gains traction and liquidity, the smart burn feature ensures a gradual increase in the asset ratio, reaching the optimal level of 40% over time.

Expediting Maturity:

The smart burn mechanism in Stablecoin 2.0 expedites the maturity process compared to traditional stablecoin models. By periodically burning a portion of the circulating tokens and replacing them with additional assets, the stablecoin continuously increases its asset coverage. This process accelerates the stabilization of Stablecoin 2.0, reducing the potential for price volatility and enhancing confidence among investors and users. The combination of the smart burn feature and gradual asset ratio increase results in a faster path to maturity, allowing Stablecoin 2.0 to establish itself as a stable and reliable medium of exchange within a relatively short timeframe of 4 years.

Market Adaptability:

The flexibility in the asset ratio and accelerated maturity of Stablecoin 2.0 contribute to its adaptability in dynamic market conditions. The smart burn feature allows the stablecoin to respond to changing market dynamics, adjusting the asset ratio to align with evolving market demands. In times of heightened volatility or increased investor interest in specific asset classes, Stablecoin 2.0 can adapt its asset composition to maintain stability and resilience. This adaptability enhances the stablecoin's appeal to investors and positions it as a reliable hedge against market fluctuations.

Investor Benefits:

The smart burn mechanism and the ability to start with a lower asset ratio provide significant benefits to investors. Firstly, a lower initial asset ratio reduces the barrier to entry, making Stablecoin 2.0 more accessible to a broader range of investors. Secondly, the accelerated maturity process allows investors to benefit from the stability and reliability of the stablecoin within a relatively short period. Moreover, the continuous rebalancing through smart burns ensures that the stablecoin remains aligned with the target asset ratio, providing transparency and trust to investors.

Stablecoin 2.0's smart burn feature and flexible asset ratio introduce a paradigm shift in stablecoin design, enabling accelerated maturity and adaptability in response to market dynamics. The combination of a lower initial asset ratio and periodic smart burns creates a robust mechanism for stability and investor confidence. This innovative approach positions Stablecoin 2.0 as a pioneering solution that addresses the limitations of traditional stablecoins, while offering unparalleled stability, liquidity, and growth potential in the evolving landscape of digital currencies.

Relationship between Holders and access to backing.

Stablecoin 2.0, with its advanced features and enhanced stability, has transformed the landscape of digital currencies. One notable aspect of Stablecoin 2.0 is the ownership structure of its asset pools, which is often facilitated through independent subsidiaries. This article delves into the reasons behind the prevalence of independent subsidiaries and explores how this model provides benefits and safeguards for the management and governance of asset pools in Stablecoin 2.0.

The utilization of independent subsidiaries for asset pool ownership in Stablecoin 2.0 ensures compliance with regulatory frameworks and mitigates potential risks. By establishing independent entities to hold the asset pools, firms can segregate the assets from their core operations, reducing the risk of commingling or misallocation. Independent subsidiaries are subject to specific regulations and oversight, ensuring that the management of the asset pools adheres to stringent guidelines and safeguards against any potential conflicts of interest.

Independent subsidiaries in Stablecoin 2.0 contribute to the transparency and accountability of asset pool management. These entities operate with a distinct legal and governance structure, often requiring regular audits and reporting. This transparency fosters trust among investors and stakeholders, as they can easily access information related to the asset pool's holdings, valuation methodologies, and performance. The separation of asset pool ownership through independent subsidiaries also establishes clear lines of accountability, as these entities have dedicated management teams responsible for their operations.

The utilization of independent subsidiaries allows for a greater focus on asset management expertise and specialization. These subsidiaries are often staffed with professionals who possess specialized knowledge and experience in managing specific asset classes. By entrusting the asset pools to these specialized entities, firms can leverage their expertise in analyzing, acquiring, and managing diverse assets. This approach enhances the effectiveness and efficiency of asset allocation strategies, optimizing the performance of the stablecoin and mitigating potential risks associated with asset management.

> "Eighty percent of success is showing up."
>
> Woody Allen

The adoption of independent subsidiaries bolsters investor protection and fosters trust in Stablecoin 2.0. The clear separation between the stablecoin-issuing firm and the independent subsidiary creates a level of insulation, shielding the assets from the potential financial or operational risks that may arise within the parent company. This separation provides an additional layer of assurance to investors that the asset pool is managed independently, ensuring the stability and integrity of the stablecoin. Investors can have confidence that the assets are held and managed by an entity solely dedicated to that purpose, safeguarding their interests, and minimizing potential conflicts of interest.

Independent subsidiaries offer flexibility and scalability for asset pool management in Stablecoin 2.0. As the stablecoin ecosystem evolves and expands, the use of independent entities provides a scalable framework for managing larger and more diverse asset pools. These subsidiaries can be structured to accommodate future growth, allowing for the addition of new assets, the implementation of advanced risk management strategies, and the adaptation to changing market conditions. The flexibility offered by independent subsidiaries ensures that Stablecoin 2.0 can meet the evolving needs of investors and maintain its stability and resilience.

The adoption of independent subsidiaries for asset pool ownership in Stablecoin 2.0 represents a sound approach that aligns with regulatory compliance, enhances transparency and accountability, leverages specialized expertise, and protects the interests of investors. By separating asset pool ownership from the stablecoin-issuing firm, independent subsidiaries provide a robust framework for managing asset pools, ensuring stability, and facilitating the long-term success of Stablecoin 2.0. This model not only addresses regulatory requirements and risk mitigation but also inspires trust and confidence among investors, fostering the continued growth and adoption of Stablecoin 2.0 in the dynamic landscape of digital currencies.

Susceptibility to Attacks

Stablecoin 2.0 has emerged as a groundbreaking digital currency solution, offering unprecedented stability and reliability in the volatile cryptocurrency market. One of the key attributes that sets Stablecoin 2.0 apart is its robust security architecture, which provides comprehensive protection against a wide range of attacks. This article explores how the unique asset architecture of Stablecoin 2.0 contributes to its enhanced security and mitigates potential vulnerabilities.

Stablecoin 2.0 employs a diversified asset architecture, comprising a wide range of asset classes. By diversifying the asset holdings across various categories such as fiat currencies, commodities, securities, and real estate, Stablecoin 2.0 reduces the risk associated with a single asset class. This diversification minimizes the impact of any potential attack targeting a specific asset class, as the stablecoin's value remains supported by the other assets in the portfolio. The resilience provided by a diversified asset architecture strengthens the stability of Stablecoin 2.0 and enhances its resistance to attacks.

Stablecoin 2.0 adopts an independent asset management model, wherein multiple independent funds manage distinct asset pools. Each fund operates autonomously, employing its own risk management strategies and investment decisions. This decentralized approach ensures that the assets are not concentrated under a single entity, reducing the vulnerability to attacks targeting a central point of control. Furthermore, the independent nature of asset management enhances transparency, accountability, and governance, which act as additional layers of protection against potential attacks.

Stablecoin 2.0 incorporates a smart burn mechanism, which dynamically adjusts the circulating supply of tokens based on market conditions. In the event of a potential attack or a significant fluctuation in the asset pool's value, the smart burn mechanism can initiate a controlled reduction in the circulating supply of stablecoins. This feature helps maintain the stability of Stablecoin 2.0 by adjusting the token supply in response to market dynamics, thereby mitigating the impact of external attacks on the stablecoin's value. The smart burn mechanism acts as a safeguard, enhancing the overall security and resilience of Stablecoin 2.0.

Stablecoin 2.0 implements rigorous monitoring and auditing processes to ensure the integrity and security of its assets. Regular monitoring of the asset pools, conducted by independent third-party auditors, helps identify any anomalies or potential vulnerabilities. This proactive approach allows for the prompt detection and resolution of security-related issues, reducing the likelihood of successful attacks. The continuous monitoring and auditing processes also provide transparency to stakeholders, fostering trust and confidence in Stablecoin 2.0.

Stablecoin 2.0 prioritizes the implementation of robust encryption techniques and secure storage protocols to protect the assets from unauthorized access and theft. The private keys associated with the asset pools are securely stored using advanced cryptographic algorithms, ensuring that they are shielded from potential breaches. Additionally, the adoption of secure storage solutions, such as hardware wallets and cold storage facilities, further enhances the security of the asset holdings. By safeguarding the keys and assets through encryption and secure storage practices, Stablecoin 2.0 fortifies its defense against attacks.

To bolster its security measures, Stablecoin 2.0 actively encourages external security researchers and experts to identify vulnerabilities. Regular security audits conducted by reputable firms provide an independent assessment of the security posture of Stablecoin 2.0. These comprehensive evaluations help identify and rectify any weaknesses, ensuring that the stablecoin remains resilient against attacks.

Stablecoin 2.0's unique asset architecture, coupled with robust security measures, establishes it as a highly secure and resilient digital currency. The diversification of asset classes, independent asset management, smart burn mechanism, continuous monitoring and auditing, encryption and secure storage, bug bounties, and security audits collectively contribute to the unparalleled protection against attacks. Stablecoin 2.0's commitment to security not only safeguards the value and integrity of the stablecoin but also instills confidence among users and stakeholders in the digital currency ecosystem. As the cryptocurrency market evolves, the advanced security architecture of Stablecoin 2.0 positions it as a trusted and resilient stablecoin solution in the face of emerging security challenges.

> "The foundation stones for balanced success are honest, character, integrity, faith, love and loyalty."
>
> Zig Zigler

Transparency

Transparency is crucial for building trust and confidence among users and investors. Stablecoin 2.0 operates on the principle of maintaining a stable value, and users rely on this stability when transacting and storing value. By providing transparent information about the underlying assets, their valuation methods, and the governance of the stablecoin, users can have greater confidence in the stability and integrity of the token.

Stablecoin 2.0 is backed by a diverse portfolio of assets managed by multiple independent funds. Transparency ensures that the asset pools are properly constituted and that the assets backing the stablecoin are accurately represented. Users can verify that the asset pools contain the expected assets and can assess their quality, ensuring that the stablecoin remains adequately collateralized.

Transparency allows users to assess the risks associated with Stablecoin 2.0. By providing information about the types of assets held in the portfolio, their risk profiles, and their performance, users can make informed decisions about the stability and safety of the stablecoin. Transparency helps users understand the potential risks and rewards of investing in or utilizing Stablecoin 2.0, enabling them to assess their risk tolerance and make informed choices.

Transparency is essential for ensuring compliance with regulatory frameworks. Stablecoin 2.0 operates within the bounds of financial regulations and may be subject to requirements such as anti-money laundering (AML) and know your customer (KYC) regulations. Transparent reporting and disclosure enable regulatory authorities to monitor and enforce compliance, protecting users and maintaining the integrity of the stablecoin ecosystem.

Transparency facilitates accountability and good governance practices within the stablecoin ecosystem. By making information about the governance structure, decision-making processes, and key stakeholders accessible, Stablecoin 2.0 fosters accountability among its participants. Transparent governance mechanisms allow users to understand how decisions are made, who is responsible, and how conflicts of interest are managed, ultimately promoting a more robust and trustworthy stablecoin ecosystem.

In summary, transparency plays a vital role in Stablecoin 2.0 by building trust, enabling verification of asset backing, facilitating risk assessment, ensuring regulatory compliance, and promoting accountability and good governance. By providing open and accessible information, Stablecoin 2.0 can foster confidence among users, protect against fraud and manipulation, and create a stable and reliable digital currency that can serve as a trusted medium of exchange and store of value.

Viability in Major Markets

Stablecoin 2.0 has emerged as a revolutionary financial instrument that combines the stability of traditional currencies with the advantages of blockchain technology. With its innovative architecture, diverse asset pools, smart burn mechanism, and enhanced transparency, Stablecoin 2.0 is poised to disrupt the financial industry and become a major market force. This essay delves into the key factors driving the potential of Stablecoin 2.0, including its stability, liquidity, versatility, regulatory compliance, and global accessibility.

Stablecoin 2.0's primary strength lies in its ability to maintain price stability, making it an attractive medium of exchange and store of value. By leveraging multi-asset pools managed by independent funds, Stablecoin 2.0 diversifies risk and minimizes the impact of price fluctuations in any single asset class. Additionally, the smart burn mechanism ensures that the stablecoin's value remains tethered to the underlying assets, facilitating stability even in volatile market conditions. This stability enhances confidence among users, merchants, and investors, paving the way for widespread adoption.

Stablecoin 2.0 offers unprecedented liquidity due to its architecture, which enables efficient asset management and redemption processes. The divided asset pools provide increased access to liquidity, as independent funds manage diversified assets with varying liquidity profiles. This structure allows for faster and more seamless conversions of stablecoin into other assets or fiat currencies, enabling users to easily enter or exit positions. Moreover, the smart burn system intelligently manages the token supply, maintaining balance and stability while optimizing liquidity within the ecosystem.

> "I've failed over and over and over again in my life and that is why I succeeded."
>
> Michael Jordan

Stablecoin 2.0's versatility is a key factor driving its potential as a major market force. Its multi-asset approach enables exposure to a wide range of asset classes, including currencies, commodities, real estate, and more. This diversification not only enhances stability but also provides users with the opportunity to tailor their investment strategies and manage risk effectively. Stablecoin 2.0's versatility extends beyond traditional financial markets, as it can also facilitate decentralized finance (DeFi) activities, such as lending, borrowing, and yield farming. This flexibility makes Stablecoin 2.0 a preferred choice for users seeking a comprehensive and adaptable financial instrument.

Stablecoin 2.0's commitment to regulatory compliance positions it favorably in the market. By adhering to financial regulations and embracing transparency, Stablecoin 2.0 mitigates risks associated with regulatory uncertainty. Compliance with AML and KYC regulations, along with transparent reporting and disclosure, ensures that the stablecoin ecosystem operates within the boundaries of legal frameworks. This regulatory compliance not only safeguards users but also fosters trust among institutional investors and regulatory authorities. Stablecoin 2.0's willingness to collaborate with regulatory bodies sets it apart from many other cryptocurrencies and positions it for widespread acceptance.

Stablecoin 2.0's global accessibility is a catalyst for its potential as a major market force. Its borderless nature allows for seamless transactions across jurisdictions, making it an ideal medium for international commerce.

Moreover, stablecoin's ability to facilitate faster and cheaper cross-border transfers compared to traditional financial systems enhances its appeal for global remittances and international trade. The inclusive nature of Stablecoin 2.0 enables individuals with limited access to traditional banking services to participate in the digital economy, thereby fostering financial inclusion on a global scale. The combination of stability, liquidity, versatility, regulatory compliance, and global accessibility positions Stablecoin 2.0 as a transformative force in the financial landscape.

Stablecoin 2.0 is poised to be a major market force due to its unparalleled stability, enhanced liquidity, versatility, regulatory compliance, and global accessibility. By leveraging diversified asset pools, a smart burn mechanism, and transparent governance, Stablecoin 2.0 addresses key challenges faced by traditional and existing stablecoins, creating a more robust and reliable financial instrument. With its potential to revolutionize cross-border transactions, facilitate DeFi activities, and empower individuals globally, Stablecoin 2.0 is set to play a significant role in shaping the future of finance and fostering a more inclusive and efficient digital economy.

Conclusion

Stablecoin 2.0 has emerged as a transformative financial instrument, combining the stability of traditional currencies with the advantages of blockchain technology. Through its innovative architecture, diverse asset pools, smart burn mechanism, and enhanced transparency, Stablecoin 2.0 has the potential to become a major market force. This chapter has explored the key features and advantages of Stablecoin 2.0, shedding light on its stability, liquidity, versatility, regulatory compliance, and global accessibility.

Stability is one of the defining characteristics of Stablecoin 2.0. By leveraging multiple independent funds managing diversified asset pools, Stablecoin 2.0 minimizes the risks associated with price fluctuations in any single asset class. This approach ensures that the stablecoin maintains its value even in volatile market conditions, enhancing confidence among users and investors. The smart burn mechanism further reinforces stability by managing the token supply and maintaining a balance between demand and supply. Stablecoin 2.0's stability makes it a reliable medium of exchange and a store of value, positioning it favorably in the market.

The liquidity of Stablecoin 2.0 sets it apart from other stablecoin variants. The divided asset pools managed by independent funds allow for increased access to liquidity. By diversifying assets with varying liquidity profiles, Stablecoin 2.0 enables users to convert their stablecoin holdings into other assets or fiat currencies more efficiently. This enhances the ease with which users can enter or exit positions, promoting a seamless user experience and attracting a broader user base. Additionally, the smart burn system optimizes liquidity within the ecosystem, ensuring a healthy balance between supply and demand.

Versatility is another key advantage of Stablecoin 2.0. The multi-asset approach allows for exposure to a wide range of asset classes, including currencies, commodities, real estate, and more. This diversification not only enhances stability but also provides users with flexibility in their investment strategies. Whether users seek traditional financial market exposure or wish to engage in decentralized finance (DeFi) activities, Stablecoin 2.0 offers a versatile and adaptable solution. This versatility positions Stablecoin 2.0 as a preferred choice for users looking for a comprehensive financial instrument that meets their unique investment objectives.

Regulatory compliance is crucial for the success and widespread adoption of stablecoins, and Stablecoin 2.0 is well-positioned in this regard. By prioritizing adherence to regulatory frameworks, including anti-money laundering (AML) and know-your-customer (KYC) regulations, Stablecoin 2.0 mitigates risks associated with regulatory uncertainty. Transparent reporting, disclosure, and collaboration with regulatory bodies build trust and confidence among institutional investors, regulatory authorities, and users. The commitment to regulatory compliance distinguishes Stablecoin 2.0 from many other cryptocurrencies and positions it favorably in the market.

The global accessibility of Stablecoin 2.0 further strengthens its potential as a major market force. Its borderless nature enables seamless transactions across jurisdictions, making it an ideal medium for international commerce. The efficiency and cost-effectiveness of Stablecoin 2.0 in cross-border transfers have the potential to revolutionize global remittances and international trade. Moreover, Stablecoin 2.0 plays a crucial role in fostering financial inclusion by providing individuals with limited access to traditional banking services the opportunity to participate in the digital economy. The inclusive nature of Stablecoin 2.0 contributes to a more equitable and accessible financial landscape.

In conclusion, Stablecoin 2.0 combines stability, liquidity, versatility, regulatory compliance, and global accessibility to position itself as a major market force. Through its innovative architecture and unique features, Stablecoin 2.0

addresses key challenges faced by existing stablecoins, offering a more robust and reliable financial instrument. As Stablecoin 2.0 continues to gain traction and adoption, it has the potential to shape the future of finance, empower individuals worldwide, and foster a more inclusive and efficient digital economy.

Chapter 19: Level Three Stablecoins (Stablecoins built based on a Stablecoin 2.0)

"Valor is stability, not of the legs and arms but courage of the soul"[44]

As the blockchain and cryptocurrency space continues to evolve, the demand for interoperability and cross-chain functionality has become increasingly apparent. Level Three stablecoins have emerged as a solution to facilitate seamless cross-chain access within the Stablecoin 2.0 ecosystem. Building upon the stability, liquidity, and versatility of Stablecoin 2.0, Level Three stablecoins offer enhanced utility and connectivity across different blockchain networks. This chapter explores the concept of Level Three stablecoins, their role in enabling cross-chain access, and the benefits they bring to the evolving digital economy.

Understanding Level Three Stablecoins

Level Three stablecoins represent a significant advancement in the realm of stablecoin technology, offering enhanced functionality and interoperability within the Stablecoin 2.0 ecosystem. These stablecoins build upon the foundation of Stablecoin 2.0, leveraging its stability, liquidity, and regulatory compliance to enable seamless cross-chain access. In this section, we delve deeper into the concept of Level Three stablecoins, exploring their features, benefits, and implications for the evolving digital economy.

---

[44] Michel de Montaigne at https://www.azquotes.com/quotes/topics/stability.html.

The Evolution of Stablecoin Technology:

Stablecoins have emerged as a vital component of the cryptocurrency ecosystem, addressing the issue of price volatility commonly associated with cryptocurrencies like Bitcoin and Ethereum. Stablecoin 2.0, with its diversified asset pools and innovative smart burn mechanisms, has introduced a new level of stability and resilience. Level Three stablecoins take this concept further by enabling cross-chain functionality, allowing for the transfer of stable value across different blockchain networks.

Cross-Chain Access and Interoperability:

Level Three stablecoins serve as gateways, providing users with the ability to move assets seamlessly between different blockchain networks. They bridge the gap between disparate chains, enabling interoperability and facilitating efficient value transfers. With Level Three stablecoins, users can leverage the stability and liquidity of Stablecoin 2.0 across multiple chains, expanding market reach and reducing reliance on single chain stablecoins. This cross-chain access unlocks new possibilities for decentralized applications, decentralized finance platforms, and other use cases that require integration across various chains.

Benefits of Level Three Stablecoins:

The integration of Level Three stablecoins within the Stablecoin 2.0 ecosystem offers several notable benefits. Firstly, it enhances liquidity by creating interconnected pools of value across different chains. Users can access and trade stablecoins seamlessly, regardless of the underlying blockchain technology, fostering deeper liquidity and reducing fragmentation in the market. Secondly, Level Three stablecoins facilitate cross-chain asset transfers, eliminating the need for complex token swaps and enhancing user experience on decentralized exchanges and lending platforms.

Moreover, Level Three stablecoins foster collaboration and interoperability between different blockchain communities. They create bridges that encourage the sharing of resources, expertise, and user bases, driving innovation and growth in the broader blockchain ecosystem. This increased collaboration leads to the development of standardized protocols and best practices for cross-chain communication, further improving the efficiency and effectiveness of value transfers.

Technical Implementation and Challenges:

Implementing Level Three stablecoins requires a robust technical infrastructure and careful attention to various challenges. Interoperability between different blockchain networks, each with its own consensus mechanisms and smart contract standards, poses technical complexities. Developers need to design and deploy efficient cross-chain bridges that ensure the secure transfer of assets while maintaining data integrity. Standardization efforts, such as the development of cross-chain communication protocols, are essential to establish seamless value transfers and data exchange across chains.

> "You can't build a great building on a weak foundation."
>
> Gordon B. Hinkley

Additionally, regulatory considerations and compliance requirements are paramount when operating Level Three stablecoins. Collaboration with regulatory authorities and adherence to global regulatory frameworks ensure transparency, anti-money laundering (AML) compliance, and investor protection. Striking the right balance between privacy and compliance becomes crucial, as cross-chain transactions involve the transfer of sensitive data across multiple networks.

Level Three stablecoins represent a significant step forward in the evolution of stablecoin technology, providing cross-chain access and interoperability within the Stablecoin 2.0 ecosystem. By leveraging the stability, liquidity, and regulatory compliance of Stablecoin 2.0, Level Three stablecoins enhance liquidity, foster collaboration, and unlock new possibilities for innovation and growth in the digital economy. While technical implementation challenges and regulatory considerations exist, the potential benefits of Level Three stablecoins are substantial. As the blockchain industry continues to mature, Level Three stablecoins are poised to become a vital market force, facilitating seamless value transfers, and driving the widespread adoption of cross-chain solutions.

Cross Chain Access

In the rapidly evolving landscape of blockchain technology, interoperability has emerged as a crucial factor for the widespread adoption and scalability of decentralized applications. Cross-chain access plays a pivotal role in enabling interoperability, allowing different blockchain networks to communicate and exchange value seamlessly. Within the context of Level Three stablecoins, cross-chain access takes center stage, empowering users to leverage the stability and liquidity of the Stablecoin 2.0 ecosystem across multiple chains.

Cross-chain access refers to the ability to transfer assets and information between different blockchain networks. It allows users to move value from one chain to another, bridging the gap between disparate ecosystems and facilitating the seamless transfer of stable value. With Level Three stablecoins, users gain the ability to navigate across various chains, regardless of their underlying consensus mechanisms, smart contract standards, or blockchain architectures.

The benefits of cross-chain access are numerous and impactful. First and foremost, it enhances liquidity by creating interconnected pools of value across different chains. Liquidity is a critical factor for the efficient functioning of financial markets, as it enables the smooth execution of trades and transactions. By enabling cross-chain access, Level Three stablecoins break down the barriers between isolated liquidity pools, fostering deeper liquidity and reducing fragmentation in the market. Users can access and trade stablecoins seamlessly, unlocking a broader range of investment opportunities and market participation.

> "One must have first of all a solid foundation."
>
> Sri Aurobindo

Furthermore, cross-chain access facilitates the efficient transfer of assets and data across different blockchain networks. Traditionally, moving assets from one chain to another requires complex token swaps or cumbersome intermediary steps, leading to inefficiencies and delays. Level Three stablecoins simplify this process by providing a standardized mechanism for asset transfers. Users can send stablecoins across chains without the need for multiple conversions, minimizing transaction costs and streamlining the user experience.

In addition to enhanced liquidity and streamlined asset transfers, cross-chain access also fosters collaboration and interoperability between different blockchain communities. It enables the sharing of resources, expertise, and user bases, creating bridges that drive innovation and growth in the broader blockchain ecosystem. Cross-chain access encourages developers and projects from various chains to collaborate, leading to the development of standardized protocols and best practices for cross-chain communication. This collaborative approach strengthens the overall blockchain infrastructure, ensuring compatibility and interoperability between different networks.

However, enabling cross-chain access is not without its challenges. Interoperability between disparate blockchain networks requires careful consideration of technical complexities. Each blockchain operates on its own set of consensus mechanisms, smart contract standards, and governance models, which can create barriers to communication and data exchange. Developers must design and deploy efficient cross-chain bridges that ensure the secure transfer of assets while maintaining data integrity. Standardization efforts, such as the development of cross-chain communication protocols, are crucial to establishing seamless value transfers and data interoperability.

Moreover, regulatory considerations and compliance requirements play a significant role in enabling cross-chain access. Collaboration with regulatory authorities and adherence to global regulatory frameworks ensure transparency, anti-money laundering (AML) compliance, and investor protection. Striking the right balance between privacy and compliance becomes crucial, as cross-chain transactions involve the transfer of sensitive data across multiple networks. Addressing these regulatory challenges is essential to build trust and confidence among users, institutions, and regulators.

> "Build your empire on a firm foundation of the fundamentals."
>
> Lou Holtz

Cross-chain access is a transformative concept within the Stablecoin 2.0 ecosystem, empowering users with seamless interoperability and access to liquidity across multiple blockchain networks. It enhances liquidity, streamlines asset transfers, fosters collaboration, and drives innovation within the blockchain industry. While technical implementation challenges and regulatory considerations exist, the benefits of cross-chain access are substantial. By enabling efficient and secure value transfers, cross-chain access paves the way for the widespread adoption and scalability of decentralized applications, propelling the growth of the digital economy.

Conclusion: Benefits of Level 3 Stablecoins

Level Three stablecoins offer a range of benefits that make them a powerful tool in the cryptocurrency ecosystem. One of the key advantages is enhanced stability. By building on the foundation of a stablecoin 2.0 system, Level Three stablecoins inherit the robust stability mechanisms and risk management features of their underlying protocol. This stability is achieved through diversified asset pools, smart burn systems, and the expertise of professional fund managers. As a result, Level Three stablecoins provide users with a reliable and predictable value store, reducing the volatility typically associated with other cryptocurrencies.

Another significant benefit of Level Three stablecoins is their ability to enable cross-chain access within the cryptocurrency ecosystem. Cross-chain interoperability is a critical requirement for the seamless transfer of assets and liquidity across different blockchain networks. Level Three stablecoins serve as a bridge between various blockchain platforms, allowing users to access and transact with different cryptocurrencies and assets. This cross-chain functionality enhances liquidity, increases market efficiency, and facilitates the integration of different decentralized finance (DeFi) protocols.

Level Three stablecoins also offer improved scalability and efficiency compared to traditional stablecoin models. By leveraging the capabilities of a stablecoin 2.0 system, Level Three stablecoins can handle a higher transaction volume, supporting faster and more cost-effective transactions. The use of advanced technology, such as layer-two solutions and optimized smart contract architectures, enables Level Three stablecoins to process transactions at a higher throughput, reducing congestion and transaction fees.

Furthermore, Level Three stablecoins promote transparency and accountability within the cryptocurrency ecosystem. The use of audited and regulated asset pools, combined with robust reporting mechanisms, ensures that users have access to accurate and up-to-date information regarding the underlying assets. This transparency builds trust among users and provides greater visibility into the stability and security of the stablecoin ecosystem. Additionally, Level Three stablecoins often employ on-chain governance mechanisms, allowing token holders to participate in decision-making processes and ensure the fair and democratic management of the stablecoin protocol.

Lastly, Level Three stablecoins contribute to the broader adoption and mainstream acceptance of cryptocurrencies. Their stability, cross-chain functionality, scalability, and transparency make them an attractive option for individuals, businesses, and institutional investors looking for a reliable and efficient means of transacting in the digital economy. The reduced volatility and increased liquidity of Level Three stablecoins make them a suitable medium of exchange, store of value, and unit of account, facilitating wider usage and acceptance in everyday transactions.

In conclusion, Level Three stablecoins offer enhanced stability, enable cross-chain access, improve scalability and efficiency, promote transparency, and contribute to the broader adoption of cryptocurrencies. As the cryptocurrency ecosystem continues to evolve, Level Three stablecoins provide a robust and versatile financial infrastructure that bridges different blockchain networks, enhances liquidity, and provides users with a reliable and efficient means of transacting. With their unique features and benefits, Level Three stablecoins are poised to play a significant role in shaping the future of digital finance.

Conclusion Part II

Throughout our extensive discussion on stablecoins, we have explored various iterations and advancements in the stablecoin landscape. From Stablecoin 1.0 to Stablecoin 2.0 and Level Three stablecoins, each iteration brings unique features and improvements to address the challenges faced by traditional cryptocurrencies. Let's summarize the key insights we've gained from our exploration.

Stablecoin 1.0 emerged as the first attempt to create a stable cryptocurrency by pegging it to a traditional fiat currency. While these stablecoins offered stability in terms of value preservation, they often lacked sufficient transparency and faced challenges related to centralization and counterparty risk. Despite their early promise, Stablecoin 1.0 models failed to gain significant market penetration due to these limitations.

The evolution of stablecoin technology led us to Stablecoin 1.3, which introduced algorithmic mechanisms to maintain stability. While these models aimed to be decentralized and autonomous, they struggled to achieve widespread adoption due to their complexity and susceptibility to market manipulations. The lack of sufficient market liquidity and trust also hindered the growth of Stablecoin 1.3.

Stablecoin 1.4 brought a new approach by anchoring stablecoins to digital assets, such as cryptocurrencies or tokenized assets. This model acknowledged the inherent volatility of the underlying assets, which required a higher risk tolerance in the short term. However, over the long term, these stablecoins demonstrated positive stability, aligning with the performance of the underlying assets. The unprecedented liquidity of Stablecoin 1.4 made it an attractive option, especially in the international market, where it gained substantial success.

One of the significant advancements came with Stablecoin 1.5, which introduced the concept of stablecoins backed by a collection of different fiat currencies. This approach aimed to leverage the stability of multiple currencies, reducing the reliance on a single currency's performance. Although Stablecoin 1.5 faced liquidity challenges in the short term, the diversification of assets and the long-term stability of fiat currencies contributed to its overall stability.

Stablecoin 1.6 introduced the concept of stablecoins backed by mutual funds. These stablecoins offered stability through a well-diversified fund, managed by experienced professionals. While this model provided stability, the liquidity of the stablecoin relied on the liquidity of the underlying fund. The required asset pool level of 50% ensured sufficient backing for the stablecoin's value, with a growth potential that allowed the fund to reach parity within seven years.

As we progressed to Stablecoin 1.7, land-based assets became the underlying collateral. While land investments offered stability, they posed liquidity challenges as the value of land cannot be quickly accessed. The asset pool ratio of around 60-62% ensured stability, and the smart burn feature allowed the stablecoin to reach maturity within a relatively short period.

The transition to Stablecoin 1.8 brought stablecoins backed exclusively by derivatives. While this model offered potential stability, the valuation of derivatives remained highly volatile. Consequently, investors needed to find a ratio that they were comfortable with, as there was no fixed asset ratio for Stablecoin 1.8. The importance of expert risk management and careful asset selection played a crucial role in maintaining stability.

With Stablecoin 1.9, stablecoins were backed by a full family of a single asset class. This approach allowed for stability through diversified assets within a specific class, reducing the reliance on a single asset's performance. The asset ratio of around 50% ensured the stablecoin's stability, with a relatively shorter time to maturity of approximately seven years.

Stablecoin 2.0 emerged as a game-changer in the stablecoin landscape. By incorporating multiple independent funds managing multi-asset pools and utilizing a smart burn system, Stablecoin 2.0 offered unprecedented stability. The smart burn system removed and reintroduced tokens into the market through the purchase of assets, ensuring stability while managing supply and demand dynamics. Additionally, the divided asset pools enhanced liquidity, even with a multitude of different assets, as each asset pool contributed to the overall liquidity of the stablecoin ecosystem.

The benefits of Level Three stablecoins became evident as they leveraged the stablecoin 2.0 system to provide cross-chain access within the cryptocurrency ecosystem. Level Three stablecoins acted as bridges between different blockchain platforms, facilitating the seamless transfer of assets and liquidity. This cross-chain functionality enhanced liquidity, increased market efficiency, and promoted the integration of various decentralized finance (DeFi) protocols. The ability to move assets and value across different blockchains efficiently opens up new possibilities for innovation and collaboration in the cryptocurrency space.

Throughout our discussions, we recognized the importance of transparency in stablecoin ecosystems. Transparency builds trust and confidence among users, allowing them to verify the stability and security of the underlying assets. Stablecoins backed by real-world assets require detailed reporting and auditing mechanisms to provide investors with accurate and up-to-date information. Furthermore, on-chain governance mechanisms empower token holders to participate in decision-making processes, ensuring fair management and democratic control of the stablecoin protocol.

The success of stablecoins in the market has been driven by their unique value propositions and the growing demand for stable and efficient digital currencies. Stablecoins offer a reliable store of value, a medium of exchange, and a unit of account, which are essential functions for wider adoption and acceptance of cryptocurrencies. The stability, liquidity, and transparency provided by different stablecoin models have attracted individuals, businesses, and institutional investors seeking secure and efficient means of transacting in the digital economy.

In conclusion, stablecoins have undergone significant evolution, addressing the challenges faced by traditional cryptocurrencies. From Stablecoin 1.0 to Stablecoin 2.0 and Level Three stablecoins, each iteration brings improvements in stability, liquidity, scalability, and transparency. These advancements have positioned stablecoins as a major force in the cryptocurrency market, bridging the gap between traditional finance and the digital economy. As the ecosystem continues to evolve, stablecoins will play a pivotal role in shaping the future of decentralized finance, providing stability, liquidity, and seamless cross-chain interoperability for a wide range of financial applications.

Part III: Regulations

Introduction to Part III

Regulations play a vital role in shaping the landscape of the crypto industry, providing a framework for its development, and ensuring the protection of stakeholders. This section explores three distinct perspectives: regulations that we have, regulations that we want, and regulations that we do not want. Firstly, we delve into the regulations that are already in place. This includes securities law, contract law, anti-fraud law, anti-money laundering law, and anti-terrorism law. These existing regulations serve as a foundation for establishing transparency, accountability, and investor protection within the crypto space. By understanding these regulations, we gain insight into the current regulatory landscape and its impact on the industry.

Moving on, we explore the regulations that we aspire to have. These desired regulations address emerging challenges and opportunities within the crypto industry. They include depository insurance for crypto, recovery protocols, international KYC clarification, protections for exchanges, ethics protocols for government officials overseeing the industry, trading bans for high-ranking politicians and judges, specialized asset classification for stablecoins, a UN protocol for handling crypto, and representation of digital citizens. These regulations aim to foster responsible growth, enhance security, and promote fair practices within the industry.

In contrast, we also examine the regulations that we don't want. This includes the classification of crypto as a security or commodity, micro-oversight of the global system, and over-regulation of the developing ecosystem. By identifying these potential pitfalls, we strive to avoid hindering innovation and stifling the growth potential of cryptocurrencies.

Finally, we explore the concept of building a better system through cooperation. Collaboration among international panels, the establishment of banking protection for cryptocurrencies, special panels to increase cryptocurrency participation, and the limitation of Central Bank Digital Currencies (CBDCs) are proposed as ways to foster innovation, create harmonious regulations, and promote a more inclusive and robust crypto ecosystem.

By examining the regulations that we have, the regulations that we want, the regulations that we don't want, and the potential for building a better system through cooperation, we aim to contribute to the ongoing dialogue surrounding the development and regulation of cryptocurrencies, with the ultimate goal of fostering a secure, transparent, and sustainable future for the industry.

Chapter 20: Regulation We Have

"If you make 10,000 regulations you destroy all respect for the law."[45]

Introduction

Regulations are a funny thing; we all hate them but most of us need them. This is very apparent in the current cryptocurrency landscape, where we do not have enough regulations to keep the system moving forward at the rate that entropy is pushing it. This puts us in a very difficult space of asking for new regulations but resisting the urge to deal with the regulatory agencies on their terms. As the cryptoverse expands, the regulatory agencies and the crypto community need to work together to build a better future, or else the future will be dystopian.

Security law is one of the biggest plugs we have seen in the development of the stablecoin industry. While, at least in the US, stablecoins do not meet the Howey test to be a security (failing the profit standard), there are still powerful groups trying to overregulate the space. The community needs a definitive stance from lawmakers to say that cryptocurrency in general, especially stablecoins are not a security unless:

1.) They are claiming that everyone will make massive profits even after the token matures; or
2.) They are using the tokens as a governance token as well as a stablecoin.

---

[45] Winston Churhchill at https://www.azquotes.com/quotes/topics/regulation.html.

If both of these factors are met, then the coin is not a security.

The remainder of the law that we have is actually quite useful for the stablecoin market. Contract law is the binding law in the area, and this is very well developed. When it comes to criminal elements, anti-fraud law is actually in a good position to defend both institutions and users within the space. Because of activities by the governments of the world in the 1980s, anti-money laundering law is another area that is quite well developed judicially. Anti-terrorism laws, specifically for funding are a difficult area- especially with governments giving billions to rogue nations; however, stablecoin can help prevent this through good KYC in the system.

Securities Law

Introduction (approx. 100 words) Cryptocurrencies have gained significant attention and adoption in recent years, but the existing regulatory landscape, particularly security law, poses challenges for the development of new models within the industry. This essay explores the current state of security law as it pertains to cryptocurrencies and analyzes its impact on stifling innovation and impeding the growth of new models. By examining key regulations and their limitations, we aim to shed light on the barriers faced by entrepreneurs, startups, and developers in creating novel cryptocurrency models.

Overview of Security Law

Security law governs the issuance, trading, and disclosure of securities, and its application in the cryptocurrency realm has significant implications. The Howey Test, established by the U.S. Supreme Court, serves as a framework to determine whether a digital asset qualifies as a security. However, this test, designed for traditional securities, can be challenging to apply to cryptocurrencies due to their unique characteristics. The absence of clear guidelines and regulatory ambiguity surrounding security classification hampers the development of new models, as entrepreneurs face the risk of inadvertently violating security laws and incurring legal consequences.

> DUMBEST QUOTE IN THE BOOK
>
> "The time has come to put scaremongering to rest and recognize that regulation can often support and advance efficient and more dynamic markets."
>
> Elizabeth Warren

Challenges and Limitations

One significant challenge lies in the broad interpretation of securities laws, which potentially classifies a wide range of cryptocurrencies as securities. This classification subjects' innovators to burdensome registration requirements and compliance costs, impeding their ability to launch new models. The lack of flexibility in regulatory frameworks restricts the creativity and experimentation necessary for the development of innovative solutions within the cryptocurrency space.

Another limitation is the inconsistency in global security law regulations. Different jurisdictions have varying approaches to classifying and regulating digital assets, leading to regulatory fragmentation and uncertainty. This inconsistency hinders cross-border collaborations and limits the global reach of new cryptocurrency models.

Moreover, the lengthy and complex registration process for securities poses a significant barrier for startups and entrepreneurs. The costs associated with compliance, legal fees, and ongoing reporting obligations can be prohibitively high, deterring innovators from entering the market and stifling the development of new models.

Impact on Innovation

The stringent security regulations and associated compliance costs inhibit innovation within the cryptocurrency industry. Startups and entrepreneurs face significant challenges in accessing capital through initial coin offerings (ICOs) or other token sales due to the complexities and costs of complying with securities laws. This limited access to funding curtails the development of groundbreaking projects and constrains the growth potential of new models.

> "Rules are often made to be broken and are too often for the lazy to hide behind."
>
> Douglas MacArthur

Additionally, the strict regulatory environment discourages experimentation with tokenized business models. Entrepreneurs are hesitant to explore novel approaches that blur traditional boundaries and may fall under security regulations, fearing potential legal repercussions. The lack of regulatory clarity stifles creativity and innovation, preventing the industry from realizing its full potential.

The Way Forward

To foster innovation and allow the development of new models, it is imperative to address the challenges posed by current security laws. Regulators should provide clear and consistent guidelines for determining whether a digital asset qualifies as a security. This would reduce legal uncertainties and enable entrepreneurs to navigate the regulatory landscape with confidence.

Furthermore, creating regulatory sandboxes and pilot programs that offer limited exemptions from securities regulations for innovative cryptocurrency projects can encourage experimentation and foster the development of new models. These programs would provide an environment where entrepreneurs can test their ideas, gain regulatory insights, and refine their models without undue legal constraints.

Collaboration between industry stakeholders, regulators, and policymakers is also crucial. Engaging in constructive dialogue to identify regulatory gaps and develop flexible frameworks can promote responsible innovation while ensuring investor protection. This cooperative approach would enable regulators to stay informed about emerging technologies and market trends, allowing for timely and appropriate regulatory updates.

Current security law presents significant challenges for the development of new models within the cryptocurrency industry. The ambiguity and complexity of regulations hinder innovation, deter investment, and limit the potential of new ideas. By addressing these challenges through clear guidelines, regulatory sandboxes, and collaborative efforts, we can strike a balance between investor protection and fostering a thriving ecosystem of innovative cryptocurrency models. Only through such measures can we unlock the full potential of this transformative technology and its contributions to the global economy.

Contract Law

Contract law, as a cornerstone of legal systems worldwide, plays a crucial role in facilitating economic transactions and ensuring parties' rights and obligations. With the rise of cryptocurrencies and their underlying technology, blockchain, the relationship between contract law and the digital asset ecosystem has become increasingly significant. This essay explores the intricate connection between contract law and cryptocurrencies, discussing the challenges, opportunities, and potential solutions that arise within this dynamic landscape.

Understanding Contract Law in the Digital Age

Contract law provides the legal framework for agreements, defining the rights and responsibilities of parties involved in a transaction. Traditionally, contracts have been documented in written or verbal form, subject to legal enforcement and dispute resolution. However, the emergence of smart contracts, which are self-executing agreements coded on a blockchain, introduces new complexities and opportunities.

Smart Contracts: Opportunities and Challenges

Smart contracts hold immense potential to streamline transactions, automate processes, and enhance efficiency in various industries. By eliminating intermediaries and relying on blockchain's transparency and immutability, smart contracts can reduce costs, increase security, and provide verifiable and enforceable agreements. However, their implementation within the legal framework raises several challenges.

One challenge lies in the interpretability and enforceability of smart contracts. Unlike traditional contracts, smart contracts are executed automatically based on predefined conditions, leaving little room for subjective interpretation. Determining the legal consequences of code-based agreements and establishing liability in the event of errors or breaches poses a significant challenge for contract law.

> "The whole duty of government is to prevent crime and preserve contracts."
>
> William Lamb

Another challenge involves jurisdictional issues and cross-border enforceability. Cryptocurrencies operate in a decentralized and globalized environment, raising questions about which legal system applies and how to enforce contractual obligations across different jurisdictions. Bridging the gap between digital assets and traditional legal systems remains a complex task.

Contract Formation and Legal Recognition

Contract formation is a fundamental aspect of contract law. In the context of cryptocurrencies, challenges arise when determining the intent of the parties and the moment of contract formation. The absence of traditional indicators such as handwritten signatures or physical documents poses difficulties in establishing the mutual agreement required for contract formation. However, blockchain technology provides a potential solution by offering a decentralized and tamper-proof record of transactional data, which can serve as evidence of contractual intent.

Moreover, legal recognition of smart contracts presents a significant opportunity for enhancing efficiency and trust within the cryptocurrency ecosystem. By acknowledging the legal validity of smart contracts, governments can provide legal certainty and enforceability to digital agreements. Legislative initiatives and regulatory frameworks that explicitly recognize and enforce smart contracts could encourage broader adoption and facilitate the growth of decentralized applications and decentralized finance (DeFi) platforms.

Dispute Resolution and Consumer Protection

Effective dispute resolution mechanisms and consumer protection are vital for maintaining trust and confidence in contractual relationships. In the cryptocurrency space, where transactions occur pseudonymously and across borders, traditional methods of dispute resolution face significant challenges. The immutable nature of blockchain transactions, while providing transparency and auditability, presents hurdles when addressing contractual disputes or instances of fraud.

In response, the concept of decentralized dispute resolution mechanisms, such as blockchain-based arbitration or smart contract escrow services, has gained attention. These mechanisms aim to provide efficient, transparent, and decentralized solutions to resolve disputes within the cryptocurrency ecosystem. Implementing such mechanisms requires collaboration between legal systems, technology developers, and industry stakeholders to ensure fair and accessible resolution processes.

Contract law plays a vital role in the cryptocurrency ecosystem, defining the rights and obligations of parties and providing a framework for trust and enforcement. The emergence of smart contracts introduces new opportunities for efficiency and automation, but also challenges related to interpretability, enforceability, and jurisdiction. By addressing these challenges through innovative legal frameworks, recognizing smart contracts, and developing decentralized dispute resolution mechanisms, we can create a legal environment that fosters the growth of cryptocurrencies and enhances trust in digital transactions. Only through collaborative efforts between legal experts, policymakers, and technologists can we navigate the complex intersection of contract law and the evolving landscape of cryptocurrency.

Anti- Fraud Law

Anti-fraud law serves as a critical safeguard against fraudulent activities, protecting investors and maintaining the integrity of financial systems. In the context of the rapidly evolving cryptocurrency ecosystem, the application of anti-fraud laws presents unique challenges and opportunities. This essay explores the intersection of anti-fraud law and cryptocurrencies, discussing the specific challenges posed by fraudulent schemes within the digital asset space and the measures needed to mitigate risks and promote trust.

The Complexities of Fraud in the Cryptocurrency Realm

Cryptocurrencies have gained popularity as a decentralized and pseudonymous means of value transfer. However, these attributes also make the ecosystem susceptible to fraudulent schemes. The absence of centralized oversight and the lack of regulatory clarity create an environment that fraudsters can exploit. Cryptocurrency scams range from fraudulent initial coin offerings (ICOs) and Ponzi schemes to pump-and-dump schemes, fake exchanges, and phishing attacks.

Regulatory Challenges and Jurisdictional Issues Anti-fraud laws are typically jurisdiction-based, posing challenges when dealing with cross-border cryptocurrency fraud. Determining jurisdiction and enforcing regulations becomes complex due to the global nature of cryptocurrency transactions and the pseudonymous nature of participants. International cooperation and harmonization of anti-fraud regulations are crucial to effectively combat fraudulent activities within the cryptocurrency ecosystem.

Enhancing Investor Protection and Education

Educating investors about potential risks and fraudulent schemes is a critical aspect of combating cryptocurrency fraud. Regulators play a pivotal role in raising awareness, issuing investor alerts, and providing guidance to help individuals make informed decisions. Implementing robust Know Your Customer (KYC) and Anti-Money Laundering (AML) procedures can also enhance investor protection by reducing the anonymity often exploited by fraudsters.

Regulatory Frameworks and Enforcement Mechanisms

Developing a comprehensive regulatory framework tailored to the unique characteristics of cryptocurrencies is essential for combating fraud. This framework should encompass registration requirements for cryptocurrency exchanges and stringent due diligence for ICOs. Effective enforcement mechanisms, such as increased regulatory oversight, monitoring suspicious transactions, and imposing penalties for fraudulent activities, are vital to deter fraudsters and protect investors.

Technological Solutions: Blockchain Analysis and Artificial Intelligence

Leveraging technological advancements can assist in detecting and preventing cryptocurrency fraud. Blockchain analysis tools can trace transactions and identify suspicious patterns, enhancing transparency and aiding law enforcement agencies in their investigations. Additionally, artificial intelligence (AI) algorithms can analyze vast amounts of data, detect fraudulent behavior, and generate risk assessments to mitigate fraud risks in real-time.

International Cooperation and Information Sharing

Cryptocurrency fraud often transcends borders, necessitating international cooperation and information sharing among regulatory bodies, law enforcement agencies, and financial intelligence units. Collaborative efforts can enable the timely exchange of intelligence, facilitate investigations, and enhance the enforcement of anti-fraud measures across jurisdictions.

Effective implementation of anti-fraud law within the cryptocurrency landscape is crucial for safeguarding investor trust and maintaining the integrity of the digital asset ecosystem. Addressing regulatory challenges, enhancing investor education, developing comprehensive frameworks, leveraging technological solutions, and fostering international cooperation are key strategies to combat cryptocurrency fraud. By striking a balance between innovation and investor protection, we can create a secure and transparent environment that fosters the growth and sustainability of cryptocurrencies while minimizing the risks posed by fraudulent activities.

Anti- Money Laundering Law

Anti-money laundering (AML) regulations play a vital role in combating illicit financial activities and preserving the integrity of the global financial system. As cryptocurrencies gain prominence, their decentralized and pseudonymous nature raises unique challenges for AML efforts. This essay explores the intersection of AML law and cryptocurrencies, examining the specific challenges posed by money laundering in the digital asset space and discussing measures to strike a balance between transparency and innovation.

The Complexity of AML in the Cryptocurrency Ecosystem

> "Its bad form to mention money laundering. Instead, you talk about asset management structures and tax beneficial schemes."
>
> John Sweeny

Cryptocurrencies offer potential avenues for money laundering due to their global accessibility, ease of cross-border transfers, and the perceived anonymity they provide. The decentralized nature of blockchain technology makes it challenging to identify the true beneficiaries of transactions, and the use of privacy-focused cryptocurrencies further complicates AML efforts. Cryptocurrency mixers and tumblers, which obfuscate transaction trails, and the potential for layering funds through complex transactions pose significant challenges for AML initiatives.

Regulatory Frameworks and the Need for Adaptation

Existing

AML regulations were primarily designed for traditional financial systems and struggle to accommodate the unique characteristics of cryptocurrencies. Adapting regulatory frameworks to address the decentralized and borderless nature of digital assets is crucial. Regulators must collaborate with industry experts to develop comprehensive guidelines and ensure that AML obligations are effectively applied across the cryptocurrency ecosystem without stifling innovation.

Enhancing KYC Procedures and Transparency

Know Your Customer (KYC) procedures are fundamental in AML efforts, allowing financial institutions to verify the identity of their customers and monitor transactions for suspicious activity. In the cryptocurrency realm, implementing robust KYC measures for cryptocurrency exchanges and wallet providers can enhance transparency and mitigate money laundering risks. Striking a balance between customer privacy and regulatory compliance is vital to maintain trust in the cryptocurrency ecosystem.

Technological Solutions: Blockchain Analytics and Data Sharing

Leveraging blockchain analytics tools can assist in identifying suspicious transactions and tracing the flow of funds within the cryptocurrency ecosystem. By analyzing patterns and monitoring addresses associated with illicit activities, law enforcement agencies and financial institutions can detect and prevent money laundering more effectively. Encouraging data sharing and collaboration between entities can enhance the efficiency and accuracy of AML efforts across borders.

International Cooperation and Regulatory Consistency

Cryptocurrencies operate in a global landscape, and effective AML measures require international cooperation and regulatory consistency. Encouraging coordination between regulatory bodies, financial intelligence units, and law enforcement agencies is crucial to address the challenges posed by cross-border money laundering. Sharing best practices, intelligence, and information can facilitate the identification of global money laundering networks and improve enforcement actions.

Educating Stakeholders and Promoting Compliance

Education and awareness play a pivotal role in combating money laundering in the cryptocurrency ecosystem. By educating stakeholders about the risks, obligations, and consequences of non-compliance, regulators can foster a culture of compliance and encourage self-regulation within the industry. Continuous engagement with cryptocurrency businesses, exchanges, and users is essential to ensure widespread adoption of AML measures.

Balancing AML efforts with the innovative potential of cryptocurrencies is a complex task. Adapting regulatory frameworks, enhancing KYC procedures, leveraging blockchain analytics, promoting international cooperation, and educating stakeholders are key strategies to mitigate money laundering risks. By striking the right balance between transparency, regulatory compliance, and privacy, we can foster a secure and trusted cryptocurrency ecosystem that upholds the principles of anti-money laundering and supports continued innovation in the digital asset space.

Anti-Terrorism Law

The intersection of anti-terrorism law and cryptocurrencies presents unique challenges and opportunities in combating terrorist financing and ensuring global security. The decentralized and pseudonymous nature of cryptocurrencies has the potential to be exploited by malicious actors seeking to fund terrorist activities. This essay examines the relationship between anti-terrorism law and cryptocurrencies, focusing on the challenges posed by illicit financing and discussing the role of Know Your Customer (KYC) procedures in preventing terrorism financing within the digital asset space.

The Threat of Cryptocurrencies in Terrorism Financing

> "You get to a point where it gets very complex, where you have money laundering activities, and terrorist support activities converging at a certain point and becoming one."
>
> Sibel Edmonds

Cryptocurrencies offer a means for terrorist organizations to bypass traditional financial systems and anonymously transfer funds across borders. The global accessibility, ease of use, and potential for anonymity within the cryptocurrency ecosystem make it an attractive avenue for illicit financing. The ability to crowdfund, solicit donations, and transfer funds with minimal oversight presents challenges for anti-terrorism efforts.

Regulatory Frameworks and Enhanced Due Diligence

Developing robust regulatory frameworks that encompass cryptocurrencies and address terrorist financing is crucial. Governments and regulatory bodies must adapt anti-terrorism laws to include digital assets, imposing stringent due diligence requirements on cryptocurrency businesses. Enhanced KYC procedures, including identity verification and transaction monitoring, play a vital role in identifying suspicious activities and preventing terrorists from exploiting the anonymity provided by cryptocurrencies.

KYC: A Critical Tool in Preventing Terrorism Financing

Know Your Customer procedures play a pivotal role in preventing terrorism financing within the cryptocurrency ecosystem. By implementing strong KYC protocols, cryptocurrency businesses can verify the identity of their customers, flag high-risk individuals or entities, and monitor transactions for suspicious activities. KYC measures act as a deterrent to terrorists seeking to exploit the decentralized nature of cryptocurrencies for illicit purposes.

Collaboration between Regulators and Cryptocurrency Businesses

Effective prevention of terrorism financing requires collaboration between regulators, law enforcement agencies, and cryptocurrency businesses. Regulators should provide clear guidance on KYC requirements, ensuring that businesses implement robust verification processes. Close cooperation between these stakeholders can facilitate the sharing of intelligence, timely reporting of suspicious activities, and the swift freezing of funds associated with terrorism.

Technological Solutions: Blockchain Analysis and Monitoring Tools

Leveraging blockchain analysis tools can aid in identifying patterns and tracing the flow of funds within the cryptocurrency ecosystem. These tools can assist in flagging suspicious transactions and addresses associated with known terrorist organizations or individuals. Collaborative efforts between technology providers, regulatory bodies, and law enforcement agencies are essential to developing effective monitoring and analysis tools to combat terrorism financing.

International Cooperation and Information Sharing

Terrorism knows no borders, and international cooperation is crucial in combating terrorism financing within the cryptocurrency realm. Sharing intelligence, best practices, and regulatory approaches between countries can help identify and disrupt global terrorist networks operating with cryptocurrencies. Timely information sharing between financial intelligence units and law enforcement agencies can facilitate the tracking and freezing of terrorist-related funds.

Education and Awareness Initiatives

Raising awareness about the risks and consequences of terrorism financing in the cryptocurrency ecosystem is essential. Education campaigns targeting cryptocurrency users, businesses, and the general public can enhance understanding and promote responsible use of digital assets. By fostering a culture of compliance and awareness, stakeholders can actively contribute to preventing terrorism financing within the cryptocurrency space.

Addressing the challenges posed by terrorism financing in the cryptocurrency ecosystem requires a multifaceted approach. Enhancing regulatory frameworks, implementing robust KYC procedures, leveraging technology for analysis and monitoring, fostering international cooperation, and promoting education and awareness are key strategies in mitigating risks. By effectively integrating anti-terrorism measures into the cryptocurrency landscape, we can strengthen security, protect global financial systems, and disrupt the funding mechanisms of terrorist organizations.

Conclusion

The intersection of law and cryptocurrencies presents a complex landscape that requires careful consideration and innovative approaches. The regulations that currently exist for cryptocurrencies, such as securities law, contract law, anti-fraud law, and anti-money laundering law, play a vital role in establishing legal frameworks and protecting stakeholders. However, they also face significant challenges in adapting to the unique characteristics of digital assets.

Securities law seeks to ensure investor protection and market integrity, but its application to cryptocurrencies is still evolving. Clear guidelines are necessary to determine when cryptocurrencies should be classified as securities, striking a balance between regulatory oversight, and fostering innovation. Contract law, on the other hand, faces challenges in verifying and enforcing smart contracts, as the decentralized nature of blockchain technology often poses jurisdictional and enforcement issues.

Anti-fraud law is crucial in combatting fraudulent activities within the cryptocurrency space, where the absence of centralized oversight and regulatory clarity can be exploited by fraudsters. Developing comprehensive regulatory frameworks and leveraging technological solutions, such as blockchain analysis and artificial intelligence, can aid in detecting and preventing fraudulent schemes.

Anti-money laundering law plays a vital role in mitigating the risks associated with money laundering and terrorist financing in cryptocurrencies. Adapting existing AML regulations to address the unique characteristics of digital assets is necessary to enhance transparency, implement robust KYC procedures, and foster international cooperation. Collaboration between regulators, financial institutions, and law enforcement agencies is crucial in sharing intelligence, detecting suspicious activities, and enforcing AML measures effectively.

Similarly, anti-terrorism law needs to adapt to the evolving landscape of cryptocurrencies to combat illicit financing by terrorist organizations. Implementing strong KYC procedures and leveraging technology for blockchain analysis can aid in identifying and preventing terrorism financing within the digital asset space. International cooperation and information sharing are vital to disrupt global terrorist networks and freeze funds associated with terrorism.

While regulations are necessary, it is equally important to strike a balance between regulatory oversight and fostering innovation. Overregulation of the cryptocurrency ecosystem may stifle its growth and hamper technological advancements. Finding the right balance requires collaboration and dialogue between regulators, policymakers, legal experts, and industry participants.

Building a better system through cooperation is paramount. International panels and collaborative efforts can facilitate the exchange of best practices, harmonize regulations, and address jurisdictional challenges in the global cryptocurrency landscape. Additionally, cooperation between traditional banking institutions and cryptocurrency platforms can provide enhanced financial services, protect customer assets, and bridge the gap between fiat and digital currencies.

The future of cryptocurrencies relies on the development of a robust regulatory framework that fosters trust, transparency, and innovation. By addressing the challenges posed by existing regulations, promoting the regulations we want, and avoiding unnecessary regulations, we can create an environment that supports the growth and adoption of cryptocurrencies while mitigating risks and protecting stakeholders.

In summary, navigating the legal landscape of cryptocurrencies requires a delicate balance between regulatory oversight and fostering innovation. By adapting existing regulations, embracing technological solutions, fostering international cooperation, and promoting education and awareness, we can build a better system that safeguards stakeholders, enhances trust, and paves the way for the continued evolution and adoption of cryptocurrencies in a secure and responsible manner.

Chapter 21: The Regulations We Want

"Good regulation should be conducive to business and customer protection."[46]

In the ever-evolving landscape of cryptocurrencies, the need for comprehensive and forward-thinking regulations becomes increasingly evident. As the global adoption of digital assets accelerates, it is essential to establish a regulatory framework that promotes innovation, safeguards investors, and ensures the integrity of the financial system. This chapter explores the regulations we want for cryptocurrencies, focusing on key areas such as depository insurance, recovery protocols, international KYC clarification, exchange protections, ethical standards for government officials, trading bans for high-ranking positions, specialized asset classification, UN protocols, and representation of digital citizens.

One of the foremost regulations we seek is the establishment of depository insurance for cryptocurrencies. Just as traditional bank deposits enjoy the protection of deposit insurance, introducing a similar mechanism for digital assets would instill confidence and provide a safety net for investors. By safeguarding against the loss or theft of cryptocurrencies held in custodial accounts, depository insurance can alleviate concerns and foster greater trust in the crypto ecosystem.

The nascent nature of cryptocurrencies poses unique challenges in terms of asset recovery. Implementing recovery protocols that enable the retrieval of lost or stolen

[46] Jamie Dimon at https://www.azquotes.com/quotes/topics/regulation.html.

digital assets would provide crucial support to individuals and businesses facing such incidents. Mechanisms like multi-signature wallets, smart contract escrows, and decentralized arbitration could enhance the ability to recover compromised funds, thereby bolstering confidence and encouraging wider participation in the crypto space.

To ensure the integrity of the global financial system and combat illicit activities, strong clarification on international Know Your Customer (KYC) standards is imperative. Harmonizing and clarifying KYC requirements for cryptocurrency transactions across borders would facilitate effective monitoring and deter money laundering, terrorist financing, and other illicit activities. By establishing a consistent and transparent framework, international KYC standards would promote regulatory compliance and enhance the legitimacy of the cryptocurrency industry.

Cryptocurrency exchanges serve as vital gateways between traditional financial systems and the digital asset realm. Consequently, robust regulations are needed to safeguard the interests of users and ensure the security of their funds. Measures such as mandatory security audits, stringent data protection protocols, and stringent operational standards can fortify exchanges against hacking attempts, safeguard user funds, and mitigate the risk of fraudulent activities.

Given the potential impact of cryptocurrencies on economies and financial systems, it is crucial to establish ethics protocols for government officials who oversee the industry. Clear guidelines and conflict-of-interest regulations can help prevent corruption and abuse of power while ensuring that public officials act in the best interest of the people they serve. Upholding transparency, accountability, and ethical conduct would foster trust and integrity within the crypto ecosystem.

) To prevent conflicts of interest and potential abuses of power, a trading ban for politicians, judges, and government workers in high-ranking positions is a necessary regulation to consider. By imposing a prohibition on cryptocurrency trading for these individuals, a level playing field is established, eliminating any potential advantage derived from insider information or influence. Implementing a "double-blind" mechanism would ensure anonymity in trading activities and strengthen the integrity of the regulatory framework.

The unique characteristics of stablecoins necessitate a specialized asset classification that distinguishes them from commodities or securities. Clear guidelines on how stablecoins should be classified and regulated would provide certainty and clarity for issuers, investors, and users. By creating a distinct category for stablecoins, regulators can address their specific features, risks, and potential systemic impacts while fostering innovation and maintaining regulatory oversight.

In an increasingly interconnected world, a unified approach to cryptocurrency regulation is essential. Establishing a United Nations (UN) protocol on how to handle cryptocurrencies would facilitate collaboration and coordination among nations, enabling the development of global standards and best practices. This protocol would serve as a foundation for international cooperation in addressing challenges such as money laundering, terrorist financing, and cross-border regulatory arbitrage.

As the crypto ecosystem expands, it is crucial to ensure that the voices and interests of digital citizens are represented in the regulatory decision-making process. Enabling mechanisms for public consultation, industry collaboration, and stakeholder engagement would help shape regulations that are inclusive, balanced, and reflective of the diverse perspectives within the digital asset community. By fostering representation, regulators can better understand the needs and concerns of digital citizens, leading to more effective and equitable regulations.

> "But what is money? A physical representation of the abstract concept of effort."
>
> Brandon Sanderson

The regulations we want for cryptocurrencies aim to strike a delicate balance between fostering innovation and safeguarding stakeholders' interests. By introducing depository insurance, implementing recovery protocols, clarifying international KYC standards, providing exchange protections, establishing ethics protocols, enforcing trading bans, creating specialized asset classifications, formulating UN protocols, and ensuring representation of digital citizens, we can cultivate a secure, inclusive, and responsible crypto ecosystem. Through collaboration between regulators, industry participants, and the global community, we can shape a regulatory framework that enables the full potential of cryptocurrencies while addressing risks and fostering trust in the digital asset landscape.

Depository Insurance for Crypto

Deposit insurance is a crucial element that can contribute to the stability and resilience of the cryptocurrency ecosystem.

By providing a safety net for users' digital assets, depository insurance can instill confidence, attract new participants, and mitigate the risks associated with holding cryptocurrencies. In this discussion, we will explore why depository insurance would help cryptocurrency become more stable.

One of the primary concerns that hinder widespread adoption of cryptocurrencies is the perceived lack of security. While cryptocurrencies offer unique advantages such as decentralization and immutability, they are also prone to hacking, theft, and loss due to the absence of traditional financial safeguards. This vulnerability has resulted in instances of significant losses for individuals and businesses, creating an atmosphere of uncertainty and mistrust.

Depository insurance would address this concern by offering protection for users' digital assets, similar to how traditional bank deposits are insured. By guaranteeing the safety and recovery of funds in the event of theft or loss, depository insurance provides a sense of security and reassurance to cryptocurrency holders. This enhanced level of protection encourages wider participation in the crypto ecosystem, attracting both individual investors and institutional players who may have been hesitant to engage due to security concerns.

Moreover, depository insurance helps mitigate the systemic risk that arises from potential security breaches or insolvency of cryptocurrency exchanges and custodial platforms. These platforms, which act as intermediaries for users to store and trade cryptocurrencies, play a critical role in the ecosystem. However, their vulnerability to cyberattacks or mismanagement poses risks to the assets held by users.

With depository insurance in place, users can have confidence that their assets are safeguarded even in the event of a platform's failure. This reduces the overall risk in the system and contributes to the stability of the cryptocurrency market. Users can be more willing to hold their assets for longer periods, knowing that their investments are protected by insurance.

Depository insurance also helps address the challenge of user confidence and trust in the cryptocurrency market. Traditional financial systems have long-established mechanisms for asset protection and compensation in case of loss. These mechanisms, including deposit insurance, have been instrumental in building trust and stability in the banking sector.

By introducing depository insurance for cryptocurrencies, a similar level of protection is extended to digital assets. This not only instills confidence in existing participants but also attracts new users who may have been wary of engaging with cryptocurrencies due to concerns about security and protection. Increased participation and trust contribute to market stability by reducing the volatility driven by panic selling or speculative behavior.

Furthermore, depository insurance acts as a catalyst for regulatory acceptance and oversight. Regulators worldwide have been grappling with the challenge of effectively regulating cryptocurrencies to protect investors and maintain financial stability. The inclusion of depository insurance in the regulatory framework for cryptocurrencies offers an additional layer of protection for users and aligns with the objectives of financial regulators.

With depository insurance in place, regulators can establish a more robust framework to oversee cryptocurrency exchanges and custodial platforms. This oversight can include requirements for risk management, security standards, and regular audits, ensuring that adequate safeguards are in place to protect users' assets. The presence of depository insurance provides regulators with assurance that users' funds are protected, reducing the potential for systemic risks and fraudulent activities.

In summary, depository insurance would significantly contribute to the stability and maturity of the cryptocurrency ecosystem. By addressing security concerns, mitigating systemic risks, building user confidence, attracting new participants, and facilitating regulatory oversight, depository insurance creates a more stable environment for cryptocurrency holders and market participants. As the cryptocurrency market continues to evolve, the introduction of depository insurance would play a crucial role in fostering trust, encouraging long-term investment, and driving the widespread adoption of cryptocurrencies as a reliable and secure asset class.

Recovery Protocols for Crypto

The emergence of cryptocurrencies has revolutionized the financial landscape, offering individuals and businesses a decentralized and secure means of transacting value. However, the unique characteristics of cryptocurrencies also present challenges, particularly when it comes to the recovery of lost or stolen digital assets. This is why the implementation of recovery protocols is vital in addressing these issues and providing users with a safety net. In this discussion, we will explore why recovery protocols are necessary for lost cryptocurrencies or those extorted by criminals.

One of the primary reasons why recovery protocols are essential is the irreversible nature of most cryptocurrency transactions. Once a transaction is recorded on the blockchain, it becomes extremely difficult, if not impossible, to reverse or retrieve the funds. This poses a significant challenge when users accidentally send funds to the wrong address or fall victim to phishing scams or fraudulent activities.

> "You will learn to manage money, or the lack of money will manage you."
>
> Dave Ramsey

Recovery protocols would provide a mechanism to retrieve lost or stolen cryptocurrencies, offering users a lifeline in such situations. These protocols could involve the cooperation of various stakeholders, including cryptocurrency exchanges, wallet providers, and blockchain forensic experts. By leveraging their expertise and technical capabilities, these parties can work together to trace and recover the lost assets.

Furthermore, recovery protocols can help combat the increasing incidents of crypto-related crimes, such as ransomware attacks and extortion. Criminals are increasingly targeting individuals and organizations with sophisticated methods to extort cryptocurrencies, capitalizing on the perceived anonymity and irreversibility of transactions. Without effective recovery mechanisms, victims may face significant financial losses and emotional distress.

By implementing recovery protocols, law enforcement agencies and cybersecurity experts can work collaboratively to identify the perpetrators and trace the flow of funds in such criminal activities. These protocols can facilitate the recovery of extorted cryptocurrencies, enabling victims to regain their assets and discouraging criminals from engaging in illegal activities within the crypto space.

Moreover, recovery protocols can contribute to the overall stability and trust in the cryptocurrency ecosystem. The fear of losing funds due to accidental transactions or criminal activities may deter potential users from entering the crypto market. By providing a safety net through recovery protocols, the perceived risk associated with using cryptocurrencies is reduced, encouraging wider adoption and participation.

Recovery protocols can also support the development of standardized best practices and security measures within the cryptocurrency industry. As incidents of lost or stolen cryptocurrencies occur, industry participants can learn from these experiences and implement enhanced security measures to prevent future occurrences. This collective learning and implementation of preventive measures can strengthen the overall security posture of the crypto ecosystem.

However, it is important to note that implementing recovery protocols for cryptocurrencies raises some challenges and concerns. The decentralized and pseudonymous nature of blockchain transactions presents technical and legal complexities in the retrieval process. Striking a balance between preserving the security and immutability of blockchain transactions while enabling recovery mechanisms requires careful consideration and collaboration between industry stakeholders and regulators.

Recovery protocols for lost or stolen cryptocurrencies are necessary to address the challenges faced by users and organizations in the crypto space. These protocols provide a means to retrieve lost funds, combat crypto-related crimes, enhance user confidence, and promote the overall stability of the cryptocurrency ecosystem. As the crypto market continues to mature, the implementation of recovery protocols will be crucial in ensuring a safer and more user-friendly environment, ultimately driving the widespread adoption and acceptance of cryptocurrencies as a reliable form of digital value.

Strong Clarification on International KYC

The implementation of Know Your Customer (KYC) requirements has become a significant aspect of regulatory frameworks in the cryptocurrency industry. KYC regulations are intended to prevent illicit activities such as money laundering, terrorist financing, and fraud by verifying the identities of individuals engaging in cryptocurrency transactions. However, there is a need for strong clarification on how KYC should work in the cryptoverse, as the current landscape lacks standardized guidelines and practices. In this discussion, we will explore the importance of clear KYC regulations and propose the concept of a digital passport for use in the crypto community.

> "The complexity of regulatory requirements provides both a barrier to and opportunity for KYC transformation."
>
> Encompass

Clear clarification on how KYC should work is vital for several reasons. First, it promotes regulatory compliance and mitigates legal risks for businesses operating in the cryptocurrency space. Without clear guidelines, companies may struggle to implement effective KYC procedures, leading to regulatory violations and potential penalties. Clear and comprehensive KYC regulations provide businesses with a framework to design robust compliance programs that meet regulatory expectations.

Second, strong clarification on KYC procedures helps protect the interests of cryptocurrency users and investors. By verifying the identities of individuals, KYC measures can help prevent fraud, identity theft, and unauthorized access to digital assets. Users can have greater confidence in the platforms they engage with, knowing that proper due diligence has been conducted to ensure the integrity and security of the ecosystem.

Third, clear KYC regulations promote transparency and trust within the crypto community. The perception that cryptocurrencies are primarily used for illicit activities or money laundering undermines the credibility and mainstream adoption of digital assets. By implementing robust KYC procedures, the industry can demonstrate its commitment to operating within legal frameworks and foster trust among users, financial institutions, and regulators.

To facilitate strong clarification and enhance KYC processes in the cryptoverse, the concept of a digital passport could be introduced. A digital passport would serve as a secure and tamper-proof digital identity document for individuals engaging in cryptocurrency transactions. This digital passport could include verified personal information, such as name, address, date of birth, and biometric data, stored on a decentralized and encrypted blockchain network.

The digital passport would enable individuals to undergo KYC processes once and use their verified identity across multiple cryptocurrency platforms. This would reduce the burden of repeatedly submitting personal identification documents for each new platform or service, streamlining the onboarding process for users while maintaining privacy and security.

Implementing a digital passport for the cryptoverse would require collaboration between regulators, technology providers, and industry participants. The digital passport could be designed to adhere to privacy regulations, ensuring that individuals have control over their personal data and how it is shared. It would also incorporate robust security measures to protect against identity theft and data breaches.

The advantages of a digital passport extend beyond KYC processes. It could enable seamless and secure peer-to-peer transactions, simplifying the transfer of assets between individuals and facilitating global interoperability. The digital passport could also support the development of decentralized finance (DeFi) applications by providing trusted and verified identities for individuals to participate in various DeFi protocols.

However, the implementation of a digital passport for the cryptoverse requires careful consideration of potential challenges and concerns. Privacy and data protection are paramount, and individuals must have control over their personal information and the ability to consent to its use. Collaboration with regulatory authorities is essential to strike the right balance between preventing illicit activities and preserving user privacy.

In conclusion, strong clarification on how KYC should work in the cryptoverse is crucial to ensure regulatory compliance, protect users, and foster transparency and trust. Introducing a digital passport concept could streamline the KYC process, enhance security, and enable individuals to use their verified identities across multiple cryptocurrency platforms. By implementing clear KYC regulations and exploring innovative solutions like the digital passport, the crypto community can promote responsible and secure participation in the digital asset ecosystem, advancing its adoption and integration with traditional financial systems.

Protections for Exchanges

The stability of the cryptocurrency system relies heavily on the integrity and security of exchanges, which serve as crucial gateways for users to buy, sell, and store digital assets. However, the history of the cryptocurrency industry has witnessed numerous exchange hacks and insolvencies, resulting in significant financial losses for users. To ensure a stable crypto system, it is essential to implement protections and depository insurance for exchanges. In this discussion, we will explore why these measures are crucial and how they contribute to a more stable cryptocurrency ecosystem.

First and foremost, protections for exchanges instill confidence and trust among users. The crypto industry is still relatively new and evolving, and many potential investors remain hesitant to enter the market due to concerns about the security of their assets. By implementing robust protections for exchanges, such as secure custody solutions and comprehensive risk management protocols, users can have greater peace of mind knowing that their funds are safeguarded against theft and loss.

Depository insurance for exchanges is another critical aspect of protecting users and ensuring the stability of the crypto system. Deposit insurance provides a safety net for users' funds in the event of an exchange's insolvency or loss of assets. It acts as a financial backstop, reimbursing users for their losses up to a certain limit, which can help mitigate the negative impact of exchange failures on the overall market sentiment.

The presence of depository insurance not only protects individual users but also contributes to the overall stability of the cryptocurrency ecosystem. In the absence of insurance, the failure of a significant exchange could have cascading effects, leading to a loss of trust, market panic, and a decline in crypto asset valuations. By providing a level of financial protection, depository insurance helps maintain market confidence and resilience, preventing potential systemic risks.

Moreover, protections and depository insurance for exchanges encourage responsible and professional practices within the industry. Exchanges that prioritize security, risk management, and compliance are more likely to attract users and build a solid reputation. They are also more likely to develop robust internal controls and operational processes to prevent unauthorized access, implement cybersecurity measures, and ensure the integrity of transactions. This promotes a higher standard of professionalism and accountability across the crypto ecosystem.

In addition to protecting users' funds, these measures also incentivize exchanges to adopt better security practices and adhere to regulatory standards. Regulatory bodies play a crucial role in enforcing guidelines and imposing compliance requirements on exchanges. By mandating protections and depository insurance, regulators can ensure that exchanges implement adequate security measures and risk management protocols. This enhances the overall resilience of the crypto system and protects users from potential vulnerabilities and fraudulent activities.

It is important to acknowledge that the implementation of protections and depository insurance for exchanges requires collaboration between industry participants, regulators, and insurance providers. It involves setting clear guidelines and standards for security practices, establishing regulatory oversight, and developing insurance products tailored to the unique risks and characteristics of the cryptocurrency market. By fostering cooperation and coordination, stakeholders can work together to create a robust framework that safeguards the interests of users and promotes a stable and secure crypto system.

In conclusion, the implementation of protections and depository insurance for exchanges is crucial for the stability of the cryptocurrency ecosystem. These measures instill confidence among users, protect against financial losses, and contribute to market resilience. By incentivizing exchanges to adopt robust security practices and adhere to regulatory standards, the industry can build a solid foundation for the continued growth and mainstream adoption of cryptocurrencies. Collaborative efforts between industry participants, regulators, and insurance providers are essential to establish a framework that prioritizes the security and stability of the crypto system.

Ethics Protocols for Government Officials overseeing the Industry.

> "I consider ethics, as well as religion, supplements to law in the government of man."
>
> Thomas Jefferson

Ethics protocols are essential for government officials who have involvement in financial or crypto matters to ensure transparency, accountability, and the integrity of the regulatory process. The cryptocurrency industry is rapidly evolving, presenting unique challenges and opportunities that require responsible and ethical decision-making. In this discussion, we will explore why ethics protocols are crucial for government officials in the financial and crypto sectors and how they contribute to a fair and trustworthy regulatory environment.

First and foremost, ethics protocols provide clear guidelines and standards for government officials, outlining expected behaviors and actions in their professional roles. These protocols establish a framework that promotes ethical conduct, integrity, and impartiality, ensuring that officials act in the best interest of the public and uphold public trust. In the context of financial and crypto matters, where significant financial interests and potential conflicts of interest exist, adherence to ethics protocols becomes even more critical.

Government officials involved in financial or crypto matters often have access to sensitive information that can impact markets and investors. Adhering to ethics protocols helps prevent insider trading, unfair advantage, and misuse of confidential information. It ensures that officials do not engage in practices that could compromise the integrity of the regulatory process or create an unfair playing field for market participants.

Ethics protocols also help address potential conflicts of interest that may arise for government officials. The crypto industry, with its rapid growth and potential for significant financial gains, may attract officials who have personal investments or affiliations with related companies. By requiring officials to disclose such conflicts and recuse themselves from relevant decision-making processes, ethics protocols help maintain transparency, impartiality, and public confidence in the regulatory system.

Furthermore, ethics protocols contribute to the prevention of corruption and bribery in the financial and crypto sectors. Government officials who are responsible for overseeing and regulating these industries must be held to the highest ethical standards to ensure that their actions are not influenced by personal gain or external pressures. Implementing ethics protocols helps create a culture of integrity and professionalism, deterring unethical behaviors and enhancing public trust.

The implementation of ethics protocols for government officials in financial and crypto matters also fosters consistency and predictability in decision-making. Clear guidelines and standards help ensure that regulatory actions are based on objective criteria and the public interest rather than personal bias or favoritism. This promotes fairness, equality, and a level playing field for all market participants, contributing to the overall stability and growth of the financial and crypto sectors.

Moreover, ethics protocols help establish accountability mechanisms for government officials. By requiring officials to adhere to ethical standards and subjecting them to disciplinary actions for violations, these protocols create a system of checks and balances. This reinforces the idea that public servants are accountable to the public and must be held responsible for their actions, promoting a culture of transparency, and preventing abuse of power.

To effectively implement ethics protocols, government agencies must provide appropriate training, resources, and oversight to ensure compliance. This includes ongoing education on ethical conduct, guidance on identifying and addressing conflicts of interest, and mechanisms for reporting potential violations. Agencies should also establish independent oversight bodies or ethics committees to review and address complaints, ensuring that ethics protocols are enforced effectively.

In conclusion, ethics protocols are essential for government officials involved in financial and crypto matters to uphold transparency, accountability, and public trust. These protocols provide clear guidelines, address conflicts of interest, prevent corruption, and promote consistent and fair decision-making. By adhering to ethics protocols, government officials contribute to the development of a trustworthy regulatory environment that fosters the growth and stability of the financial and crypto sectors.

Trading Ban for Politicians, Judges and Government Workers in High-Ranking Positions (Double Blind)

Implementing a stock/security/crypto trading ban for politicians is a crucial measure to ensure the integrity of the regulatory process and prevent conflicts of interest. As public officials, politicians hold positions of power and influence that can significantly impact financial markets. Allowing them to engage in trading activities directly could create a fertile ground for insider trading, market manipulation, and abuse of power.

By imposing a trading ban, politicians are prevented from leveraging their privileged positions for personal gain through advantageous market knowledge. This safeguard helps maintain a level playing field for all investors, ensuring fairness and market integrity. It also reinforces the principle that public servants should prioritize the public interest above their personal financial interests.

However, it's important to note that a trading ban doesn't necessarily mean politicians are completely restricted from participating in financial investments. To address concerns of potential financial exclusion, an alternative solution can be the establishment of government-run double-blind funds. These funds operate in a transparent and independent manner, managed by financial experts who are not directly affiliated with politicians or subject to their influence.

Politicians can invest their funds in these double-blind funds, but with important restrictions in place. The values of these investments could be diminished by the inflation rate, ensuring that the politicians' financial gains are relatively limited. This approach aims to strike a balance between allowing politicians to invest their personal wealth while minimizing the potential for conflicts of interest or unfair advantage.

By investing in double-blind funds, politicians are subject to the same market risks and fluctuations as other investors. This diminishes their ability to benefit from insider knowledge or manipulate markets for personal gain. It also reinforces transparency and accountability, as the investment decisions and performance of the double-blind funds can be independently monitored and audited.

Moreover, the establishment of double-blind funds can enhance public trust and confidence in the political system. It demonstrates a commitment to transparency, ethical conduct, and responsible financial management by ensuring that politicians' investments are held to the same standards as those of the general public.

While implementing a trading ban for politicians and promoting the use of double-blind funds is a significant step in preventing conflicts of interest, it is equally important to establish robust monitoring and enforcement mechanisms. Independent oversight bodies should be established to monitor politicians' financial activities, investigate potential violations, and impose appropriate sanctions if necessary. This helps ensure compliance with the trading ban and strengthens accountability within the political system.

In conclusion, a trading ban for politicians in stocks, securities, and cryptocurrencies is a necessary measure to preserve the integrity of the regulatory process and prevent conflicts of interest. By directing politicians towards government-run double-blind funds, where their investments are subject to the same market risks as other investors, transparency, fairness, and accountability are maintained. This approach strikes a balance between allowing politicians to engage in financial activities while safeguarding against potential abuse of power and insider trading.

Specialized Asset Classification for Stablecoins (not commodity or Security)

Cryptocurrencies have emerged as a unique and innovative financial asset class, presenting governments around the world with a regulatory challenge. Instead of attempting to force cryptocurrencies into existing asset classifications, it would be more efficient and effective for governments to establish a specialized classification for cryptocurrencies. This tailored approach recognizes the distinct characteristics and functionalities of cryptocurrencies, fostering regulatory clarity, innovation, and growth in the digital asset space.

Firstly, cryptocurrencies possess distinct qualities that set them apart from traditional asset classes like commodities or securities. Unlike commodities such as gold or oil, cryptocurrencies do not have physical form and their value is derived from their underlying technology and network effects. Additionally, cryptocurrencies are not easily fit into the traditional definition of securities, as they often lack a centralized issuer and do not necessarily represent ownership or dividend rights. By recognizing these unique attributes, governments can avoid imposing ill-fitting regulations that may stifle innovation or hinder the potential benefits that cryptocurrencies offer.

A specialized asset classification for cryptocurrencies would provide regulatory clarity and certainty for market participants. Cryptocurrency projects, investors, and service providers often struggle with the ambiguity and inconsistency of regulatory frameworks across different jurisdictions. This uncertainty hampers the development of new business models and investment opportunities. By establishing a clear and distinct classification, governments can provide a transparent regulatory environment that facilitates compliance and encourages responsible growth in the cryptocurrency industry.

Moreover, a specialized classification would enable governments to implement targeted regulations that address the specific risks associated with cryptocurrencies. Cryptocurrencies, due to their decentralized nature and pseudonymous transactions, present unique challenges related to consumer protection, anti-money laundering (AML), and financial stability. By having a dedicated classification, governments can design regulations that specifically address these challenges without unnecessarily burdening other traditional asset classes. This approach allows for more efficient and effective oversight, enhancing investor protection and mitigating the risks associated with cryptocurrencies.

Furthermore, a specialized classification would foster innovation and technological advancements in the crypto space. Cryptocurrencies and the underlying blockchain technology have the potential to revolutionize various sectors, including finance, supply chain management, and identity verification. By creating a regulatory environment that recognizes the distinct nature of cryptocurrencies, governments can encourage experimentation and innovation. This can attract investment, talent, and businesses to their jurisdictions, positioning them as leaders in the evolving digital economy.

> "Ethics must be reintroduced to public servants to restore people's faith in government. Without such faith, democracy cannot flourish."
>
> Walter Cronkite

Additionally, a specialized classification can facilitate international cooperation and coordination in regulating cryptocurrencies. Given the global nature of cryptocurrencies and their borderless transactions, a harmonized approach among governments is essential to effectively address regulatory challenges. By establishing a standardized classification, governments can more easily collaborate and share best practices in regulating cryptocurrencies. This promotes consistency, reduces regulatory arbitrage, and encourages cross-border innovation and investment.

It is important to note that a specialized classification does not mean a lack of regulation or oversight. Rather, it signifies a regulatory framework tailored to the unique characteristics and risks of cryptocurrencies. Governments can still establish and enforce regulations related to consumer protection, AML, investor disclosure, and market integrity within this specialized classification. The aim is to strike a balance between fostering innovation and protecting the interests of market participants and the wider economy.

In conclusion, governments worldwide should consider the advantages of creating a specialized asset classification for cryptocurrencies. This approach recognizes the unique characteristics of cryptocurrencies, provides regulatory clarity, targets specific risks, fosters innovation, and promotes international cooperation. By embracing a tailored approach, governments can establish a regulatory framework that balances innovation with investor protection, paving the way for responsible growth and the realization of the potential benefits that cryptocurrencies and blockchain technology offer.

UN protocol on how to handle crypto.

The establishment of an international protocol, under the auspices of the United Nations (UN), to protect the rights of crypto companies and crypto users would bring numerous benefits to the cryptocurrency ecosystem. Such a protocol would provide a globally recognized framework for safeguarding the interests of all stakeholders involved in the crypto industry, fostering trust, stability, and legal certainty.

Firstly, an international protocol would address the regulatory fragmentation and inconsistency that currently exists across different jurisdictions. Cryptocurrency companies and users often face challenges navigating a patchwork of regulations that vary from country to country. This lack of harmonization creates uncertainty and hampers the development of innovative solutions and cross-border transactions. With a UN protocol in place, governments could collaborate and establish common standards, enhancing regulatory clarity and facilitating global interoperability for cryptocurrencies.

Secondly, the protocol would help protect the rights of crypto companies and users by establishing clear guidelines for compliance, security, and privacy. The crypto industry is susceptible to risks such as hacking, fraud, and data breaches. An international protocol could outline best practices and security measures to ensure the protection of user funds and personal information. This would enhance the trust of individuals and businesses in the crypto ecosystem, leading to increased adoption and participation.

Additionally, a UN protocol would promote transparency and accountability within the crypto industry. It could establish guidelines for disclosure requirements, audit procedures, and reporting standards. This would create a more transparent environment, reducing the potential for fraudulent activities and promoting investor confidence. By setting clear expectations for companies operating in the crypto space, the protocol would contribute to the overall integrity and legitimacy of the industry.

Furthermore, an international protocol would facilitate cross-border cooperation and information sharing among governments. Cryptocurrencies operate on a global scale, transcending national boundaries. A UN protocol would encourage collaboration among countries, enabling them to jointly combat money laundering, terrorist financing, and other illicit activities associated with cryptocurrencies. By sharing intelligence and cooperating in investigations, governments could effectively mitigate the risks posed by malicious actors while safeguarding the rights and interests of legitimate crypto businesses and users.

Moreover, a UN protocol would provide a platform for dispute resolution and legal recourse. Currently, the absence of a standardized international framework makes it challenging to resolve disputes involving cross-border transactions or conflicts between crypto companies and users. A UN protocol could establish mechanisms for arbitration and mediation, ensuring fair and efficient resolution of disputes. This would enhance legal certainty and provide a reliable framework for resolving conflicts, thereby boosting investor confidence, and encouraging participation in the crypto market.

In conclusion, the development of an international protocol under the UN to protect the rights of crypto companies and users would be highly beneficial for the cryptocurrency ecosystem. It would promote regulatory harmonization, enhance security and privacy standards, foster transparency, and accountability, facilitate cross-border cooperation, and provide mechanisms for dispute resolution. By establishing a globally recognized framework, the protocol would create a more stable and trustworthy environment for the crypto industry to thrive, driving its widespread adoption and realizing its potential as a transformative force in the global economy.

Representation of Digital Citizens

In the rapidly evolving digital landscape, the emergence of cryptocurrencies has given rise to the concept of "digital citizens." These are individuals who actively participate in the cryptoverse and engage in various digital activities, including cryptocurrency transactions, decentralized finance, and digital asset ownership. As the cryptoverse operates independently of traditional nation-states, there is a growing need to recognize and protect the rights of digital citizens separately from the rights guaranteed by national governments. In this context, it is essential to explore the reasons why digital citizens should have distinct rights in the cryptoverse.

Firstly, the cryptoverse transcends geographical boundaries and traditional jurisdictional frameworks. Cryptocurrencies and decentralized technologies operate on a global scale, enabling individuals from different countries to engage in borderless transactions and interactions. As digital citizens actively participate in this decentralized ecosystem, they require rights that are not constrained by national boundaries. Recognizing their distinct rights would ensure that they have access to a fair and inclusive digital environment that respects their individual liberties and protects their interests, regardless of their nationality or geographic location.

Secondly, the cryptoverse offers unique opportunities and challenges that require tailored rights for digital citizens. This digital realm is characterized by features such as decentralization, transparency, and pseudonymity. The underlying principles of cryptocurrencies, such as financial sovereignty and privacy, call for specific protections that may differ from those guaranteed by nation-states. By recognizing the distinct rights of digital citizens, we can address the unique risks and concerns they face in this rapidly evolving digital ecosystem, including privacy protection, data security, and safeguarding against digital fraud and theft.

Moreover, digital citizens deserve rights that enable them to fully participate in the cryptoverse and exercise control over their digital assets and identities. As individuals engage in various activities such as token ownership, decentralized governance, and participation in decentralized applications (DApps), they require clear and enforceable rights that empower them to manage their digital assets and identities with confidence. These rights may include the ability to securely store and control their private keys, access, and transfer their digital assets, and participate in the decision-making processes that govern decentralized platforms.

Furthermore, recognizing distinct rights for digital citizens can foster innovation, collaboration, and the growth of the cryptoverse. By providing a clear legal framework that respects and protects the rights of digital citizens, governments and regulatory bodies can instill confidence in individuals and businesses to actively participate in this transformative space. This, in turn, can attract investment, talent, and entrepreneurial activities, leading to the development of innovative solutions and the advancement of decentralized technologies. By recognizing the unique rights of digital citizens, we can unleash the full potential of the cryptoverse and create a thriving ecosystem that benefits individuals and society as a whole.

Lastly, as the cryptoverse evolves and intersects with traditional systems, there is a need to ensure harmonization between the rights of digital citizens and the regulatory frameworks of nation-states. This calls for constructive dialogue and collaboration between governments, international organizations, and the digital community. By acknowledging the distinct rights of digital citizens, governments can work towards creating a regulatory environment that promotes innovation, safeguards individual rights, and ensures compliance with broader legal and ethical principles.

Digital citizens actively participating in the cryptoverse require distinct rights separate from those guaranteed by nation-states. The global nature of the cryptoverse, unique opportunities and challenges it presents, the need for individuals to exercise control over their digital assets and identities, and the fostering of innovation and growth all emphasize the importance of recognizing and protecting the rights of digital citizens. By doing so, we can establish a framework that respects individual liberties, promotes inclusivity, and supports the continued development and maturation of the cryptoverse as a transformative force in our increasingly digital world.

Conclusion

In conclusion, the chapter on "The Laws We Want" explores the regulatory improvements and enhancements that are needed to foster a stable, secure, and inclusive cryptoverse. Throughout this chapter, we have highlighted several key areas where regulatory interventions can play a crucial role in shaping the future of cryptocurrencies and decentralized technologies.

Firstly, the implementation of depository insurance for cryptocurrencies would provide a vital layer of protection for digital assets held by individuals and institutions. This insurance would enhance confidence in the security and custodial practices of crypto service providers, mitigating the risks associated with hacking, theft, or loss of funds. By safeguarding the interests of users and encouraging responsible custody practices, depository insurance would contribute to the overall stability of the crypto ecosystem.

Secondly, the establishment of recovery protocols for lost or extorted cryptocurrencies is crucial in addressing the growing concern of funds becoming inaccessible due to accidental loss, theft, or ransomware attacks. These protocols would provide a framework for retrieving lost funds and facilitating the return of assets to their rightful owners. By providing a mechanism to address such incidents, recovery protocols would instill trust and promote the adoption of cryptocurrencies by assuring users that their assets are not irretrievably lost.

Furthermore, the need for strong clarification on international Know Your Customer (KYC) standards is essential to combat money laundering, terrorist financing, and other illicit activities. A clear and consistent framework for KYC procedures would enhance regulatory compliance, reduce fraudulent activities, and foster trust within the crypto industry. It would also facilitate cross-border transactions and ensure the global interoperability of cryptocurrencies while maintaining the necessary safeguards to protect against illicit activities.

Protections for exchanges, both in terms of regulatory oversight and depositary insurance, are critical to maintaining the integrity and stability of the crypto ecosystem. By establishing robust regulatory frameworks and implementing measures to protect against security breaches and operational risks, exchanges can provide a secure environment for users to trade and transact with confidence. These protections would not only enhance market integrity but also foster broader adoption and participation in the cryptoverse.

Moreover, the implementation of ethics protocols for government officials overseeing the industry is essential to prevent conflicts of interest, corruption, and undue influence. By establishing clear guidelines and standards of conduct, governments can ensure that public officials act in the best interests of the community and avoid any impropriety that may undermine the integrity of the crypto sector. Ethics protocols would promote transparency, accountability, and good governance practices, thereby creating an environment conducive to the healthy development of the industry.

A trading ban for politicians, judges, and government workers in high-ranking positions, coupled with a double-blind investment approach, would further strengthen the integrity of the cryptoverse. This measure would prevent potential conflicts of interest and ensure that individuals in influential positions cannot abuse their authority for personal gain. By implementing such a ban, we can foster trust, eliminate potential avenues for manipulation, and enhance the credibility of the crypto industry.

Specialized asset classification for stablecoins, separate from commodities or securities, is vital to recognize the unique characteristics and functions of these digital assets. Stablecoins play a crucial role in facilitating efficient and stable transactions within the cryptoverse. By establishing a clear and distinct asset classification, we can provide regulatory clarity, enable innovation, and ensure that stablecoins can operate within appropriate regulatory frameworks.

Furthermore, the development of a UN protocol to handle cryptocurrencies would provide a globally recognized framework for protecting the rights of crypto companies and users. This protocol would foster cooperation among nations, promote regulatory harmonization, and establish mechanisms for addressing cross-border challenges. By facilitating international collaboration and information sharing, the protocol would help combat illicit activities while preserving the rights and interests of legitimate participants in the cryptoverse.

Lastly, the representation of digital citizens in the regulatory decision-making process is crucial to ensure inclusivity, fairness, and representation in shaping the future of cryptocurrencies. Digital citizens, who actively engage in the cryptoverse, possess unique perspectives and insights that can inform effective policymaking. By involving digital citizens in regulatory discussions and governance processes, we can foster a sense of ownership, empower diverse voices, and promote the development of regulations that cater to the needs and aspirations of the crypto community.

In conclusion, the laws we want aim to establish a robust regulatory framework that fosters trust, stability, and innovation in the cryptoverse. By implementing depository insurance, recovery protocols, strong KYC standards, protections for exchanges, ethics protocols for government officials, trading bans for influential individuals, specialized asset classifications, UN protocols, and representation for digital citizens, we can create an environment that encourages responsible growth, safeguards user interests, and promotes the long-term sustainability of cryptocurrencies and decentralized technologies. It is through collaborative efforts, constructive dialogue, and adaptive regulation that we can strike a balance between innovation and protection, unleashing the full potential of the cryptoverse for the benefit of individuals, businesses, and society as a whole.

Chapter 22 Regulations We Don't Want

"I am a strong believer in the free market. I am strong believer in capitalism. But I am also a strong believe that there are certain common goods- our air, our water, making sure that people are safe- that require some regulation."[47]

Chapter 22 delves into the realm of regulations that, while well-intentioned, could have adverse effects on the development and growth of cryptocurrencies and decentralized technologies. In this chapter, we critically examine four regulatory aspects that may hinder innovation, impede market efficiency, and stifle the potential benefits of the cryptoverse.

One of the contentious issues discussed is the classification of cryptocurrencies as securities. The debate revolves around whether cryptocurrencies should be treated as traditional investment instruments subject to the same regulatory requirements as securities. While some argue that such classification could provide investor protections and market stability, others believe that it may impose burdensome compliance measures, impede accessibility, and limit the transformative potential of cryptocurrencies.

Similarly, the classification of cryptocurrencies as commodities has raised concerns within the crypto community. Commodities, being physical goods or raw materials, differ significantly from digital assets. Applying the same regulatory framework designed for physical commodities may not adequately address the unique characteristics and functionalities of cryptocurrencies,

---

[47] Barack Obama at https://www.azquotes.com/quotes/topics/regulation.html.

potentially stifling innovation and hindering the growth of the cryptoverse.

Another area of concern is the potential for micro-oversight of the global system governing cryptocurrencies. While regulatory oversight is essential for maintaining market integrity, excessive micro-level regulations may create unnecessary complexity, hamper market efficiency, and limit the agility and adaptability of decentralized technologies. Striking the right balance between oversight and fostering innovation is crucial to ensure the sustainable growth of the cryptoverse.

Furthermore, overregulation of the developing system can have unintended consequences. Cryptocurrencies and decentralized technologies are still in their nascent stages, rapidly evolving and exploring new frontiers. Imposing excessive regulations at this early stage could stifle innovation, deter investment, and impede the development of transformative solutions. It is vital to foster an environment that encourages experimentation, promotes responsible growth, and allows the technology to mature organically.

In this chapter, we aim to critically analyze the potential drawbacks and unintended consequences associated with these regulatory approaches. By exploring the implications of classifying cryptocurrencies as securities or commodities, as well as the impact of micro-oversight and overregulation, we can gain a comprehensive understanding of the challenges posed by these regulatory paths.

While the need for regulation in the cryptoverse is undeniable, it is crucial to strike a balance that fosters innovation, protects investors, and ensures market integrity without stifling growth or impeding technological advancements. Through thoughtful analysis and discussion, we can explore alternative regulatory approaches that address legitimate concerns while preserving the fundamental principles of decentralization, privacy, and financial inclusivity.

Classification of crypto as a security

The classification of all cryptocurrencies as securities has been a subject of intense debate within the crypto community and regulatory circles. While the intent behind such classification may be to provide investor protections and regulatory oversight, it could have devastating consequences for the industry and potentially infringe upon the human rights of the digital community. In this discussion, we will explore the reasons why such a classification would be detrimental and why it should be avoided.

> "We should favor innovation and freedom over regulation."
>
> George Allen

First and foremost, treating all cryptocurrencies as securities fails to recognize the fundamental differences between traditional securities and digital assets. Securities typically represent ownership or debt in a company or organization and are subject to specific regulatory requirements. Cryptocurrencies, on the other hand, encompass a wide range of digital assets with varying functionalities, including mediums of exchange, store of value, and utility tokens for accessing decentralized networks. Applying securities regulations to all cryptocurrencies overlooks their unique characteristics, hindering their potential for innovation and growth.

By classifying cryptocurrencies as securities, burdensome regulatory requirements designed for traditional securities would be imposed on the entire industry. These requirements include costly registration processes, extensive disclosure obligations, and ongoing reporting and compliance measures. Such onerous obligations would disproportionately affect smaller projects and startups, limiting their ability to innovate and compete in the market. It could create a regulatory barrier that stifles entrepreneurship and inhibits the development of new decentralized solutions, hindering technological progress and the potential benefits it can bring to society.

Moreover, the classification of cryptocurrencies as securities could infringe upon the human rights of the digital community. The right to access and participate in the digital economy is increasingly recognized as a fundamental human right. By subjecting cryptocurrencies to securities regulations, individuals may face barriers to entry, limited access to financial services, and reduced opportunities for financial inclusion. This hampers the ability of individuals, particularly those in underserved regions or marginalized communities, to exercise their economic rights and participate fully in the digital revolution.

Furthermore, the broad classification of cryptocurrencies as securities fails to acknowledge the decentralization and community-driven nature of many blockchain projects. Cryptocurrencies often operate on decentralized networks governed by consensus mechanisms, where decisions are made collectively by the community. Unlike traditional securities, there is no centralized issuer or controlling entity. Imposing securities regulations on these decentralized networks undermines their fundamental principles and stifles their ability to operate as intended. It restricts the autonomy and self-governance of digital communities, impeding their ability to innovate, experiment, and develop decentralized solutions.

It is important to recognize that the crypto industry is already subject to regulatory frameworks in many jurisdictions. Governments and regulators have taken steps to address concerns such as fraud, market manipulation, and investor protection through targeted regulations that focus on specific aspects of the crypto ecosystem. This approach allows for tailored regulations that address legitimate concerns without stifling the industry as a whole. Blanket classification of all cryptocurrencies as securities disregards this nuanced approach and imposes a one-size-fits-all solution that is ill-suited for the diverse and evolving nature of the cryptoverse.

Instead of a broad securities classification, a more appropriate approach would involve developing a regulatory framework that distinguishes between different types of cryptocurrencies based on their characteristics and functionalities. This approach would allow for targeted regulations that address specific risks associated with certain types of tokens, while still fostering innovation, encouraging investment, and safeguarding the rights of the digital community.

In conclusion, classifying all cryptocurrencies as securities would be devastating for the industry and a violation of the digital community's human rights. It disregards the unique characteristics of cryptocurrencies, imposes onerous regulatory burdens, hampers innovation and financial inclusion, and undermines the principles of decentralization and community governance. Instead, a nuanced and targeted regulatory approach that recognizes the diversity of digital assets is necessary to strike a balance between investor protection and fostering innovation in the cryptoverse. By embracing regulatory frameworks that are tailored to the specific risks and functionalities of different cryptocurrencies, we can promote a thriving and sustainable industry that respects the rights and aspirations of the digital community.

Classification of crypto as a commodity

The classification of all cryptocurrencies as commodities, rather than recognizing their true nature as currencies, would have significant detrimental effects on the industry and hinder the growth and adoption of cryptocurrencies. While the classification of cryptocurrencies as commodities may stem from the desire to subject them to existing regulatory frameworks, it fails to acknowledge the unique characteristics and functionalities of digital currencies. In this discussion, we will explore why such a classification would be damaging and how it overlooks the fundamental aspects of cryptocurrencies.

First and foremost, cryptocurrencies fundamentally function as currencies rather than commodities. They are designed to serve as mediums of exchange, units of account, and stores of value. Unlike commodities, which are physical goods or raw materials, cryptocurrencies exist purely in a digital form and operate on decentralized networks. Treating cryptocurrencies as commodities fails to capture their primary purpose as digital currencies and undermines their potential as a transformative force in the global financial system.

By classifying cryptocurrencies as commodities, they would be subject to regulatory frameworks that were primarily developed for physical goods. This approach overlooks the unique characteristics of cryptocurrencies, such as their decentralized nature, cryptographic security, and borderless transactions. Applying commodity regulations to digital currencies would impose unnecessary burdens on the industry, hindering innovation, stifling market efficiency, and impeding the adoption of cryptocurrencies as viable mediums of exchange.

Moreover, the classification of cryptocurrencies as commodities neglects the distinct features of decentralized networks that underpin many cryptocurrencies. Unlike commodities, which are typically controlled by centralized entities, cryptocurrencies operate on blockchain technology, enabling peer-to-peer transactions and eliminating the need for intermediaries. This decentralization ensures transparency, immutability, and security, which are not inherent features of commodities. Treating cryptocurrencies as commodities would disregard the revolutionary potential of decentralized technologies and restrict their ability to transform traditional financial systems.

Additionally, the classification of cryptocurrencies as commodities fails to consider the implications for financial inclusivity and access. Digital currencies have the potential to provide financial services to unbanked and underbanked populations, allowing them to participate in the global economy. By treating cryptocurrencies as commodities, access to these digital currencies could be limited by regulatory barriers, hindering financial inclusion, and impeding the empowerment of individuals who are currently excluded from traditional financial systems.

Furthermore, the classification of cryptocurrencies as commodities overlooks the evolving nature of the crypto industry. Cryptocurrencies are not static assets; they are dynamic and constantly evolving. The crypto ecosystem encompasses a diverse range of tokens, including utility tokens, security tokens, and stablecoins, each serving different purposes and exhibiting unique characteristics. Treating them all as commodities fails to recognize this diversity and inhibits the development of innovative token models and use cases that can drive the adoption of cryptocurrencies in various sectors.

Instead of classifying all cryptocurrencies as commodities, a more nuanced regulatory approach is needed. This approach should consider the specific functionalities and characteristics of different types of cryptocurrencies and develop targeted regulations accordingly. Such an approach would allow for the development of a regulatory framework that fosters innovation, ensures market integrity, and protects investors without stifling the inherent potential of cryptocurrencies.

In conclusion, the classification of all cryptocurrencies as commodities would be damaging to the industry and overlooks the fact that cryptocurrencies function primarily as currencies rather than physical goods. Such classification would subject cryptocurrencies to regulatory frameworks ill-suited to their unique characteristics, hindering innovation, limiting financial inclusion, and impeding the transformative power of decentralized technologies. A more balanced and nuanced regulatory approach that recognizes the true nature of cryptocurrencies is necessary to foster an environment conducive to their growth, adoption, and positive impact on the global financial system.

Micro-oversight of the global system

Government micro-management of the crypto industry, characterized by excessive regulatory oversight and control, not only hampers the growth and development of the digital community but also violates the principles of the United Nations (UN) protocols regarding human expression and freedom. In this discussion, we will explore the reasons why such micro-management is detrimental and how it infringes upon the rights of digital citizens.

> "Self-regulation stands in relation to regulation the way self-importance stands in relation to importance."
>
> William Buiter

One of the key strengths of cryptocurrencies and blockchain technology lies in their decentralized and community-driven nature. These technologies operate on peer-to-peer networks, empowering individuals and communities to participate in a trustless and transparent ecosystem. Excessive government micro-management disrupts this decentralized nature and stifles the creativity, innovation, and self-governance that define the crypto community.

Micro-management of the crypto industry is often driven by a desire to mitigate risks such as fraud, money laundering, and market manipulation. While it is important to address these concerns, imposing overly strict regulations without considering the unique characteristics of cryptocurrencies can have unintended consequences. Excessive regulatory burdens create barriers to entry for startups and small businesses, limiting competition and stifling innovation. It also hampers the ability of the industry to adapt quickly to technological advancements and market demands.

Furthermore, micro-management of the crypto industry undermines the principles of freedom of expression and individual sovereignty that are enshrined in the UN protocols. Digital citizens have the right to express themselves and engage in economic activities without unnecessary government intervention. Excessive regulation curtails this freedom by imposing restrictions on how individuals can transact and participate in the crypto ecosystem. It inhibits the ability of individuals to exercise their economic rights and stifles the potential of decentralized technologies to empower individuals and foster economic inclusion.

Moreover, government micro-management often results in a lack of regulatory clarity and inconsistency across jurisdictions. The absence of clear and consistent regulations creates uncertainty for businesses and investors, discouraging their participation in the crypto industry. This lack of clarity also hampers the ability of digital citizens to engage in cross-border transactions and global collaborations, impeding the growth and development of the global crypto ecosystem.

In contrast, a more balanced regulatory approach that encourages industry self-regulation and fosters collaboration between governments and the crypto community can achieve better outcomes. This approach respects the decentralized nature of cryptocurrencies and allows for innovation and experimentation, while still addressing legitimate concerns such as consumer protection and market integrity.

The UN protocols emphasize the importance of fostering an open, inclusive, and accessible digital environment that respects human rights. Government micro-management of the crypto industry directly contradicts these principles by impeding the free expression of digital citizens. It is crucial to ensure that regulations strike the right balance between addressing risks and preserving the inherent benefits of decentralized technologies.

To uphold the UN protocols and promote a thriving crypto industry, governments should focus on creating a regulatory framework that encourages responsible behavior, enhances transparency, and protects the rights of digital citizens. This can be achieved through collaboration between governments, industry stakeholders, and the crypto community to develop best practices, codes of conduct, and standards that ensure the integrity and stability of the crypto ecosystem.

Additionally, governments should invest in education and awareness initiatives to empower individuals with the knowledge and skills necessary to navigate the crypto landscape safely. By promoting digital literacy and providing accessible resources, governments can foster an environment where individuals can make informed decisions and engage with cryptocurrencies in a responsible manner.

Government micro-management of the crypto industry not only hampers the growth and development of the digital community but also violates the UN protocols regarding human expression. Excessive regulatory burdens stifle innovation, hinder economic inclusion, and curtail the freedom of digital citizens. A balanced regulatory approach that respects the decentralized nature of cryptocurrencies, encourages self-regulation, and promotes collaboration between governments and the crypto community is essential to foster a vibrant and inclusive crypto ecosystem. By upholding the principles of the UN protocols, governments can create an environment that empowers individuals, drives innovation, and harnesses the transformative potential of decentralized technologies for the benefit of society as a whole.

Conclusion

In conclusion, over-regulation of a developing system, such as the crypto industry, can have detrimental effects and hinder its true potential. While regulatory frameworks play a vital role in ensuring consumer protection, market integrity, and stability, an excessive and burdensome regulatory environment can stifle innovation, discourage entrepreneurship, and impede the growth of the industry.

Over-regulation often stems from a cautious approach that seeks to apply traditional regulatory models to emerging technologies. However, this approach fails to recognize the unique characteristics and benefits of the crypto ecosystem. Cryptocurrencies and blockchain technology offer novel solutions to long-standing problems in the financial sector, promoting transparency, decentralization, and financial inclusivity. By imposing excessive regulations, governments risk stifling these innovations and forcing the industry into an old-world framework that may not be suitable for the emerging digital economy.

Moreover, over-regulation can create a perception that the traditional financial system is inherently superior to the crypto industry. This perception is misguided as it disregards the shortcomings and inefficiencies of the traditional system. The crypto industry has the potential to address these shortcomings, such as high transaction costs, slow settlement times, and limited financial access for marginalized populations. By over-regulating the crypto industry, governments perpetuate the notion that the old world is better, even when evidence suggests otherwise.

Additionally, over-regulation hampers the ability of entrepreneurs and innovators to experiment and develop new models within the crypto space. The dynamic and rapidly evolving nature of the industry requires flexibility and adaptability to accommodate emerging technologies, business models, and market demands. Excessive regulatory burdens create barriers to entry and compliance costs that disproportionately affect startups and small businesses, limiting competition and stifling innovation.

Furthermore, over-regulation in a developing system may lead to regulatory arbitrage, where businesses relocate to jurisdictions with more favorable regulatory environments. This can result in a brain drain of talent, capital flight, and the loss of economic opportunities for the home country. It is crucial for regulators to strike a balance between fostering innovation and safeguarding against risks, allowing for experimentation and growth within a well-defined regulatory framework.

A more balanced approach to regulation is needed—one that encourages responsible behavior, protects consumers, and safeguards against financial crimes, while still allowing room for innovation and growth. Such an approach would involve engaging with industry stakeholders, seeking their input, and collaborating to develop regulatory frameworks that strike the right balance between oversight and innovation.

Additionally, regulators should prioritize education and awareness campaigns to help users understand the risks and benefits of engaging with cryptocurrencies. By empowering individuals with knowledge and resources, they can make informed decisions and protect themselves against fraud and scams without stifling the industry's progress.

In conclusion, over-regulation of a developing system, like the crypto industry, hampers its true formation and growth. It perpetuates an old-world mentality that overlooks the potential of emerging technologies and fails to address the limitations of the traditional financial system. A balanced regulatory approach that encourages innovation, fosters competition, and protects consumers is crucial to unleash the transformative power of cryptocurrencies and blockchain technology. By embracing this approach, governments can support the evolution of a more inclusive, transparent, and efficient financial ecosystem for the benefit of individuals and society as a whole.

Chapter 23: Building a Better System Through Cooperation

"Teamwork divides the task and multiples the success."[48]

In an increasingly interconnected and digital world, cooperation at the international level is crucial to foster a thriving and sustainable crypto ecosystem. Recognizing the need for collaboration and shared responsibility, this chapter delves into key aspects of cooperation that can enhance the development, stability, and inclusivity of the crypto industry. From international panels to banking protection for cryptocurrencies, special panels for increased participation, and limitations on central bank digital currencies (CBDCs), this chapter explores the opportunities and challenges associated with international cooperation.

Collaboration among nations through international panels is vital for establishing common standards, best practices, and regulatory frameworks that facilitate the global growth of the crypto industry. These panels serve as platforms for open dialogue, knowledge exchange, and policy coordination, enabling countries to harmonize their approaches and address cross-border challenges. By sharing experiences, insights, and expertise, international panels foster a deeper understanding of the unique dynamics and potential risks associated with cryptocurrencies, leading to more informed and effective regulatory decisions.

One area that necessitates international cooperation is the establishment of banking protection for cryptocurrencies. Given the global nature of digital assets, ensuring access to

---

[48] Unknown at https://www.greatexpectations.org/resources/life-principles/cooperation/quotes-about-cooperation/.

reliable and secure banking services is crucial for the stability and growth of the crypto industry. International collaboration can facilitate the development of frameworks that provide banking institutions with clear guidelines and risk management practices to safely engage with cryptocurrencies. This cooperation can help mitigate the risk of financial exclusion and foster greater confidence among businesses and individuals looking to participate in the crypto economy.

To encourage broader participation in the crypto industry, the establishment of special panels dedicated to enhancing inclusivity and accessibility is essential. These panels can bring together industry leaders, policymakers, and representatives from marginalized communities to identify barriers to entry and develop strategies for addressing them. By actively engaging with underrepresented groups, these panels can help bridge the digital divide and ensure that the benefits of cryptocurrencies are accessible to all, regardless of geographical location, socioeconomic background, or technological literacy.

While CBDCs hold promise for advancing digital payments and financial inclusion, international cooperation is vital to establish appropriate limitations and safeguards. Collaborative efforts can help prevent the monopolization of digital currencies by central banks, maintain competition, and ensure a level playing field for all participants in the crypto ecosystem. Cooperation can also facilitate the development of interoperability standards that enable seamless transactions between CBDCs and other digital assets, fostering a more interconnected and efficient global financial system.

Cooperation at the international level is a fundamental pillar for building a better system in the crypto industry.

Through international panels, banking protection, special panels for increased participation, and limitations on CBDCs, nations can work together to foster innovation, enhance financial inclusion, and address potential risks. By embracing collaboration, governments, industry stakeholders, and global organizations can create a harmonized regulatory landscape that promotes trust, stability, and responsible growth in the crypto ecosystem. This chapter explores the opportunities and challenges presented by international cooperation and highlights the importance of collective action in shaping the future of the crypto industry.

International Panels

> "Coming together is a beginning. Keeping together is progress. Working together is success."
>
> Henry Ford

As the crypto industry continues to grow and evolve, the need for effective regulatory frameworks becomes increasingly apparent. To ensure balanced and forward-thinking regulations, it is essential to involve industry stakeholders in the regulatory process. This article explores the significance of international panels consisting of crypto company owners and experts in crafting regulations for the industry. By harnessing the expertise and insights of those directly involved in the field, international panels offer a unique opportunity to drive efficiency, foster innovation, and ensure the development of regulations that align with the needs of the crypto community.

1. In-depth Industry Knowledge: International panels composed of crypto company owners possess extensive firsthand knowledge of the challenges,

opportunities, and intricacies of the crypto industry. Their deep understanding of the technology, market dynamics, and regulatory landscape allows them to provide valuable insights and expertise in shaping regulations. This industry-specific knowledge enables panels to identify potential pitfalls, propose practical solutions, and develop regulations that are both effective and feasible.

2. Real-world Experience: Crypto company owners bring valuable real-world experience to the table. They have firsthand experience navigating the complexities of the crypto ecosystem, including issues related to security, privacy, governance, and user protection. This practical experience equips them with a comprehensive understanding of the industry's nuances, allowing them to provide meaningful input into regulatory discussions. By incorporating their experiences and perspectives, international panels can ensure that regulations are not only theoretically sound but also grounded in the practical realities of the crypto landscape.
3. Industry-wide Representation: International panels provide a platform for diverse representation from across the crypto industry. By bringing together stakeholders from various sectors, such as exchanges, wallet providers, blockchain developers, and decentralized finance (DeFi) platforms, panels can capture a comprehensive range of perspectives. This inclusivity helps to avoid biases and ensures that the regulatory framework considers the unique characteristics and needs of different stakeholders. It also fosters collaboration and consensus-building,

leading to more robust and widely accepted regulations.

4. Rapid Adaptability: The crypto industry is known for its fast-paced and dynamic nature. Regulations that fail to keep up with technological advancements and evolving market trends can stifle innovation and hinder industry growth. International panels, comprising crypto company owners, provide an agile and adaptable approach to regulation. Their firsthand experience and close proximity to emerging trends enable panels to respond quickly to market developments and propose timely regulatory adjustments. This agility helps to foster an environment that supports innovation while addressing emerging risks and challenges effectively.
5. Global Perspectives and Harmonization: The crypto industry is inherently global, transcending national borders and jurisdictions. Regulations developed in isolation can create fragmentation, regulatory arbitrage, and hinder the global adoption of cryptocurrencies. International panels offer an opportunity for global perspectives to converge, facilitating harmonization of regulations across countries. By working together, panel members can identify common principles, standards, and best practices that transcend national boundaries, fostering regulatory consistency and minimizing regulatory gaps.
6. Public Confidence and Trust: The active involvement of industry stakeholders in the regulatory process enhances public confidence and trust in the regulatory framework. When regulations

are perceived as fair, transparent, and well-informed, it instills confidence among users, investors, and businesses. International panels, with their expertise and industry insights, can contribute to the development of regulations that strike a balance between consumer protection and fostering innovation. This involvement helps to build trust between regulators, industry participants, and the general public, ultimately promoting the healthy growth of the crypto ecosystem.

7. Efficient Regulatory Process: Incorporating international panels of crypto company owners can streamline the regulatory process and make it more efficient. Panels can proactively engage with regulators, providing timely feedback, insights, and suggestions. By involving industry stakeholders from the outset, regulators can benefit from their expertise, avoiding the need for extensive revisions and amendments later. This collaborative approach ensures that regulations are well-informed, practical, and readily implementable, saving time and resources for all parties involved.

International panels comprising crypto company owners play a crucial role in the development of efficient and effective regulatory frameworks for the crypto industry. Their in-depth industry knowledge, real-world experience, and global perspectives contribute to the crafting of regulations that are balanced, forward-thinking, and conducive to innovation. By embracing the expertise and insights of those directly involved in the field, regulators can create an environment that fosters growth, protects users, and maintains the integrity of the crypto ecosystem. The inclusion of international panels in the regulatory process is a step toward empowering industry stakeholders and ensuring that regulations keep pace with the dynamic nature of the crypto industry.

Banking protection for cryptocurrencies

The cryptocurrency industry has experienced remarkable growth and widespread adoption in recent years. As the industry matures, it is essential to recognize the importance of protecting cryptocurrencies on par with traditional financial institutions. This article delves into the significance of providing robust protection measures for cryptocurrencies and highlights how such measures can foster industry growth. By establishing a secure and trustworthy environment, where cryptocurrencies are protected like other financial assets, we can inspire confidence, attract institutional investors, and unlock the industry's full potential.

1. Enhancing Investor Confidence: Protecting cryptocurrencies as financial institutions bolsters investor confidence. Traditional financial systems have established regulatory frameworks, oversight mechanisms, and investor protection measures to instill trust. By extending similar safeguards to cryptocurrencies, such as robust security standards, transparency requirements, and insurance coverage, we can enhance investor confidence in the asset class. This increased confidence will attract more institutional investors who seek regulated environments and, in turn, provide the industry with substantial capital inflows.
2. Mitigating Risks and Fraud: The cryptocurrency industry has been vulnerable to fraud, hacks, and scams due to its decentralized and nascent nature.

Treating cryptocurrencies as financial institutions enables the implementation of stringent anti-fraud measures, such as know-your-customer (KYC) procedures, transaction monitoring, and anti-money laundering (AML) protocols. These measures can help identify and prevent illicit activities, mitigating risks for investors and fostering a more secure ecosystem. Protecting cryptocurrencies with robust security protocols and regulatory oversight minimizes the occurrence of fraud and contributes to the long-term stability and credibility of the industry.

3. Facilitating Institutional Adoption: Institutional adoption of cryptocurrencies has the potential to drive significant industry growth. However, many institutional investors remain hesitant due to concerns about security, regulatory uncertainty, and lack of investor protection. Treating cryptocurrencies as financial institutions addresses these concerns by establishing clear regulatory frameworks, compliance standards, and custodial requirements. This provides institutional investors with a familiar and regulated environment, instilling confidence in their participation. Increased institutional adoption not only injects substantial capital into the industry but also contributes to mainstream acceptance and further market development.
4. Access to Financial Services: Treating cryptocurrencies like other financial institutions ensures equal access to financial services for crypto-related businesses and individuals. Many traditional financial institutions have been reluctant

to engage with the crypto industry due to regulatory uncertainties and compliance risks. By extending protections to cryptocurrencies, such as banking services, access to loans, and merchant processing, the industry gains access to essential financial infrastructure. This integration bridges the gap between traditional and crypto financial systems, facilitating seamless interactions and enabling the industry to thrive.

5. Regulatory Clarity and Market Stability: Regulating cryptocurrencies as financial institutions provides much-needed clarity and stability to the market. Clear regulatory frameworks help businesses understand their obligations, compliance requirements, and the legal boundaries within which they operate. This clarity encourages responsible and sustainable growth, attracting more participants to the industry. Moreover, regulatory oversight curbs excessive volatility and manipulative practices, ensuring a more stable market environment that is conducive to long-term investment and innovation.
6. Safeguarding User Funds: Treating cryptocurrencies as financial institutions necessitates robust security measures to safeguard user funds. By imposing strict operational standards, custodial requirements, and insurance coverage, the industry can protect users' assets against hacks, theft, or mismanagement. This heightened level of protection provides peace of mind to crypto users, encouraging wider adoption and increased usage. When users have confidence that their funds are secure, they are more likely to engage with

cryptocurrencies for various financial transactions, further driving industry growth.

Protecting cryptocurrencies on par with traditional financial institutions is crucial for the sustained growth and development of the industry. By enhancing investor confidence, mitigating risks, facilitating institutional adoption, ensuring equal access to financial services, providing regulatory clarity, and safeguarding user funds, the industry can thrive in a secure and regulated environment. Treating cryptocurrencies as financial institutions not only fosters growth but also strengthens the industry's credibility, mainstream acceptance, and long-term sustainability.

Special Panels to increase cryptocurrency Participation.

> "Cooperation is the thorough conviction that nobody can get there unless everybody gets there."
>
> Virginia Burden

In the realm of cryptocurrencies, fostering widespread participation and inclusivity is crucial for the industry's growth and development. Local panels for individuals serve as a powerful mechanism to enhance cryptocurrency participation at the grassroots level. This article explores the significance of establishing local panels and how they can empower individuals to engage with cryptocurrencies, contribute to the industry, and promote financial inclusion.

1. Education and Awareness: Local panels play a vital role in educating and raising awareness about cryptocurrencies within communities. By organizing workshops, seminars, and informational

sessions, these panels can disseminate knowledge about blockchain technology, digital assets, and the benefits of cryptocurrency adoption. Through educational initiatives, individuals can gain a deeper understanding of cryptocurrencies, dispel misconceptions, and make informed decisions about their participation. Increased knowledge empowers individuals to explore the potential of cryptocurrencies, contributing to a more inclusive and informed crypto community.

2. Addressing Local Concerns: Cryptocurrency adoption faces unique challenges at the local level, including regulatory hurdles, cultural perceptions, and limited access to digital infrastructure. Local panels can act as platforms for individuals to voice their concerns, discuss challenges, and provide insights specific to their communities. These panels enable dialogue between industry experts, policymakers, and community members, facilitating the identification of localized solutions and the formulation of policies that address regional needs. By actively addressing local concerns, these panels create an environment conducive to crypto participation, paving the way for broader adoption.
3. Building Trust and Confidence: Local panels foster trust and confidence in cryptocurrencies by establishing direct connections between individuals and industry experts. Participants can engage with knowledgeable professionals who can address their questions, clarify doubts, and provide guidance on various crypto-related matters. Building trust through personal interactions helps demystify cryptocurrencies and dispel skepticism, encouraging

more individuals to participate in the industry. This trust-building process is particularly important for those who may be unfamiliar with digital assets or have concerns about the legitimacy of the crypto ecosystem.

4. Encouraging Entrepreneurship and Innovation: Local panels can serve as incubators for entrepreneurship and innovation within the crypto space. By bringing together individuals with diverse backgrounds and skill sets, these panels create an environment that nurtures creative thinking and collaboration. They provide opportunities for aspiring entrepreneurs to showcase their ideas, receive feedback, and connect with potential investors or mentors. By encouraging entrepreneurship and innovation, local panels not only contribute to the growth of the crypto industry but also foster economic development and job creation within their communities.
5. Supporting Local Crypto Projects: Local panels can actively support and promote local crypto projects, thereby stimulating economic growth and community engagement. They can facilitate connections between developers, investors, and community members, enabling the development of decentralized applications (DApps), blockchain-based solutions, and crypto-focused initiatives that cater to local needs. Supporting local projects fosters a sense of ownership and community involvement, strengthening the sustainability and resilience of the crypto ecosystem.

Local panels for individuals play a crucial role in increasing cryptocurrency participation by promoting education, addressing local concerns, building trust, encouraging entrepreneurship, and supporting local projects. By establishing these panels, communities can actively engage with the crypto industry, shape its development, and reap the benefits of financial inclusion and technological innovation. Empowering individuals through local panels contributes to the overall growth and vibrancy of the crypto ecosystem, fostering a more inclusive and participatory future.

Limiting CBDCs

Central Bank Digital Currencies (CBDCs) have been a subject of growing interest among governments and financial institutions worldwide. While CBDCs present potential benefits, it is essential to consider the implications of their widespread adoption. This article argues for the limitation or elimination of CBDCs, advocating for private cryptocurrencies to become the norm. By exploring concerns related to government corruption and cybersecurity vulnerabilities, it becomes evident that embracing private cryptocurrencies offers a more secure, efficient, and inclusive future.

1. Government Corruption and Control: One of the primary concerns associated with CBDCs is the potential for increased government control over financial transactions. CBDCs grant governments access to real-time data, enabling extensive monitoring and control over individuals' financial activities. This concentration of power raises valid concerns about privacy, individual liberties, and the potential for abuse of authority. In contrast, private cryptocurrencies offer decentralized systems that empower individuals with greater financial

autonomy and control, free from the influence of centralized authorities.

2. Cybersecurity Vulnerabilities: CBDCs, being digital in nature, face significant cybersecurity risks. The large-scale implementation of CBDCs would make them attractive targets for hackers and malicious actors. Breaches in centralized systems could compromise the financial security and personal information of individuals on a massive scale. In contrast, private cryptocurrencies leverage advanced cryptographic technologies and decentralized networks, providing enhanced security measures. By embracing private cryptocurrencies, individuals can have greater confidence in the security and integrity of their financial transactions.
3. Efficiency and Innovation: Private cryptocurrencies have demonstrated their potential for driving efficiency and innovation in financial transactions. Blockchain-based cryptocurrencies offer fast, secure, and borderless transactions, eliminating the need for intermediaries and reducing transaction costs. The decentralized nature of private cryptocurrencies promotes competition, fosters innovation, and encourages the development of new financial products and services. Embracing private cryptocurrencies as the norm enables individuals and businesses to benefit from technological advancements, leading to a more dynamic and efficient financial ecosystem.
4. Financial Inclusion and Empowerment: Private cryptocurrencies have the potential to significantly enhance financial inclusion, particularly in regions

with limited access to traditional banking services. Individuals without access to formal banking systems can participate in the crypto economy and engage in cross-border transactions seamlessly. The borderless nature of private cryptocurrencies enables greater financial empowerment for individuals in underserved communities. Limiting or eliminating CBDCs encourages the development of a diverse range of private cryptocurrencies, catering to the specific needs and preferences of individuals and promoting financial inclusion on a global scale.

5. Market Competition and Consumer Choice: A diverse ecosystem of private cryptocurrencies fosters market competition and empowers individuals with a range of options to choose from. By allowing private cryptocurrencies to flourish, individuals can select currencies that align with their values, priorities, and desired functionalities. Competition among cryptocurrencies encourages innovation, price stability, and consumer-oriented services. Restricting the dominance of CBDCs and supporting the growth of private cryptocurrencies ensures a more open and competitive market environment.

Limiting or eliminating CBDCs in favor of private cryptocurrencies offers numerous advantages in terms of security, efficiency, financial inclusion, and individual empowerment. By leveraging the decentralized nature of private cryptocurrencies, individuals can have greater control over their financial transactions while benefiting from innovative technologies and market competition. It is essential to recognize the potential risks associated with CBDCs, including government control and cybersecurity vulnerabilities. Embracing private cryptocurrencies as the norm will lead us towards a more secure, efficient, and inclusive future, where individuals have the freedom to transact and engage in the crypto economy with confidence and autonomy.

> "Nip the shoots of arbitrary power in the bud, is the only maxim which can every preserve the liberties of any people."
>
> John Adams

Conclusion

In conclusion, the chapter on cooperation at the international level highlights the significance of embracing collaborative efforts to shape the future of the cryptocurrency industry. Through the establishment of international panels, banking protections for cryptocurrencies, special panels to increase participation, and the limitation of Central Bank Digital Currencies (CBDCs), we can foster an environment conducive to growth, security, and innovation.

The concept of international panels comprising crypto company owners and industry experts holds immense potential. By involving key stakeholders in the regulatory process, we can ensure that regulations are comprehensive, practical, and in line with industry needs. This collaborative approach enables the industry to grow while maintaining a balance between innovation and responsible practices.

Furthermore, providing banking protections for cryptocurrencies is essential for the industry's stability and legitimacy. Cryptocurrency users must have access to robust and reliable banking services, including custody solutions, lending platforms, and payment gateways. By extending traditional banking protections to cryptocurrencies, we can bridge the gap between traditional financial systems and the emerging crypto ecosystem, enhancing confidence and facilitating broader adoption.

Special panels dedicated to increasing cryptocurrency participation are crucial in addressing barriers to entry and fostering inclusivity. These panels can focus on issues such as education, accessibility, and regulatory frameworks tailored to specific regions or demographics. By promoting awareness and providing resources, we can empower individuals and businesses to actively participate in the crypto economy, driving economic growth and financial empowerment.

Regarding CBDCs, limiting their scope or even eliminating them altogether can have significant benefits for the industry and its stakeholders. The concerns surrounding government control and potential cybersecurity vulnerabilities associated with CBDCs underscore the importance of maintaining a decentralized and secure ecosystem. Embracing private cryptocurrencies as the norm allows for greater financial autonomy, efficiency, and innovation, while safeguarding individual privacy and protecting against cyber threats.

In summary, international cooperation and collaboration are pivotal for the development and sustainability of the cryptocurrency industry. By bringing together industry experts, regulators, and stakeholders, we can craft effective regulations, ensure banking protections, foster inclusivity, and strike a balance between innovation and security. Through these collective efforts, we can pave the way for a vibrant, inclusive, and secure crypto ecosystem that benefits individuals, businesses, and economies worldwide.

Conclusion Part III

This extensive discussion on regulations surrounding cryptocurrency has shed light on various aspects of the industry, ranging from existing regulations to desired changes and the importance of international cooperation. As the cryptocurrency market continues to evolve and gain traction, it is crucial to strike a balance between regulatory oversight and fostering innovation, while also safeguarding the interests of users and the overall stability of the financial system.

The first section explored the regulations that are already in place, including securities law, contract law, anti-fraud law, anti-money laundering law, and anti-terrorism law. These regulations play a vital role in mitigating risks, ensuring transparency, and maintaining the integrity of financial transactions. However, it is essential to continually evaluate and update these regulations to keep pace with the evolving nature of the cryptocurrency industry.

The second section highlighted the regulations that the industry desires. From depository insurance for crypto and recovery protocols to strong clarification on international Know Your Customer (KYC) standards and protections for exchanges, these proposed regulations aim to enhance security, stability, and trust within the cryptocurrency ecosystem. Additionally, ethics protocols for government officials and specialized asset classifications for stablecoins would promote responsible governance and facilitate the growth of the industry.

On the other hand, the third section addressed the regulations that the industry does not want. Classifying all cryptocurrencies as securities or commodities can impose unnecessary burdens and stifle innovation. Micro-oversight and overregulation can impede the development of the industry and discourage participation. Striking the right balance is crucial to foster growth, maintain market integrity, and prevent undue restrictions that hinder the potential benefits of cryptocurrencies.

The fourth section emphasized the importance of international cooperation and collaboration. By establishing international panels, ensuring banking protections, and limiting Central Bank Digital Currencies (CBDCs), the global community can create a harmonized regulatory environment that facilitates innovation, inclusivity, and secure financial systems. Collaboration among countries, regulators, and industry participants is essential to avoid fragmentation, foster trust, and establish consistent standards that protect the rights of digital citizens.

Throughout this discussion, it became evident that the cryptocurrency industry is at a critical juncture, where the right regulatory framework can either foster its growth or stifle its potential. Balancing regulatory oversight with the need for innovation and user protection is a complex task that requires careful consideration and collaboration among various stakeholders.

It is crucial for regulators and policymakers to approach cryptocurrency regulations with an open mind, embracing the transformative potential of digital currencies while addressing legitimate concerns such as financial stability, consumer protection, and security. A well-crafted regulatory framework should promote innovation, protect investors, prevent illicit activities, and provide clarity to industry participants, enabling them to operate in a secure and transparent manner.

Furthermore, it is imperative to recognize the global nature of the cryptocurrency industry and the need for international cooperation. Harmonizing regulations, sharing best practices, and establishing cross-border collaboration mechanisms will create a level playing field, facilitate market growth, and enhance consumer confidence in the global cryptocurrency ecosystem.

In conclusion, the future of cryptocurrencies and the regulatory landscape in which they operate will heavily influence their potential impact on the global financial system. By carefully balancing regulations, fostering innovation, and promoting international cooperation, we can harness the benefits of cryptocurrencies while safeguarding against potential risks. It is a collective effort involving regulators, industry participants, and users to shape a regulatory framework that supports the growth, stability, and inclusivity of the cryptocurrency industry. With the right approach, cryptocurrencies have the potential to revolutionize finance, empower individuals, and contribute to a more inclusive and prosperous global economy.

Conclusion

If I were to create a world, I would create a world where what is is. I would create a world where people knew what they saw, and what they saw was what they knew. The lies and misconceptions that plague the world in this age would be washed away by the realities that would give the people a better life. Now as we conclude this text, look back at the quote at the start of the book from the Mad Hatter, does my statement here or his statement there look madder in the world we live in. We have the ability to define the world as someplace where things can be better and reality cannot be a state of suffering broken by moments of pleasure, it can be a lifetime of helping others to have a better life and to be helped yourself so that you can enjoy your life.

Sometimes, I think that when we look at the world, we think that the world has to be as it is now. With nations hating one another, societies hating one another, and all of us saps living in the world being played against one another. Do you hate the average person in China or are you just worried that their government will do something stupid. If you are reading this book outside of the United States, do you hate the people in America, or do you just dislike the idiots who we elect to represent us. One of the greatest sins that this world faces is the idea that we hate based on who a person is- racism should be dead, yet it is alive and well in this world. Oh, what a world it would be if we changed the world just a little where people were not judged on where they come from.

One of the reasons that we face so many problems in this world is because we are conditioned to lose the childlike zeal we had when we were young. Too often we hear 'you cannot do that,' eventually we quit asking the world 'why not?' and just take our licks like everyone else. The sloth we feel in the world is condition, and we need to try to break it. If we want a better world and a better life, we need to take the time to challenge the world as it is, to create something better and to work to help others.

Likewise, sometimes our listlessness in this world is because we are looking for something to fill a hole in our lives. Money, power, control is all lauded as virtues in this world, yet I see many people who have them that are so miserable. The people that I see that are happy are the people who have good lives, who have families supporting them, and who are loving other people rather than trying to fill the emptiness in their lives with material goods.

Control is one of the greatest myths that I see people suffering because of. We all want control over the world we live in, yet control is something that cannot be gained. It is often said, "the way to make God laugh is to make a plan," I wonder if this is true, or whether God is sad when we choose a plan that is different from His plan and try to do things that only benefit ourselves and do not work toward the benefit of others.

This leads to the idea of greed and envy. Greed is wanting something you did not work for; envy is not wanting someone else to have something they did work for. Both can destroy the happiness in a person's life. When investing, we need to be sure that greed does not cloud our judgement and envy does not provoke us. This is why speculation and investing are two different things. When we let our vices take over us, regardless of if we are religious or not, we have major problems.

And that brings us to pride. Sometimes we are so focused in our own lives that we do not realize that there are problems outside of our lives. In our minds, the government cannot have as much control as they do, our job is not important as it is, and no one can love us as much as we love others. So why am I talking about this at the end of a stablecoin book, because stablecoin is a way to make the world a better place. It is a way to stop the inflation that is robbing us, it is a way to give people access to what they have worked for, it is a way to keep things private in our lives to stop envy. It is safe from the greed which plagues many other endeavors because it is not about getting more for free' but it is about keeping and using what we have, however we choose to use it. Once we give up the idea that we are in total control, we become more in control than we have ever been before. Whether you call it fate, serendipity, or God's will, when we work for the betterment of the world, we find that we can be happier, end up with more in the long run, and can leave the stress and anxiety at the door to a better life.

Glossary

1 Minute average- the average value of all public trades of a given asset in the last 1 minute.

10 Minute average- the average value of all public trades of a given asset in the last 10 minute.

15 Minute average- the average value of all public trades of a given asset in the last 15 minute.

Accessibility (Ability)- the ability of a document of presentation to be accessed by persons who are differently abled.

Accessibility (Crypto)- The ability of a member of the crypto community to access their assets at their discretion.

Active Reserve- The holdings of a company in a traded portfolio, bringing more value to the currency ecosystem.

Aggregate Market Value- The valuation taken from a weighted average of all trade on all markets within a system.

Air-gapped computer- A computer that is physically disconnected from the internet so that they data on the computer cannot be accessed remotely. In an air-gapped computer, there is no wife, Bluetooth, or direct internet connection. All interfaces are directly connected and also non-connected to the internet in any way.

Air-Gapped Storage- storage of information on an air-gapped computer to ensure that the data on the system cannot be accessed except select connected terminals.

Alt-Coin- Generally any digital currency that is not bitcoin. In this text, the term is used concurrently with 2nd wave cryptos. The technical difference between 2nd wave cryptos and altcoin is that Ethereum is defined as a 1st wave crypto, though in common parlance it is considered an altcoin.

Alternate Accounts- Accounts created by people on social media in order to continue attacking a webpage, even after the initial account has been banned. (See Dummy Accounts)

American Candle- A candle that expands or contracts related to the value of trades in its course.

American Option- a put or call option allowing the investor to enforce the option at any time up until the final execution date.

Anchors- an attachment to another currency (generally the parent chain) which allows the value of the underlying currency to be related to the parent currency until such a point where the underlying currency "drops anchor" and moves its valuation to another method, commonly the valuation of a fully developed asset pool.

Anti-Trust- anti-monopoly laws that were created to prevent one person or corporation from becoming too powerful and forcing others out of commercial space. (Clayton and Sherman Acts)

Anti-Whale Technology- technology developed to prevent large holders within the system from manipulating the currency through large trades.

Arbitrage- the process of triangulating currency. When the currency is exchanged on multiple exchanges, the value on each exchange may be fractionally different. As such, there may be opportunities to buy on one exchange, convert to another currency, then sell the currency to convert back to the original currency at a favorable rate to increase the value of the holding. This was popular in the 1990s and is making a comeback in crypto trading.

Ashe School of Thought- during a tiered ICO, when a tier sells out quickly, the following tier should be for a smaller amount of coin/token. The thought behind this is that the smaller tier levels will allow the coin/token to find its market value more quickly and the larger (later) launches can be set at the true market value. This mode is used to prevent day trading a speculation during the early release.

Asset Value- The totality of the values of different asset pools of a cryptocurrency.

Asset Value Product- GDP of an asset system.

Automated Buy Order- An order telling the computer to buy when the stock reaches a specific price.

Automated Sell Order- An order telling the computer to sell when the stock reaches a specific price.

Automated Trading- a method of participating in financial markets use a program that executes pre-set rules for entering and exiting trades.

Banked Sick Days- The ability for workers to put their unused sick days into a bank for other workers who run out of PTO due to extended illness.

Bargained For exchange- one of the basic principles of contract law. If a person does not have the ability to bargain for their position, then it is not a contract which the parties freely entered into.

Batches- Groups of coins or tokens released together during an ICO.

Bear- A person who believes that the market or a specific asset will be in the decline and plays the market to make money off of that belief.

Bear Market- A market where the majority of the stocks are in a sell-off.

Bilateral Contract- a contract in which both parties exchange a promise to perform.

Bitcoin Effect- where the value of a coin gets high enough that one unit is too high for most normal uses.

Black Zone crypto- a crypto that is preforming above expectations.

Blacklist- the process of declaring a person or company should not be worked with by members of the community.

Boiler Room- A group of brokers who are pushing a product on the market through calls and emails.

Borg Coins- Coins with the expressed purpose of assimilating all other coins on the market.

BPSAA- Blockchain Privacy, Security & Adoption Alliance, an association of cryptocurrencies working to improve the market through better practice in the field.

Brokered Exchanges- These etchings allow a third-party broker to use a custodial wallet to conduct internal trades within the community. The trading community, therefore, is not as large as the total market and there can be fluctuations in price and market valuation.

Bulge (American Candle)- a space that shows many trades at this value in a candle.

Bull- A person who believes that the market or a specific asset will rise and is buying to take advantage of that movement.

Bull Market- A market where a majority of the assets are rallying.

Burn- removing a set number of coins from the system in order to increase the value of the remaining coins.

Burning Coins- The process of removing coins from the system in order to increase the value of the remaining coins in the system. Burns can be conducted either in a simple or complex fashion.

Buying the Dip- When stock or crypto is in a downward trend and the holder believes that the decline will continue then rebound, the holder sells the coins on the way down then buys at a lower rate to increase their holdings in the stock or crypto.

Call Option- A type of option where the holder of the option has the opportunity on or before a specific date to buy up to a specific amount of asset or security for a set price.

Candles- See Japanese Candles

Charity Shuffle- The process of giving out amounts of an initial currency offering to increase the use of the crypto in a given market. This works by benefiting the crypto (for having more users) and the charities (as a donation).

Chasing a Stock- The process of "avenging down" if there is little likelihood of the stock rebounding.

Choice of Law Clause- the portion of a contract that dictates what type of law will be used to resolve disputes.

Choice of law clauses should have the area that the court is in and the jurisdiction of the law to be used. (ex. The court would be in the United States, but the parties agree to be judged under U.K. law).

Circulating supply- the tokens that are available for trade in the market or in the hands of holders (non-project).

Clearing House Rate- the published rate of an assets value by a clearing house.

Click-Through Agreement- a Semi contract where a person can read the agreement or just click “ok” to accept the agreement. They are tenuous as a contract at best.

Clone Accounts- accounts that are created to mimic a person’s account so that the blame for any attack falls on that person. *Alternate Definition*- an account that mimics the personal account of another person to get their friends and family to open malicious links.

Closely Held Supply- the tokens that are held by initial team members and limited private sale buyers at the beginning of cryptocurrencies life cycle. These will generally be locked.

Coin- a type of cryptocurrency that has its own blockchain.

Cold Storage- The term for maintaining crypto in a wallet that cannot be accessed for normal trade or usage. Generally used in the staking process.

Cold Wallet- See Cold Storage

Commodity, Hard- Precious metals and commodities taken from the earth.

Commodity, Soft- Generally agricultural commodities or lumber.

Comparable function method- Compares a crypto to other similar cryptos.

Conditional Burns- burns that are set up by a series of cognitions, whether acute or chronic to reduce the number of coins. These burns are not static and are at the behest of the person who controls the crypto, or the community.

Conditional Staking- a system based on burns, where coins taken out of the system are placed in an active wallet to increase the value of the entire ecosystem. The coins are used to purchase stocks, bonds, property, other crypto, currencies, or other assets to act as a holding to increase the inherent value of the system.

Controller of a Crypto- The person or company who is responsible for the system management of a cryptocurrency.

Corporate Corpus- The structure of a company including all subsidiaries.

Corporate Plan- the physical location of the company, including buildings and other assets.

Creator of a Coin- the person or persons who created the concept behind a given cryptocurrency.

Critical Mass- the point at which a standing burn system empties the black wallet into active wallets to increase the value of the system. This is done so that the black wallet is not a target for hacks.

Cross Reality Adoption- the idea that a cryptocurrency must be accepted in the physical world to break the digital barrier.

Crypto- a short-hander term for cryptocurrencies including, but not limited to, coins, tokens, and NTFs.

Crypto Bill of Rights- A list of the rights that the crypto community agrees are needed for a fair global economy.

Crypto centric investment strategy- An investment strategy generally focused on cryptocurrencies.

Crypto Xenophobia- the fear that other cryptos from other block chains are destructive to the system.

Cryptocurrency- digital currencies, based off on or existing on a blockchain, used for transactions in the digital and real-world space.

Crypto diversity- The concept that the more cryptos that get created, the faster the system will be able to find the problems within specific models and create solutions to those problems.

Cryptonomics- the economics of the cryptos space, taking the best aspects of the field of economics, IT, and business finance and combining them into a functional system.

Cryptophobia- the irrational fear that cryptocurrencies are all frauds and have no inherent value, despite evidence to the contrary.

Cryptoverse- All digital space dealing with blockchain and crypto technology (including metaverse).

Custodial Wallet- A Crypto wallet where the owner allows the company holding their assets to hold the private key.

Dark Pool Trades- Trades made off market behind closed doors.

Dead Wallet- a 2nd wave crypto concept where coins were burned to a wallet that the owner “lost” the key. This was used to reduce the number of coins or tokens without deleting coins. This system was effective till dead wallets started to activate again, which caused a market loss in late 2021.

Decentralized Finance- the concept that the economic system is more efficient when it is run by the market, rather than the Keynesian system where the market is manipulated by central banks or governments.

Defi- See decentralized finance.

Derivative- a security that is based on the underlying value of another asset.

Derivative Variable- A variable that is used in the determination of the valuation of a system.

Desktop Wallet- a crypto wallet maintained on a laptop or desktop, which can be air gapped from the cryptoverse.

Determinative Variable- A valuation of a system based on another variable.

DGP- See digital gross product.

Diamond Hands- Traders who stick with a currency for the long term, ignoring the short-term rises and falls.

Digital Gross Product- the amount of commerce that a given digital system generates each year.

Diminishing Valuation Model- plan where a company’s asset pool to declines steadily.

Discounted Cash Flow Analysis- finds the present valuate of future cash flows (expected) using an agreed upon discount rate.

Dummy Accounts- Secondary accounts with no filler that are used by media attackers to continue attacking even after first account is banned. (see alternate accounts).

Dumps- The process of getting rid of all crypto that one person holds.

Echo Chamber- the effect which occurs when a social media network continuously repeats the same message repeatedly allowing members to think that it is true.

Economics- a liberal art that looks for patterns in the economic system and asserts grounded theory based on those observations.

Emic- Researching something from the inside.

Equal Distribution- In a re-release, a process where coins are returned to the holders and are distributed equally to everyone holding the coins, regardless of how many coins they are holding.

Equal Release (ICO)- An ICO where all batches are the same size.

Equitable Distribution- In a re-release, a process where coins are returned to the holders and are distributed based on the amount of coin that they are holding on a given partner platform.

Equitable Release (ICO)- An ICO where the batch size is dictated by the speed at which the previous level sold out (generally based on either the Ashe or Smithmyer school of thought).

Etic- Researching something from the outside/

European Option- An option that can only be executed at the end of the time period.

Exchanges- Platforms that allow the trade and exchange of currencies, stocks, cryptos, commodities, and other assets.

Extrusion Account- a clone, dummy, or alternate account created by a company to promote company propaganda on social media designed to convince people it was not the company itself but one of the communities.

Family of Coins- A group of coins (regardless of platform) that work together to increase the value of an ecosystem.

Fear of Missing Out- See FOMO.

Fiat Currency- Central bank-issued currency which is only backed on the faith and credit of the government for which it is minted.

Five-Tiered ICO- A ICO where the crypto is released in five "batches."

First Wave Crypto- Bitcoin or Ethereum.

Flat Top- (American Candle) a space that shows many trades at the top value in a candle

FOMO- Fear of Missing Out, something that drives people to get in on fads.

Forward- An agreement where a person has the right to purchase stock at the end of a period of time for a set amount of money. Forwards cannot be traded on the market and are settled at once.

Fractal Reserves- a group term for any reserves the company has, this can be a simple set of reserves, a liquidity pool, or an asset pool.

Fungible Tokens- Tokens where each token is similar to any other token in the system.

Future- A derivative where a person has the obligation to purchase or sell a given amount of coin on a specific date for a specific amount. These are more restrictive than options and can also be traded on the open market.

Gas Fees- The amount that a user pays for a transaction for using the chain to confirm the transaction (these can get very steep on Ethereum)

Gas Fee Bleed- transaction term referring to the loss over multiple transactions from gas fees.

General ICO- When a coin is introduced onto the market in one "batch" allowing the market to dictate the price at a later time.

GETs- See Global Exchange Tokens

General Burns- When a controller of a coin announces that a portion of the reserve will be removed from the market to increase the value of each individual coin/token. In most general burns the controller asks that the holders burn a like percentage.

Ghosts in the Machine- lines of code that interact with each other in an unforeseeable manner, which can cause odd things to happen.

Ghost picture- The grey outline picture when someone does not fully build out their social media account.

Ghost Profile-a social media account with no data in it, used for posting anonymously.

Global Exchange Tokens- Tokens that are designed to allow traders and users to bridge different blockchains via the existence of partnerships within the system.

Goldilocks Zone- the idea area when the value of a coin stays at a reasonable level.

Green Zone crypto- a crypto that is preforming within expectations.

Growth Valuation Model- plan where a company's asset pool to grows steadily.

Hard Fork- A change in a blockchain system that all coins moving forward must use to maintain their status on the system.

Hardware Wallet- A plug in device which contains your crypto wallet- one of the most secure ways to maintain your wallet unless it is stolen.

Holder (Stock)- A person who possesses a stock. The holder has all the legal rights to the stock.

Holder (Derivative)- A person who has control over the execution of a derivative before the time on the derivative expires.

Hourglass Candle (American Candle) a space that shows many trades at the top and bottom values in a candle.

Hot Wallet- A wallet in a cryptosystem that is active (alternate meaning- A wallet in a system where the key has been stolen.

Howey Test- a four-part test which identifies if an asset is a security "an investment of money in a common enterprise with the expectation of profit and the work is to be derived from the efforts of others."

ICO- See Initial Coin offering.

Infused Value- the value added to an asset by community activity.

Insider Trading- The process of a person acting on information, not generally available to the public, to make trades leveraging the special access to information to make a profit.

Internal Token- a Token that is used by a company for internal matters and not traded on the open market.

Into the Wild- When crypto is released from the compete for control of its creator/controller to the marketplace.

IPO- See Initial Public Offering

Inherent Value- The value of a crypto within an ecosystem that has a holdings system. The holding system can either have percentile holdings (where the amount of the system is derivative of the market cap of each crypto in the system) or joint holdings, where the cryptos in the ecosystem have joint membership in the ecosystem (better for system management)

Initial Coin Offering- When a crypto (both coins and tokens) is offered to the public through an exchange or a brokerage.

Initial Public Offering- When a stock is first offered to the public through an exchange or a brokerage.

Japanese Candle- a way of following the market where the stock is shown by the high bid, the low bid, the high average, the low average, and the average in one infographic for some time. So named because it looks like a candle. On most platforms, ascending candles are green and descending candles are red.

Joint Holding Value- A system where a Stablecoin 2.0 ecosystem is valued based on the number of coins in the system and the value of the holdings shared by the family of coins.

KIROS Development Plan- The portion of the KIROS white pages looks at how the token will be distributed and developed.

KIROS White Paper- The operating plan for the KIROS Global Exchange Token Program

Lemmings- People in the market that follow personalities on the market and do what they tell them (if you are invested in a system where a guru tells you what to do, you are a lemming). Lemmings can cause runs and rallies on the market.

Limited Tokens- Tokens to which only a limited number are created.

Lipstick Politician – A politician that will say anything about a subject that they do not know, then do exactly what their donors tell them to do.

Liquidity pool- a pool of assets maintained to ensure liquidity in the market, can be used to loan to exchanges during times of low liquidity.

Live currency- a currency on a blockchain.

Locked Investment- When a purchaser buys a token in an early sale as has a period which they must hold the token before they can sell the token.

Longs- People who buy a stock for the long term, these people tend to hold stocks for weeks, months, or years, and in crypto people who hold cryptos for days, weeks, months, or years.

Macro-transactional dynamics- the relationship between transactions and value at the macro level.

Margins- the amount of money that a person has in their brokerage account, basically a trader borrows from the brokerage the amount of money their account has as a net value and uses the borrowed amount to make trades that they do not have enough liquid value to trade.

Market Cap, Total- the total market cap is the value of the project in a fiat currency.

Market Cap, Tradable- the tradable market cap is the total market cap divided by the number of tokens in circulation or number of tokens that could be in circulation.

Market Cap, Company- the total value of a company in the market.

Market Value- The valuation of what people are playing for crypto on a market. This can be a single market value or an aggregate market value.

Market valuation Analysis- Analysis based on market value.

Market Volume- The amount of trade that is being done in a specific coin on a given exchange or across multiple exchanges.

Meso-transactional dynamics- the relationship between transactions and value at the meso level.

Meta Macro-transactional dynamics- the relationship between transactions at the exchange level.

Mickey Bucks- A currency offered by Disney in its parks that was pegged to the dollar. Rubes could buy the currency for use in the park. As time went on, fewer and less stores in the park would accept the Mikey bucks turning the holders into bag holders for a novelty currency.

Micro-transactional dynamics- the relationship between transactions and value at the micro level.

Mining- the process of conducting calculations on your computer, rig, or validating transactions for payment in coin or tokens.

Mobile Wallets- A wallet that is stored on your cell phone.

Monocurrency Reserves- holdings where only one coin is maintaining its reserves (common in Stablecoin 1.0).

Monopoly- the process of a company having control over the means of production for all of one type of product. This can be illegal or legal in certain jurisdictions.

Multicurrency Reserves- When multiple cryptocurrencies work together to build holding which functions as a value base for all the coins in that group.

Multilateral Contract- A contract where multiple parties have interests, duties, and expectations.

Multilevel Portfolio- A portfolio with more than one type of asset in the portfolio (precious metals AND stocks, Commodities, AND bonds, etc.)

Multiple method analysis- valuation based on how similar assets trade.

Mutual Fund- a financial vehicle that pools assets from shareholders to invest in securities like stocks, bonds, money market instruments and other assets.

Nakamoto Coefficient- a way to quantify the decentralization of the blockchain, looking at the number of nodes need to crash a subsystem.

National Favoritism – The process of a country giving preferential treatment to crypto or company that is founded or operates within its national borders.

Native Token Pool- A pool of the chain's native tokens for gas fees.

Net present value (NPV)- A method of valuation analysis looking at the future cash flows of an investment against the value of investments in other methods as measured by the discount rate.

NFT- See Non-Fungible Token

Niche Token- A token **only** usable for a specific purpose or community.

Non-Custodial Wallets- a wallet where the owner holds the private key.

Non-Fungible Token- A token that is unique in its existence and the chain verifies that the holder has exclusive rights to that token.

Ogres- sight attackers with a specific goal when they attack the site, generally malicious.

Oligopoly- The process of a person or company obtaining the purchase rights for all of one type of good.

One wallet one vote system- a system where governance only allows one vote per wallet, no matter how many coins are in the wallet.

Option- A type of derivative where a person buys the "option" to sell (put) or buy (call) at a certain price on or before a named time. This option can then be traded on the open market with the option being executed when the holder calls in the option, or the option can expire rendering it worthless.

Optional Staking – The concept that a group of people, by action or by consent, can put away a portion of their holdings to increase the reserve of crypto.

OTC Trades- See over-the-counter trades.

Over the Counter Trades- Trades that are done person to person and not handled by any exchange.

Overminting- when too many coins are minted causes inflation of the value.

Paper Hands- A person who sells at the first sign of a decline in the coin value.

Paper Wallet- a printed wallet (generally on a cart) that allows you to access your wallet through an app using a barcode or QR code.

Pariah Crypto- a crypto that deals with criminal, pornographic, or drug related subject (any taboo)

Patterns (Stock)- movements in the market that traders look for when they want to know if there will be a market reversal.

Percentile Holding Value- The value of the holding reserve assigned to specific crypto in a family of coins related to the percentile value of the holdings of the ecosystem in relation to the market cap of the specific crypto.

Pegs- an attachment for a crypto that is associated to the value of another asset, generally through a formal technical attachment.

Pins- An attachment for a crypto that is associated to the value of another asset, though no formal technical attachment is created.

Poachers- persons who wait until a domain name comes up to buy then hold it hostage against the original user.

Pre-Planned Release- this is a cycle in a tiered release where the number of coins in each "batch" is released at a specific price in a specific volume. This is generally the easiest system to use for a tiered ICO, but usually has the most difficulty finding the real market rate.

Presale- a third party sale where individuals buy sale tokens which can be exchanged for the currency on sale.

Private Coins- coins which cannot be traced via the distributed ledger, which some would say are not cryptocurrencies.

Professional assault accounts- Blackhat accounts used specifically for attacking other companies accounts. Can be set up through a third party.

Proof of Stake- The ability of crypto or an ecosystem to prove that their holdings of the system (Stablecoin 1.0 or Stablecoin 2.0) have a value that can be measured or verified. This is needed to give an actual value calculation for a coin.

Proof of Utility- The ability to prove that a coin has a use for a process. Example: when a GET is used to trade coins across blockchains without paying gas fees on both chains.

Protected Information- Information that should not be released to the public, includes names and addresses of traders.

Pump and Dump- The process of running the value of crypto or stock up, only to sell the crypto or stock quickly causing the rate to fall. Organized schemes to do this are illegal in most markets.

Purchase Futures- Futures that create the obligation for a party to purchase a set amount of assets at a given time.

Pure market crypto analysis- an analysis which is just looking at the spot rate.

Put Option- in an option system, a put option allows you to "put" the goods on the market, selling them at a predetermined amount, which may be above or below the market rate (usually you do not execute a call option below the market rate). New investors must realize that they are only buying the option to do this action, not the actual stock spoke about in the option.

Reality Testing- The process of questioning an assumption to allow the person holding the assumption to accept challenges to the idea.

Red Zone crypto- a crypto that is preforming below expectations.

Relative Supply Dynamics- the estimation of the relevance of a crypto project on the global economy in general.

Remedial Security- security that is brought in to fix a problem already known to be in existence. This can be a good sign for a crypto that recently has had problems knowing that they are fixed and that the project is aware of what can happen.

Rental Society- A society where more people rent and least goods than own them. This system is highly conditional on government support.

Reserves- The amount of coin or token that a system keeps out of play (or in controlled play) to maintain the stability of the system. This reserve can be active or passive and generally is directly related to the holdings.

Resistance- an indicator where many people have offered to sell stock at a given price that can cause problems during a rally as the resistance line tends to eat up all purchases, not letting the stock or crypto move past that point.

Reversal (Stock)- When a stock is moving one way, then following a pattern begins to move the other direction. A declining stock would reverse to an ascending stock.

Reversal (Wrestling)- When one wrestler is in a hold and moves out of the hold to take control of the match by putting the other wrestler in a hold.

Rug Pulls- The process of pumping up a coin and then devaluing the coin quickly either by the creators cashing out or some other means. Should be illegal in all asset markets.

Scalping- the process of buying then quickly selling during a bull run.

Schrodinger's Box- A box with a cat in it where the cat could be alive and dead at the same time. If the cat were dead, we could not know it is dead because, in the process of opening the box, the cat would be killed. It represents a theoretical concept with only one outcome, where the people are afraid of the outcome (which is predestined) and thus are unwilling to take the steps that are needed to fix it. Fiat currency is dead, but we are unwilling to make changes to the system to improve the lives of people around the world because we may kill the fiat currency, which is dead.

Schrodinger's Wallet Fallacy- The idea that the coin in the wallet can be active and inactive at the same time. The use of a dead wallet implies that the coin is not destroyed but is inactive. As a coin that can be accessed may be accessed, then the coin cannot be active and inactive at the same time.

Second Wave Crypto- Any crypto built upon the principles of smart contract technology after Ethereum.

Seed Capital- investment, generally for equity or options on the token, used to fund the initial stages of the company.

Sell Futures- Futures where a person has an obligation to sell a given number of assets at a given price at a given time.

Sell to Close- the process of selling all of one's holdings of a specific asset.

Server Room (Slang)- A digital boiler room.

Short Trade- The process of borrowing an asset from a holder, then selling the asset hoping that the price of the asset goes down so that the seller is then able to buy back the asset before the end of the term, at a lower rate, and pocket the difference.

Single Market Value- The value of crypto on a single exchange or market.

Slush Fund- money that is put into a reserve fund for use that may not be directly corporate related.

Smithmyer School of Thought- the school of thought that during a tiered ICO, when a tier sells out quickly, the next tier should be larger. The philosophy behind this is that more people will be able to get involved with the coin at a lower rate to increase the dynamic holdings in the Stablecoin 2.0 system.

Social Contract- The agreement between people in a society to give a government a specific list of privileges and duties in exchange for specific restrictions being placed on individual rights. The idea behind a social contract is that the government may only function in the matter prescribed and can never increase its own power or gain rights of its own.

Soft Fork- an increase in efficiency for a specific blockchain that each coin or token on the chain has the option of participating in.

Spot Rate- the value that a stock or crypto is trading at any given moment.

Speculation Pool- a pool designed to grow in value rapidly to fuel a boom market for an asset.

Spiderweb Bridge- A concept that coins can be interconnected by the system, rather than code, and the interconnectivity of the system allows for crypto ecosystems to develop that are not based on one specific blockchain.

Squatters- people who buy websites they have no user for a hold them hostage against projects that have the same or a similar name.

Stability Pool- A pool designed to maintain a paired value for a crypto.

Stable Valuation Model- plan where a company's asset pool stays relatively the same.

Stablecoin- tokens or coins backed by an asset or system of assets that are established to maintain value relative to a fiscal point. This fiscal point can be anything that has a value, which in relation to something else has some form of stability.

Stablecoin 1.0- a System where the value of crypto is based on its relationship to a fiat currency.

Stablecoin 1.1- a system where the value of crypto is based on its relationship to a single asset or family of assets.

Stablecoin 2.0- A system where the value of the crypto is based on the combined/communal holdings of a crypto ecosystem. This system can include stocks, bonds, property, and other real-world or digital assets withstanding value.

Stability, Long-Term- The ability to maintain a general direction on the charts for a period of over 6 months.

Stability, Negative-- factoring in at a generally stable negative relationship between the price and the asset to which it is being compared (or pinned, pegged, or anchored).

Stability Neutral- factoring in at a generally stable relationship between the price and the asset to which it is being compared (or pinned, pegged, or anchored).

Stability, Positive -- factoring in at a generally stable positive relationship between the price and the asset to which it is being compared (or pinned, pegged, or anchored).

Stability, Short-Term- The ability to maintain a general direction on the charts for a period of under 6 months.

Staking- The process of a member of the community (including the creator) holding a portion of their coins/tokens in cold storage to decrease the number of coins/tokens in play and increase the value of the system.

Standard NPV Crypto Assessment- Poor analytical method where a spot NPV takes over an estimated time frame.

Standing Burn- a burn that is on every transaction. An idea replacing "gas fees" as a taxation system for the crypto ecosystem. It should never be more than 2% on any given trade.

Standing Burn System- A system defining the standing burn for crypto. Examples are "The standing burn will be between 0.5% and 1.5%" or "The burn rate will be set on a five-minute basis by [X] controlling company.

Starter Burn- a period of higher-than-normal crypto burning to ensure that people are no "pump and dumping" during the initial ICO.

Static Reserve- a reserve of coins or tokens that are kept in a separate account for the purpose of managing the liquidity of the cryptosystem.

Stock Swap- the process of exchanging one stock for another, usually used in mergers and/or acquisitions.

Subsidiary Map- a listing and organizational structure of all subsidiaries and holdings of a single company.

Support Line- The area where many buyers have placed an order setting a lower limit for a stock value. A support line tends to "eat up" all the sales which allow the price to maintain at that level.

Swaps (Derivatives)- The process of exchanging future value for present value in the derivative market. This can fall under other types of equity, but crypto only really looks at present value v. future value currently.

Swaps (Stocks)- The process of swapping one stock for another (usually in amounts along with the same value) during a stock transaction. This method is usually used in mergers.

Tag- a person's online name.

Tax Revenue- any funding that exists after the basic functions of the tax have been fulfilled.

Ten-Tiered-ICO- An initial coin offering system where the coins are broken up into ten "batches" for release, looking for the market value of a specific coin or token.

Terran- a non-political term for a human being born on earth (so all of us).

Third Wave Crypto- Any coin or token based on the Stablecoin 2.0 platform, with a standing burn, willing to participate in the spiderweb bridge.

Tiered ICO- an initial coin offering where the coin is offered in "batches" allowing the coin to find market equilibrium for price. The two most common forms are five and ten tiered ICOs.

Tipple Wallet- the wallet into which a smart burn system feeds to maintain equilibrium within the system.

Token- a form of cryptocurrency that exists on a blockchain that is not its own. Unlike a coin, a token is an overlay that increases the functionality within a set of defined parameters and bases its value on that procedural increase.

Tokens, Corporate- tokens used by a corporation to facilitate inter-office work.

Tokens, Government- any token issued by a government.

Tokenomics- the practice of assessing the value of a cryptocurrency based on the valuation, the distribution, the application, the reserves, and the community of the coin.

Total Value- The aggregated value of the quadratic method.

Trade Confirmation- a digital document that confirms that a trade has been completed.

Transactional Value- The value of a currency is determined by the amount of "real world" or digital world goods that the token or coin can purchase.

Transactional Volatility- The amount of value change between one trade and another.

Transitional Value- The value of a currency is determined by the amount of value-added to a process by the information passing through the block operating on the coin/tokens chain.

Trolls- malicious posters who simply attack people or sites for the enjoyment of other's misery.

Trust (monopoly)- a group of companies working together to maintain a monopoly on a system.

Trust (Systems)- the amount of faith the users have in a system.

Uniasset System- A system that maintains the inherent value in a portfolio by using a single asset (such as gold.

Unilateral Contract- a contract where one party's completion can only be done by an action.

Unilevel Portfolio- a Portfolio that only has one family of assets to manage the inherent value of the system (precious metals, stocks, bonds, etc.)

Utility Token- a token used for a specific value purpose. (internal tokens are used within a company, subset tokens are used within a specific project.)

Variable Burn Rate- a burn rate that is set between two points (usually percentage points), which can be changed by the controller of the coin to increase or decrease the burn rate to help with coin growth. In a Stablecoin 2.0 system, this process should be managed by an AI.

Vested (cryptocurrency)- when a holder has the full ability to trade the coins that they hold, generally in relation to a locking system.

Virtual Private Network- A screening system used to help protect the identity of a person using a digital system.

The volume of Trade- The number of coins or tokens that are being traded in a given segment.

Utility of Trade- The relationship between the volume of trade and the price point for crypto. Crypto with a low volume and a high rate of movement are less dependable than crypto with a high volume of trade and a low movement level when looking for indicators and patterns.

Underminting- when a currency does not create enough tokens for the forecasted use of the coin.

Valley- a space that shows few trades at this value in a candle.

VPN- See virtual private network.

Warrants- a type of crypto derivative (based on a stock derivative) where coins are released from a named reserve via a per sale of the coin, where the buyer buys the right to the crypto at the present price for a later date. These coins are coins released from the reserve.

Web Wallet- A crypto wallet accessible from a website.

Whale- A trader who purchases or sells large volumes and large values of coins.

Whitelist- a group of companies or people with whom a system is willing to work.

White Paper- A document showing the creation, development, and distribution plan of a cryptocurrency.

Wide Bottom (American Candle) a space that shows many trades at the bottom value in a candle.

Made in United States
Orlando, FL
04 November 2026

90218961R00348